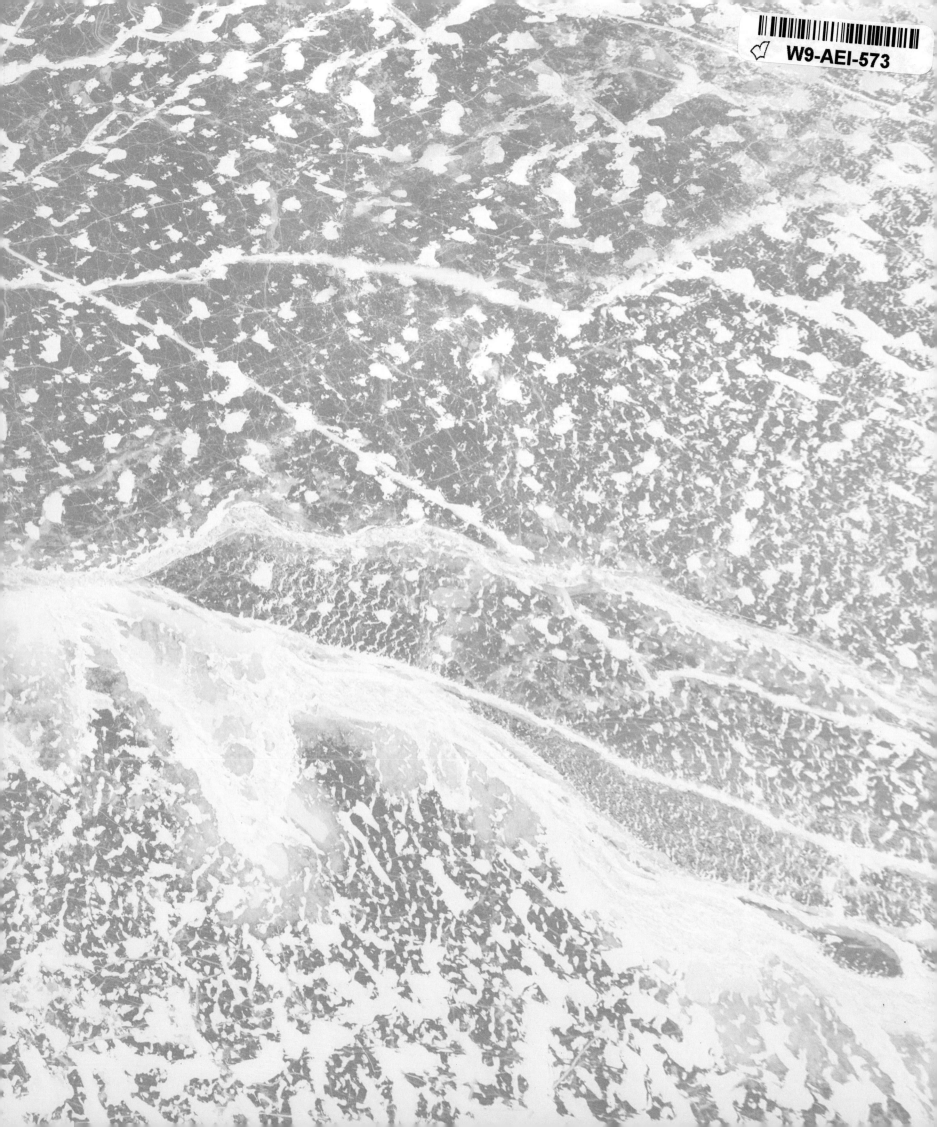

W9-AEI-573

NATURAL
WONDERS
OF THE WORLD

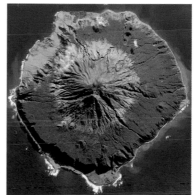

DK SMITHSONIAN

NATURAL WONDERS
OF THE WORLD

CONTENTS

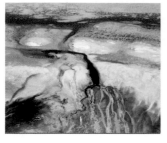   

DK | Penguin Random House

DK LONDON

Senior Editor Peter Frances
Project Editors Gill Pitts, Miezan van Zyl
Editors Claire Gell, Frankie Piscitelli, Kaiya Shang
US Editors Jill Hamilton, Kayla Dugger
Indexer Elizabeth Wise
Jacket Editor Claire Gell
Jacket Design Development Manager Sophia MTT
Managing Editor Angeles Gavira Guerrero
Associate Publisher Liz Wheeler
Publishing Director Jonathan Metcalf

Senior Art Editor Ina Stradins
Project Art Editors Francis Wong, Steve Woosnam-Savage, Mik Gates
Senior Graphics Co-ordinator Sharon Spencer
Picture Researcher Liz Moore
Senior Jacket Designer Mark Cavanagh
Pre-production Producer Gillian Reid
Senior Producer Anna Vallarino
Managing Art Editor Michael Duffy
Art Director Karen Self
Design Director Phil Ormerod

Illustrators Adam Benton, Peter Bull, Dominic Clifford, Dynamo Ltd, Arran Lewis, Sofian Moumene, Michael Parkin
Cartographer Ed Merritt

DK INDIA

Senior Editor Dharini Ganesh
Editors Riji Raju, Sonia Yooshing
Assistant Editor Priyanjali Narain
Picture Researchers Ashwin Raju Adimari, Surya Sankash Sarangi
Jackets Editorial Coordinator Priyanka Sharma
Managing Editor Rohan Sinha
Managing Jackets Editor Sreshtha Bhattacharya
Pre-production Manager Balwant Singh

Senior Art Editor Vaibhav Rastogi
Art Editor Debjyoti Mukherjee
Assistant Art Editors Yashashvi Choudhary, Simar Dhamija, Vaishali Kalra
Managing Art Editor Anjana Nair
Picture Research Manager Taiyaba Khatoon
Jacket Designer Dhirendra Singh
Senior DTP Designer Harish Aggarwal
DTP Designers Mohd Rizwan, Anita Yadav
Production Manager Pankaj Sharma

Smithsonian

Established in 1846, the Smithsonian—the world's largest museum and research complex—includes 19 museums and galleries and the National Zoological Park. The total number of artifacts, works of art, and specimens in the Smithsonian's collections is estimated at 154 million, the bulk of which is contained in the National Museum of Natural History, which holds more than 126 million specimens and objects. The Smithsonian is a renowned research center, dedicated to public education, national service, and scholarship in the arts, sciences, and natural history.

First American Edition, 2017
Published in the United States by DK Publishing
345 Hudson Street, New York, New York 10014

Copyright © 2017 Dorling Kindersley Limited
DK, a division of Penguin Random House LLC

17 18 19 20 21 10 9 8 7 6 5 4 3 2 1
001-297088-October/2017

All rights reserved.
Without limiting the rights under the copyright reserved above, no part of this
publication may be reproduced, stored in or introduced into a retrieval system,
or transmitted, in any form, or by any means (electronic, mechanical,
photocopying, recording, or otherwise) without the prior written permission of
the copyright owner.
Published in Great Britain by Dorling Kindersley Limited

A catalog record for this book is available from the Library of Congress
ISBN 978-1-4654-6417-0

DK books are available at special discounts when purchased in bulk for sales
promotions, premiums, fund-raising, or educational use. For details contact:
DK Publishing Special Markets, 345 Hudson Street, New York, New York 10014
SpecialSales@dk.com

Printed and bound in China

A WORLD OF IDEAS:
SEE ALL THERE IS TO KNOW
www.dk.com

Contributors and consultant

Jamie Ambrose is an author, editor, and Fulbright scholar with a special interest in natural history. Her books include DK's *Wildlife of the World*.

Robert Dinwiddie is a science writer who specializes in Earth science, the Universe, the history of science, and general science. When not writing, he enjoys exploring the world's volcanoes, glaciers, coral reefs, and other natural phenomena.

John Farndon has written extensively on Earth sciences and natural history. His books include *How the Earth Works*, the acclaimed *The Oceans Atlas*, and *The Practical Encyclopaedia of Rocks and Minerals*. He has been shortlisted a record six times for the Royal Society's Young People's Science Book Prize.

Tim Harris has been fascinated by all aspects of the natural world since a young age. He has studied glaciers in Norway and has written books about natural history for children and adults, including the award-winning *Migration Hotspots: The World's Best Bird Migration Sites*.

David Summers is a writer and editor with a training in natural history filmmaking. He has contributed to books on a range of subjects, including natural history, geography, and science.

Editorial consultant Dr. Don E. Wilson is Curator Emeritus, National Museum of Natural History, Smithsonian. The author of more than 250 scientific publications and 25 books, he is an elected fellow of the AAAS and an honorary member of the American Society of Mammalogists.

Half-title page Tristan da Cunha Island, South Atlantic Ocean
Title page avalanche wall near Mount Everest, Nepal
Preface Mount Wanaka Mountains, New Zealand

PREFACE

We live on a planet that is covered in amazing features, ranging from the red sands of the Sahara Desert to the blue ice of glaciers such as Vatnajökull, and from the grandeur of mountain chains like the Himalayas to the simple beauty of a bluebell wood. You might think that your list of things to see and do is complete, but once you start to look through the pages of this book you will realize there is so much more to discover and be inspired to go travelling once again.

However, you don't have to leave the comforts of home to enjoy the wonders of our world. Through this book you can see Earth's most active and dangerous volcanoes without taking any risks. If you don't have a head for heights but want to know what the view from the top of Angel Falls looks like, then turn to the entry on South America's Guiana Highlands. Even if you can't swim, you'll be able to find out what it would be like to dive inside the Great Blue Hole off the coast of Belize. And if you don't have time to trek through the Congo, you can still marvel at the wildlife in its rainforest. This book will also take you to places where access is severely limited—only a few people each year are given permission to explore the Hang Son Doong cave system in Vietnam but you will see what it is like to stand in its vast interior.

While the book is filled with fantastic photographs, it isn't just a celebration of the beauty of our natural world. It reveals the geology—and sometimes physics, chemistry, botany, and zoology—behind different features. Even the most complicated processes are explained in simple terms. And incredibly detailed artworks, constructed from satellite terrain data and imagery, reveal aspects of the natural wonders that cannot be seen with the naked eye. By viewing from above, going underneath, and looking inside, the artworks explain major geological formations and the processes behind them, including rift valleys, river canyons, island arcs, and stratovolcanoes.

This book doesn't just focus on geological wonders—the key animals and plants that live in, on, or around the different features are included, and the world's most important forests and grasslands are given their own entries. And if you want to know how a supercell storm forms, then turn to the section on extreme weather. In all, more than 240 of Earth's natural wonders are covered in detail and a further 230 are described in the directory. But rather than reading from cover to cover, this is a book that you will pick up, put down, and then return to time and time again as you try to decide where next in the world you wish to travel to.

Lake in the rift
Kenya's Lake Magadi is one of many lakes in East Africa's vast Great Rift Valley, formed as blocks of Earth's crust shift downward into the underlying mantle. The unusual colors come from minerals washed into the lake.

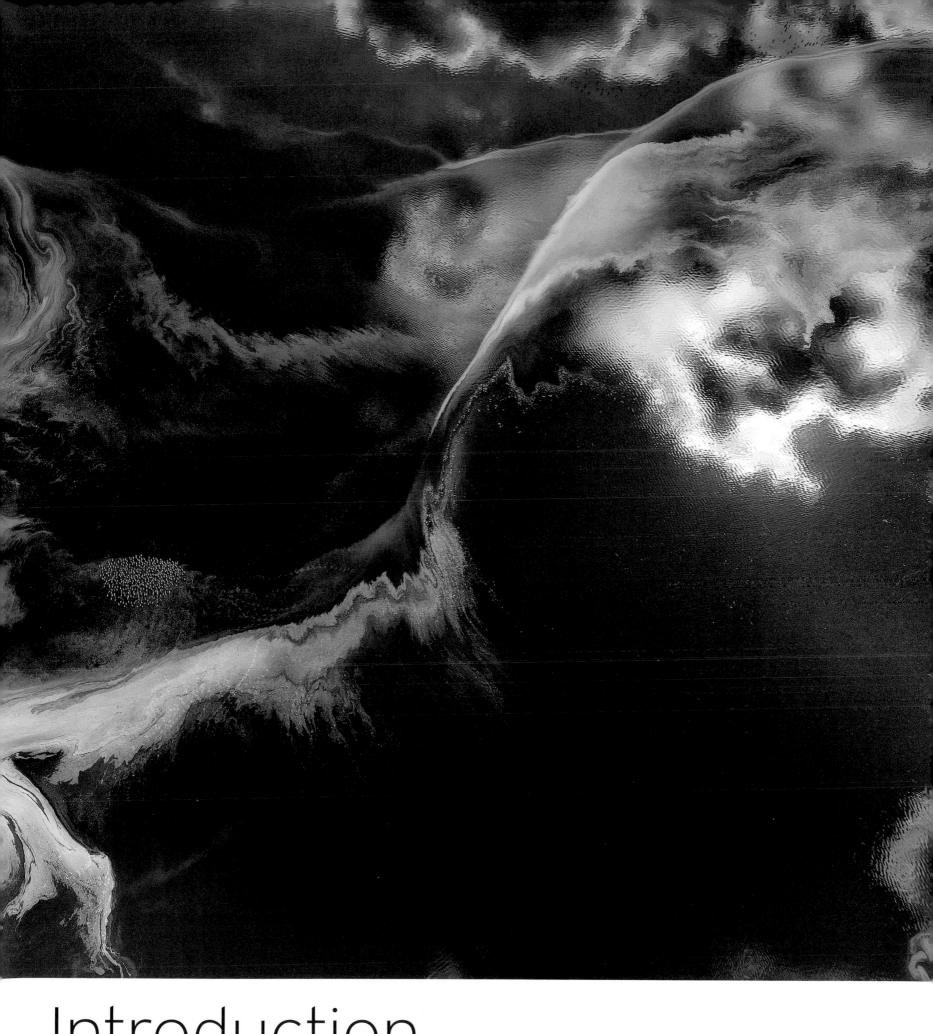

Introduction

EARTH'S STRUCTURE

Our planet has a diverse surface, with a huge variety of landscapes and materials, ranging from water and gases, to living matter, soil, ice, and rock. However, its interior is much less varied, consisting mainly of just rock, metal, and some water.

Internal structure

Much of what is known about Earth's structure has been established from the study of seismic waves created by earthquakes. Internally, our planet has three main layers: the core, mantle, and crust. The core is made mainly of iron, with some nickel, and is itself double-layered—a liquid outer core enclosing a solid inner core.

Surrounding the core is the mantle, which consists mainly of solid (but in parts, deformable) silicate rock. Its top part is called the upper mantle and has two layers. The uppermost of these, sometimes called lithospheric mantle, is solid and brittle and is fused to the crust. Below it is a more deformable layer, the asthenosphere. Below the upper mantle is a region called the transition zone, now known to contain significant amounts of water "locked into" its rocks. Beneath this is the thickest of all of Earth's internal layers, the lower mantle.

The outermost layer is the crust. This is solid, and there are two different kinds. Thick continental crust, made up of many different rock types, forms the land surface, while thin, relatively dense oceanic crust—containing just a few rock types—lies under the oceans. Both types are fused to the underlying lithospheric mantle, forming a combined rigid shell, the lithosphere.

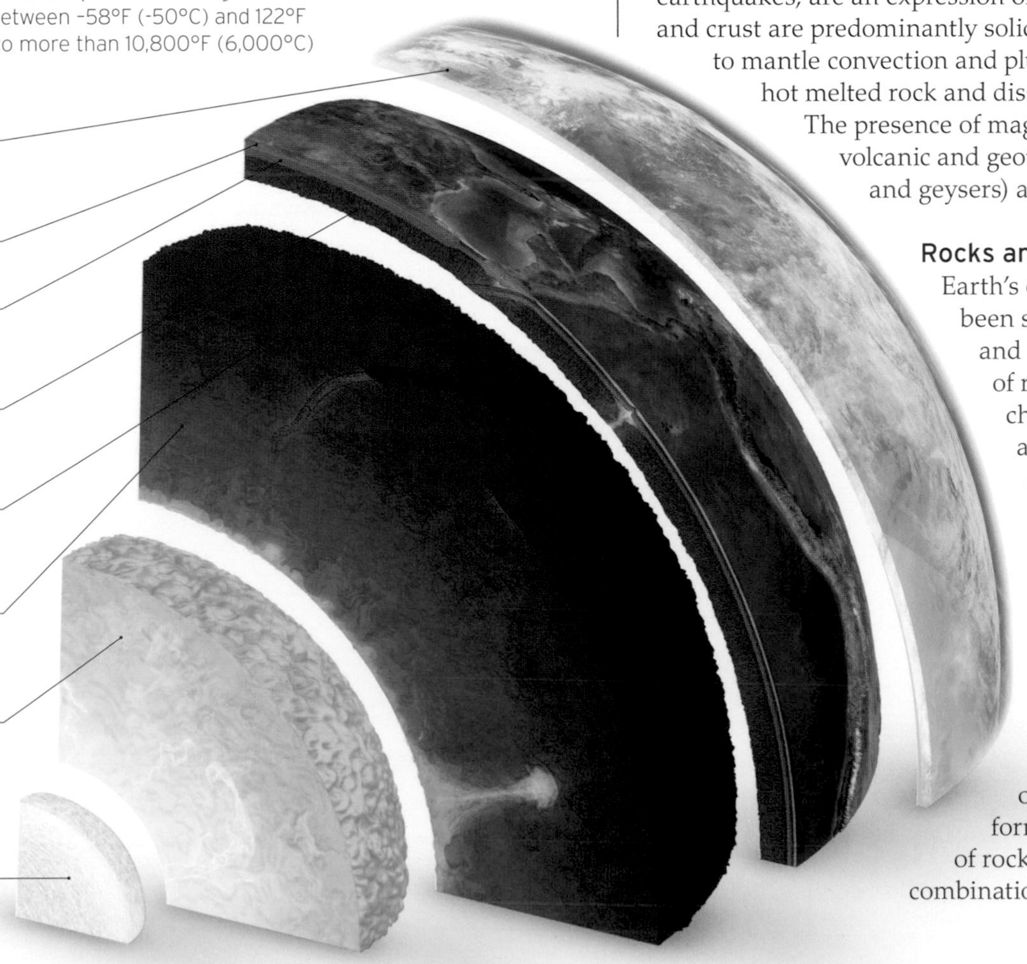

Mantle convection

A slow, gradual circulation of rock occurs in the mantle. This circulation is called mantle convection and is driven by heat flowing out of Earth's core. The mantle is also thought to contain plumes of rising, hotter, semisolid or liquid rock, called mantle plumes. Where they penetrate into the crust, these create surface hotspots (see p.13). Earth holds a colossal amount of internal heat energy, much of it generated during the planet's original formation and still trapped inside. This energy is continually added to by the radioactive decay of unstable isotopes of various chemical elements scattered throughout

▷ **WHOLE-EARTH CONVECTION**
The overall pattern of convection in Earth's mantle is not fully understood. One hypothesis, shown here, suggests that a series of convection cells slowly carry rock from the bottom of the mantle to the top and then back again.

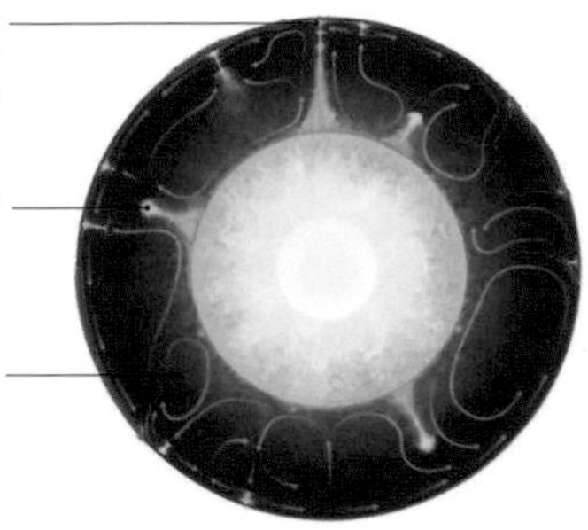

hotspot, where extra-hot material reaches crust

mantle plume

convection cell

the interior. Earth's energy is constantly trying to escape, and mantle convection and mantle plumes, as well as phenomena such as earthquakes, are an expression of this. Although our planet's mantle and crust are predominantly solid, as a result of processes connected to mantle convection and plumes both contain collections of hot melted rock and dissolved gases, called magma. The presence of magma in the crust is associated with volcanic and geothermal activity (such as hot springs and geysers) at Earth's surface.

Rocks and the rock cycle

Earth's crust, the part of our planet that has been shaped into a myriad of landscapes and physical features, is made up mostly of rocks. Rocks are assemblages of chemical substances called minerals, and are of three main types: igneous, sedimentary, and metamorphic. Igneous rocks are the result of magma from deep within the planet being erupted at the surface or injected into the crust and then cooling and solidifying. Sedimentary rocks are formed by the deposition and cementation of mineral or rock particles created by the weathering or breakdown of other rocks. Metamorphic rocks are formed by the alteration of other types of rocks by high temperature, pressure, or a combination of the two. A variety of processes

▽ **EARTH'S LAYERED STRUCTURE**
Each layer from the crust inward has a temperature and density higher than the layer surrounding it. The temperature ranges between -58°F (-50°C) and 122°F 50°C) at the surface to more than 10,800°F (6,000°C) in the inner core.

Atmosphere
more than 620 miles
(1,000 km) thick

Crust

Lithospheric mantle
25–70 miles
(40–120 km) thick

Asthenosphere
155–210 miles
(250–340 km) thick

Transition zone
155 miles
(250 km) thick

Lower mantle
1,385 miles
(2,230 km) thick

Liquid outer core
1,370 miles
(2,200 km) thick

Solid inner core
790 miles
(1,275 km) radius

▽ SHAPING THE LAND
The flow of rivers and streams over Earth's surface, causing erosion, is part of the rock cycle and the water cycle, and one of the major processes sculpting surface landscapes.

occurring within Earth's crust and surface continually change crustal rocks from one type into another, in a never-ending succession of events known as the rock cycle. The various components of the rock cycle, which include, for example, volcanism and numerous erosional and depositional processes, are extremely important in shaping Earth's surface landscapes.

Earth's atmosphere
The atmosphere is part of our planet's overall structure and has several layers. Air circulates only in the lowest layer, which is called the troposphere, and it is only in this layer that weather occurs—taking the form of, for example, winds, precipitation (rain, hail, and snow), and changes in temperature, pressure, and humidity. Weather is driven largely by a combination of energy coming from the Sun and from Earth's rotation. The averaged-out weather in a region over a long period of time is termed that region's climate. Both weather and climate, and the connected processes of the water cycle (which include, for example, precipitation, the flow of water over Earth's land surface in streams and rivers, and the evaporation of water

The mantle transition zone is thought to **contain more water** than **all the oceans**

from the sea) are other vital factors in shaping landscapes. Changes in Earth's climate, in particular a persistent warming caused by increased levels of carbon dioxide in the atmosphere, are having a widespread effect on surface landscapes. For example, a general retreat and shrinkage of glaciers is occurring worldwide.

◁ HURRICANE
One of the more extreme forms of weather, a hurricane originates from solar warming of the surface of a tropical ocean. This creates a region of low pressure, with dense clouds above it and winds circulating around it.

PLATE TECTONICS

Numerous features and events at Earth's surface, from volcanic activity to earthquakes and mountain-chain formation, can be explained on the basis that our planet's outer shell is split into fragments called tectonic plates, which slowly move around on the surface.

Earth's plates

Earth's crust is fused to the top layer of the underlying mantle (see p.10), forming a shell-like structure called the lithosphere. The lithosphere is split into several chunks, called tectonic plates, which slowly move around on Earth's surface, driven by convection in the mantle. There are seven large and numerous medium-sized and smaller plates. As they move, Earth's continents are slowly shuffled around. The rate of movement is small, but over tens or hundreds of millions of years, it can produce major rearrangements. During Earth's long history, continents carried by different plates have collided from time to time, pushing up mountains, or combined to form supercontinents. Conversely, large landmasses have sometimes split into smaller chunks in a process called rifting.

Plate boundaries

The science of plate tectonics largely revolves around the study of what happens near the edges of plates as they move relative to each other. These plate boundaries

upwelling magma solidifies to add new material to plate edges

plate movement

SPREADING RIDGE

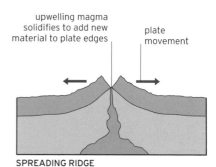

◁ **TECTONIC PLATES**
Earth's plates have irregular shapes and a wide range of sizes. They cover Earth's surface like a jigsaw puzzle. Here, several plates are visible (their boundaries highlighted in orange).

△ **DIVERGENT BOUNDARY**
At a divergent boundary, plates move apart (top). This type of boundary can be seen at the Silfra fissure in Iceland (bottom), where the Eurasian and North American Plates are separating.

are dynamic places where important landscape-altering events take place, such as volcanic activity, mountain-building, island formation, rifting, and earthquakes.

Plate boundaries fall into three main classes. The first, called divergent boundaries, occur where two plates are moving apart. At this type of boundary, material continuously welling up from the mantle fills the gap and creates new plate. Divergent boundaries are found extensively on the ocean floors, in the form of mid-ocean spreading ridges. But they also occur, among other places, in Iceland and in East Africa, along what is called the Great Rift Valley (see pp.184–87). They are always associated with volcanic activity.

The second class of boundaries, convergent boundaries, are found where two plates move toward each other. At these boundaries, all or some of one plate moves down, or subducts, under the other and is destroyed. Where both plates carry continental crust, mountains form as chunks of crust are forced together. This process accounts for the origin of many mountain chains, such as the Himalayas. Other instances involve a plate carrying oceanic crust subducting the other plate. Boundaries of this type are characterized by features such as a deep-sea trench along the line of the boundary; a chain of volcanoes, always on the side of the plate that is not subducting; and the frequent occurrence of major earthquakes.

The third class of boundaries, called transform boundaries, occur where plate edges push past each other, without new plate being created or existing plate destroyed. These boundaries are also a site and source of earthquakes. Examples can be found in California (along the famous San Andreas Fault, see pp.30–31), in New Zealand's South Island, and elsewhere, including extensively on the ocean floors.

Hotspot volcanism

Although two of the three main types of plate boundary are frequently linked to volcanism, not all volcanoes develop at plate boundaries. Some appear in the middle of plates. This type of volcanism is usually accounted for by the presence of mantle hotspots—locations at the top of Earth's mantle that appear to be the source of peculiarly large amounts of energy. The movement of a plate across a hotspot (whose position is fixed) can, over a long period of time, create a chain of volcanic features at the surface. This explains, for example, evidence of much ancient volcanic activity in a line to the southwest of Yellowstone National Park (due to movement of the North American Plate over this hotspot). It is also regarded as the best explanation for the formation of some chains of volcanic islands, including the Hawaiian Islands (see pp.316–19).

Plates move at a rate of between ¼in (7 mm) per year– one-fifth the rate that fingernails grow–and 6 in (150 mm) per year–the rate human hair grows

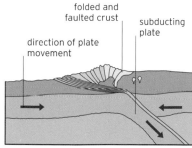

SUBDUCTION ZONE

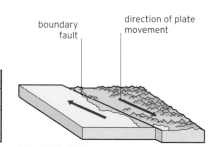

SLIDING PLATES

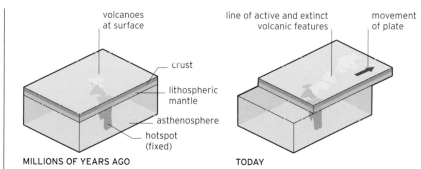

MILLIONS OF YEARS AGO · TODAY

△ **CONVERGENT BOUNDARY**
At this type of boundary, one plate moves under, or subducts, another and is gradually destroyed (top). If both plates carry continental crust, mountains are pushed up (bottom).

△ **TRANSFORM BOUNDARY**
At a transform boundary, two plates push past each other in opposite directions (top). This type of boundary may be visible as a linear fault running across the landscape (bottom).

△ **VOLCANIC CHAINS**
Where a plate moves over a mantle hotspot, it can lead to a chain of volcanic features at the surface. The Hawaiian Islands formed in this way through movement of the Pacific Plate over a hotspot currently located under the volcano Kilauea (bottom) on the Big Island of Hawaii.

EARTH'S PAST

The landscape features we see on Earth today are the result of events and processes stretching back more than 4 billion years. It is only in the last few hundred years, however, that scientists have pieced together some of the main details of the story.

Earth's age and origins

A precursor body of what is now planet Earth formed around 4.55 billion years ago, from collisions of smaller bodies in a disk of material spinning around the Sun. When it was perhaps 40 million years old, the proto-Earth is thought to have experienced one final collision, leading to the formation of what we now call Earth and its natural satellite, the Moon.

The early Earth was probably born in an extremely hot, molten state. Heavier materials, mainly iron, sank to the center, with lighter materials forming layers around this. For around the first 150 million years, no solid crust formed because there were continuing impacts from comets and asteroids, while high volcanic activity continually reworked the surface. Around 4.37 billion years ago, the oceans had begun to form, through condensation of water released into the atmosphere by ancient volcanoes. By 4 billion years ago, the first pieces of continental crust had also formed. Movements of early tectonic plates caused landmasses to collide and merge, gradually forming the ancient cores of today's continents.

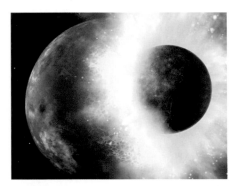

◁ **ANCIENT COLLISION**
A collision between a proto-Earth and a Mars-sized body, some 4.51 billion years ago, is thought to have led to the formation of the Earth-Moon system.

△ **CRYSTAL EVIDENCE**
A zircon crystal, from Jack Hills, Australia, has been dated at 4.375 billion years old and is the oldest known material of terrestrial origin.

The geological timescale

The fact that Earth is extremely ancient, and that the sedimentary rock layers (strata) of its continental crust were laid down in sequence over a vast period of time, first became clear to scientists between the 17th and 19th centuries. They noticed that many rock strata contain fossils—the remains of ancient, apparently extinct, animals and plants. They divided the fossil-bearing rock strata—and thus Earth's history as a life-bearing planet—into three eras: the Paleozoic (meaning ancient life), Mesozoic (middle life), and Cenozoic (recent life). Later, these eras were subdivided into geological periods. In rock strata deeper and older than the fossil-bearing rocks, there seemed at first to be no life. These rock strata were later subdivided into three extremely long time

▷ GEOLOGICAL TIME

Over the enormous span of Earth's history, times during which there was no life or only the simplest life forms have lasted longer than the eras where multicellular life-forms exist.

KEY

Cenozoic 66–0 MYA
Proterozoic 2,500–541 MYA
Mesozoic 252–66 MYA
Archean 4,000–2,500 MYA
Paleozoic 541–252 MYA
Hadean 4,550–4,000 MYA

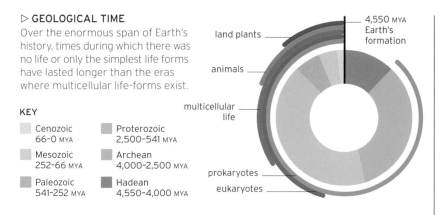

land plants
animals
multicellular life
prokaryotes
eukaryotes
4,550 MYA Earth's formation

intervals called eons—the Hadean followed by the Archean and Proterozoic eons. It is now known that simple organisms called prokaryotes (single-celled bacteria-like organisms) first evolved during the Hadean. Some 2 billion years then passed before more complex single-celled organisms called eukaryotes evolved (see pp.16–17).

Agents of change

Geologists have argued for centuries about the nature of the processes, operating through Earth's past, that have brought about the varied landscapes seen today. In the late 18th century, support grew for the argument that Earth's features were mostly formed by slow, gradual, and ongoing changes over time. It was clear, for example, that erosion had shaped the planet everywhere. This view contrasted with the opposing doctrine, which proposed that a series of cataclysms was responsible for most of the Earth's surface features. Today, it is agreed that most landforms on Earth now are the result of gradual processes. Nevertheless, there have been catastrophes: for example, an asteroid or comet impact about 67 million years ago (MYA) is thought to have caused fires, massive earthquakes, worldwide darkness, and possibly wiped out the dinosaurs.

Changes in climate and sea level

Earth's climate has changed quite drastically over long periods of time. At one extreme, there have been times when there were no ice caps and temperate deciduous forests grew at the poles. At the other, it seems probable that Earth was at least once completely frozen over. Similarly, sea level has fluctuated dramatically. For much of the past 540 million years, it has been higher than it is today. But, for the last 2.5 million years or so, sea level has mostly been lower, since during this time Earth has been in an ice age with much of its water locked up in polar ice sheets. Over the past 17,000 years or so, Earth has been in a warming period within this ice age, called an interglacial, and sea level has risen again. Since the late 18th century, the planet has warmed by around 1.8°F (1°C), due to increased atmospheric carbon dioxide, and sea level has risen by about 12 in (30 cm).

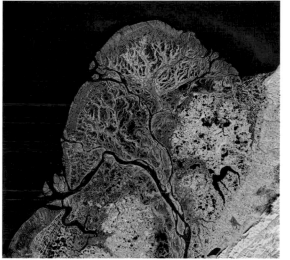

◁ **SLOW BUILDUP**
Deltas, such as the Yukon Delta seen here, are examples of a landform built up slowly over a long period of time.

▽ **ERODED LANDFORM**
Horseshoe Bend in Arizona exhibits the effects of 75 million years of erosion. The rock layers in its walls range from around 180 to 200 million years old.

20269221

LIFE ON EARTH

Earth's land surface, oceans, and atmosphere teem with life, which has colonized every corner of our planet. This is remarkable given that when Earth formed, it was just a dead rock. Life has drastically altered the appearance of our world and the chemistry of the atmosphere.

Early life on earth

Life on Earth probably originated around 4.25 billion years ago, from precursor chemicals. The oldest definite evidence of life is around 3.7 billion years old and suggests that simple single-celled organisms existed in oceans at that time. These microbes were already able to photosynthesize—that is, to use light energy to manufacture sugars from water and dissolved carbon dioxide, releasing oxygen in the process. By around 1.9 billion years ago, photosynthesizing organisms had released significant quantities of oxygen into the atmosphere. This was important to further evolution of life, because some of the oxygen (consisting of O_2 molecules) was converted into ozone (O_3 molecules), which formed a protective layer around the planet, filtering out harmful incoming ultraviolet rays. The presence of oxygen in the atmosphere, and dissolved in the oceans, also made animal life possible. By 1.2 billion years ago, multicellular organisms made from eukaryotic cells (more complex than the earlier prokaryotic cells of the simple microbes) had evolved.

◁ **OLDEST SIGNS OF LIFE**
The oldest fossils known are 3.7 billion-year-old traces of stromatolites–layered structures formed by microbes living on an ancient seafloor–within this rock found in Greenland.

Evolution, spread, extinctions

Between around 525 and 500 million years ago (MYA), during a time in Earth's history called the Cambrian Period, a great proliferation and diversification of life with mineralized shells and skeletons occurred in the oceans. Today, this is known as the Cambrian explosion. By 490 MYA, the first plants had appeared on land, and by 420 MYA, there were also animals on land. From then onward, life continued to spread and diversify rapidly, but there were also extinctions.

One major wave of extinctions occurred at the end of the Permian Period, around 252 MYA. The exact cause is not known, although a series of massive volcanic eruptions is one suspect. Another occurred at the end of the Cretaceous Period, 66 MYA, when more than 75 percent of all species died out, including most dinosaurs. However, one dinosaur group survived and evolved into birds.

▷ **EXTINCT FORM**
Tylocidaris sea urchins, of which this is a fossil example, lived between about 140 million and 40 million years ago, but then died out.

Kingdoms of life today

Life on Earth is extremely diverse. Estimates of the number of Earth's current species range from 2 million to 1 trillion, of which only about 1.6 million have been identified and described so far. Various schemes have been put forward for dividing known life forms into kingdoms. A consensus scheme, proposed by scientists at the Smithsonian Institution in 2015, defines seven kingdoms. Two consist solely of

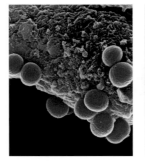

△ **ARCHAEA**
Many of these simple, single-celled lifeforms live in extreme environments, such as acidic hot springs.

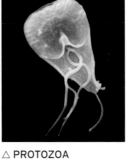

△ **PROTOZOA**
Some protozoa cause disease. This one, *Giardia lamblia*, causes a type of human intestinal infection.

△ **PLANTS**
There are over 300,000 known species of plants. They are the basis of most ecosystems, especially on land.

Life has been found inside rocks 12 miles (19 km) below Earth's surface

prokaryote organisms—bacteria and archaea. The other five are based on eukaryotic cells: protozoa (single-celled life forms such as amoebae); chromista (algae of various sorts, including diatoms and water molds); and three more familiar groups: fungi, plants, and animals. The various life-forms interact with Earth, affecting landscapes in many ways. For example, plants cover a significant proportion of the land surface, and their roots help keep soil from being carried away by rainwater. Life has also changed the chemistry of Earth's atmosphere, making it uniquely oxygen-rich. Without life, Earth would have evolved and would appear and work quite differently.

▷ **RIVER CROSSING**
One of the most awe-inspiring natural sights on Earth is the annual migration of vast numbers of wildebeest, zebras, and gazelles–here about to cross a river–across the Serengeti ecosystem.

FRANKLIN COUNTY LIBRARY
906 NORTH MAIN STREET
LOUISBURG, NC 27549
BRANCHES IN BUNN,
FRANKLINTON, & YOUNGSVILLE

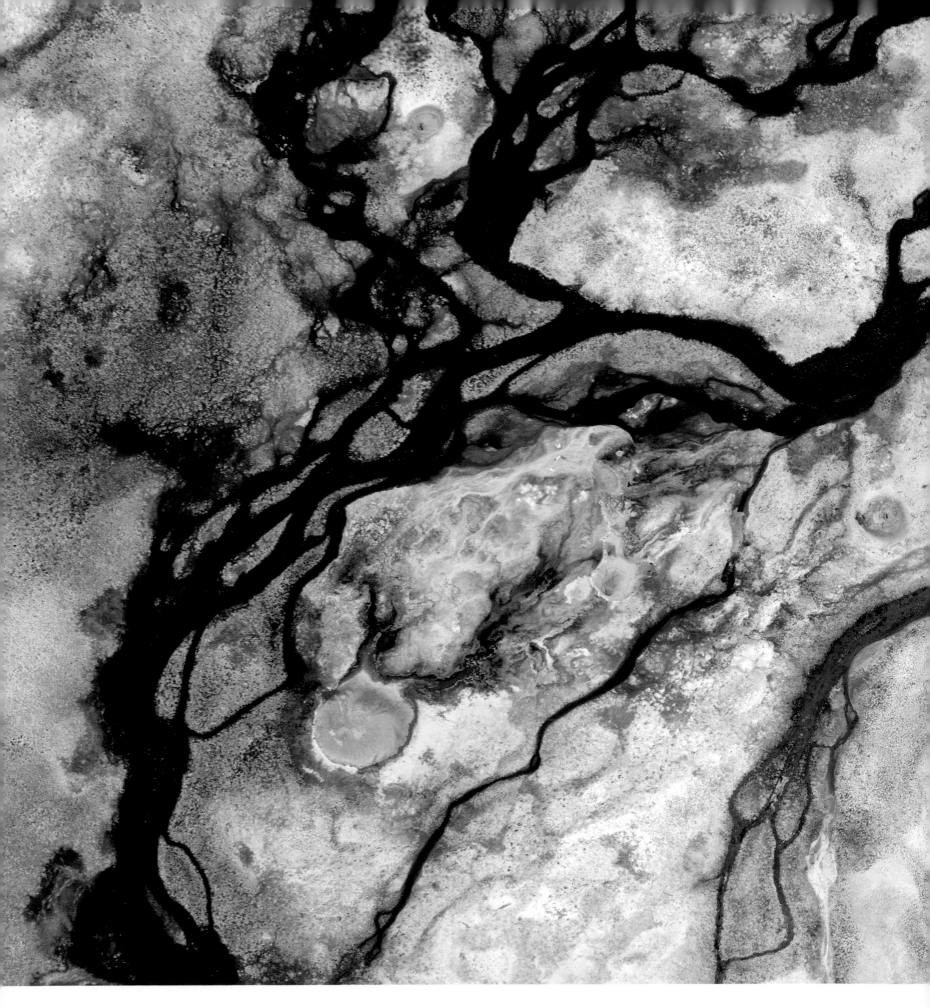

Heating system
Powered by a vast plume of hot, semisolid rock, or magma,
many miles below the surface, Yellowstone National Park
is the site of spectacular geothermal features, including
geysers, hot springs, and pools of boiling mud.

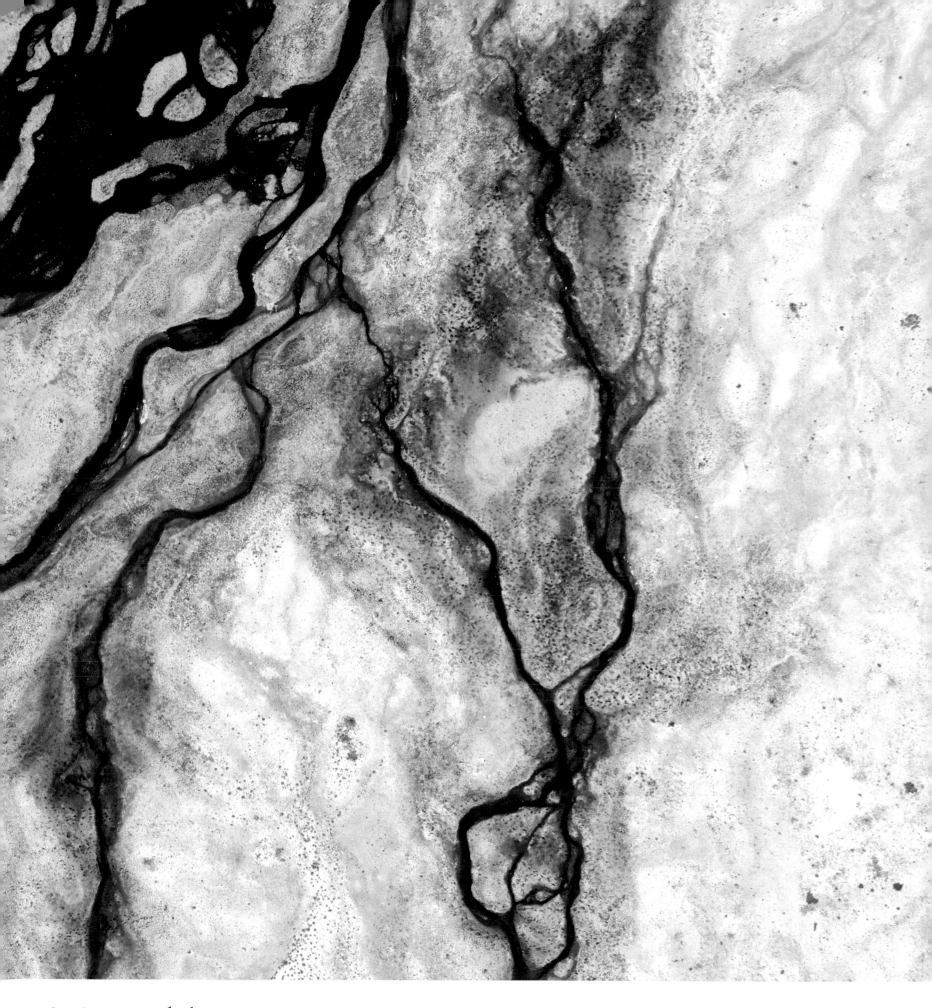

North America

MOUNTAINS, LAKES, AND PRAIRIES

North America

North America is the third largest continent. Geographically, it is also considered to include Greenland. Most of the continent lies on a single tectonic plate, with small parts of Mexico and California lying on the Pacific Plate to the west, which borders the North American Plate along the infamous San Andreas Fault.

Unusually, North America has a vast and ancient lowland heart, almost completely surrounded by younger belts of fold mountains thrown up by continental collisions in the past. This lowland heart, with its Great Plains and Great Lakes, is drained almost entirely by one of the world's great river systems, the Mississippi–Missouri, which channels water south into the Gulf of Mexico. The Appalachian range in the east

is relatively old and has been worn low over time. But the Western Cordillera, which includes the Alaska, Cascade, and Coast ranges and the inland Rocky Mountains, running down the western limb of the continent, is geologically young and still high—so high that these mountains block the flow of moist air from the Pacific. This gives much of the interior a markedly continental climate with bitter winters and hot summers punctuated by tornadoes, and in the south the lack of rain creates large deserts. Only to the north are there no mountains, and the expanse of the Canadian Shield stretches north into the bitter wastes of the Arctic.

KEY DATA

▲ **Highest point** Denali, Alaska: 20,308 ft (6,190 m)

▼ **Lowest point** Death Valley, California: -282 ft (-86 m)

● **Hottest record** Death Valley, California: 134°F (57°C)

● **Coldest record** Northice, Greenland: -87°F (-66°C)

CLIMATE

Most of North America lies in the temperate zone, with a subtropical south and an Arctic north. The coasts are moist, but the interior is drier.

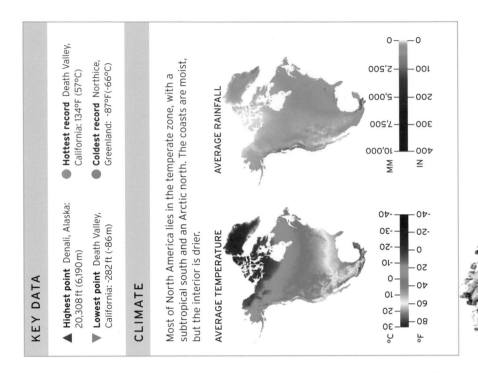

AVERAGE TEMPERATURE

AVERAGE RAINFALL

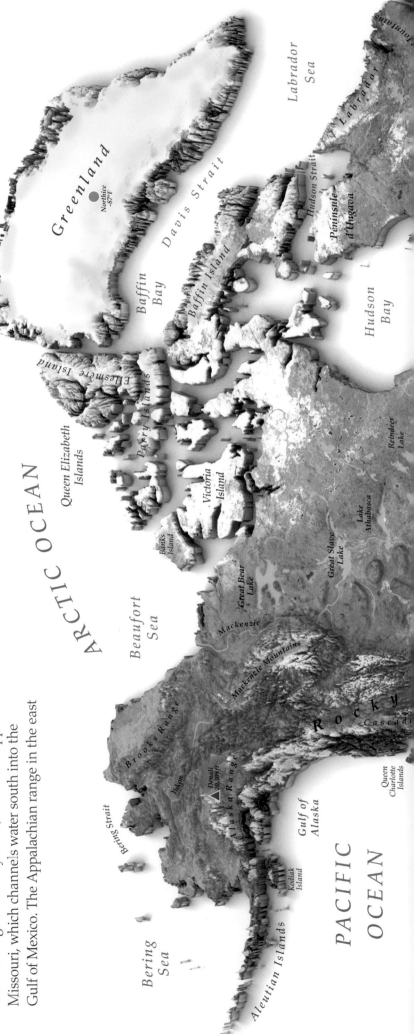

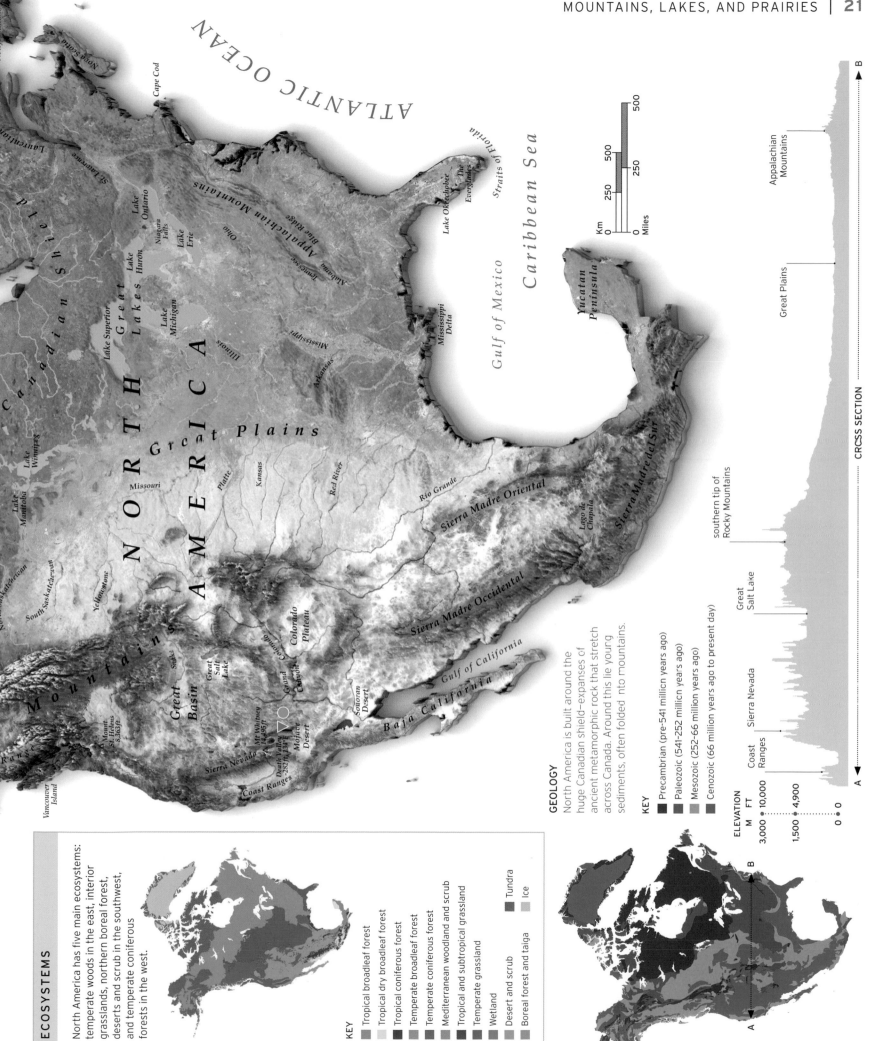

ATLANTIC OCEAN

Newfoundland

Nova Scotia

Cape Cod

Laurentian

St. Lawrence

Lake Ontario

Niagara Falls

Lake Erie

Lake Huron

Appalachian Mountains

Blue Ridge

Ohio

Tennessee

Alabama

Lake Okeechobee

The Everglades

Straits of Florida

Gulf of Mexico

Caribbean Sea

Yucatan Peninsula

Canadian Shield

Lake Superior

Great Lakes

Lake Michigan

Lake Winnipeg

Lake Manitoba

North Saskatchewan

South Saskatchewan

Yellowstone

NORTH AMERICA

Great Plains

Missouri

Platte

Kansas

Illinois

Mississippi

Arkansas

Red River

Río Grande

Mississippi Delta

Sierra Madre Oriental

Lago de Chapala

Sierra Madre del Sur

Colorado Plateau

Sierra Madre Occidental

Gulf of California

Baja California

Great Basin

Great Salt Lake

Colorado

Grand Canyon

Sonoran Desert

Mojave Desert

Snake

Mountains

Range

Vancouver Island

Mount St. Helens 8,365 ft

Sierra Nevada

Mt Whitney 14,495 ft

Death Valley -282 ft

Coast Ranges

GEOLOGY

North America is built around the huge Canadian shield—expanses of ancient metamorphic rock that stretch across Canada. Around this lie young sediments, often folded into mountains.

KEY

- Precambrian (pre–541 million years ago)
- Paleozoic (541–252 million years ago)
- Mesozoic (252–66 million years ago)
- Cenozoic (66 million years ago to present day)

ELEVATION

M	FT
3,000	10,000
1,500	4,900
0	0

CROSS SECTION

B

Appalachian Mountains

Great Plains

southern tip of Rocky Mountains

Great Salt Lake

Sierra Nevada

Coast Ranges

A

ECOSYSTEMS

North America has five main ecosystems: temperate woods in the east, interior grasslands, northern boreal forest, deserts and scrub in the southwest, and temperate coniferous forests in the west.

KEY

- Tropical broadleaf forest
- Tropical dry broadleaf forest
- Tropical coniferous forest
- Temperate broadleaf forest
- Temperate coniferous forest
- Mediterranean woodland and scrub
- Tropical and subtropical grassland
- Temperate grassland
- Wetland
- Desert and scrub
- Boreal forest and taiga
- Tundra
- Ice

Km
250 500

Miles
250 500

△ **VAST SLAB**
Parts of the Canadian Shield—a huge slab of tough and very ancient volcanic and metamorphic rocks—are exposed by erosion in areas such as riverbanks.

◁ **SCULPTED PEAKS**
One of the world's greatest mountain ranges, the Rockies run all the way along the western edge of North America. They have been heavily sculpted by glaciers.

THE SHAPING OF NORTH AMERICA

North America began life as part of the great, ancient landmass geologists call Laurentia, which journeyed across the world, colliding with and separating from other continents. The continent we know today came into being less than 200 million years ago.

Ages of rock

Underlying much of North America is a large shield of ancient rock that once formed the core of Laurentia. In the U.S., it is covered by a blanket of more recent sedimentary rock. But in Canada, where the shield is exposed on the surface, geologists have found some of the world's oldest rocks—survivors from the dawn of Earth's story, up to 4.4 billion years old.

About 750 million years ago, this Laurentian heartland formed part of the supercontinent of Rodinia, but after Rodinia broke up Laurentia drifted almost to the South Pole before drifting north again. For a while, Laurentia was alone, but eventually it was drawn into the regathering of continents and became part of a new supercontinent—Pangaea.

About 480 million years ago, the plate Laurentia sat on crunched into neighboring ocean plates, causing the Appalachian Mountains to start rising. It took 250 million years for the mountains to reach

their maximum height. At first, they were probably as tall and majestic as the Himalayas. In the southwest, rocks in Nevada and Utah were twisted by the same tectonic forces.

America's backbone

About 200 million years ago, Laurentia split from Pangaea to drift northwest, opening up the Atlantic Ocean and creating North America. The new continent's east coast settled down as it moved away, and the Appalachians were worn gradually lower and lower by millions of years of erosion and weathering. This is why the Appalachians are smaller than the Rocky Mountains even though they are older.

The west, on the other hand, became a titanic battleground between opposing tectonic plates, as the North American Plate plowed into and rode over the oceanic crust of the Pacific Plate, forcing it to subduct, or descend into the mantle. The Rocky

The Appalachians are over 480 million years old

KEY EVENTS

480 million years ago
The Appalachians start to rise. They continue rising for another 250 million years to Himalayan heights, but have since been lowered by erosion.

80-55 million years ago
The rising Rocky Mountains cause the surrounding land to uplift. As a result, the Western Interior Seaway, a large inland sea, dries out.

18 million years ago The Greenland ice sheet forms around this time. However, it may be even older than this, and it is likely the ice sheet has receded and advanced many times since then.

11,000 years ago The last of the ice sheets covering much of North America start to recede, relieving most of the continent of its covering of ice.

1.5-1 billion years ago Most of the land comprising the crust of central and eastern North America today forms.

200 million years ago
Tectonic activity splits Laurentia apart from Europe, allowing the land that will become North America to drift northwest.

26-22 million years ago The Earth's climate becomes cooler, favoring the spread of grasslands. The Great Plains first appear, covering a swathe of flat land that used to be the Western Interior Seaway.

15,000 years ago,
During the last ice age, some glaciers start to melt, and water pools to form the Great Lakes.

10,000 years ago Ice melts and sea levels rise, causing the land bridge connecting Alaska and Russia to submerge.

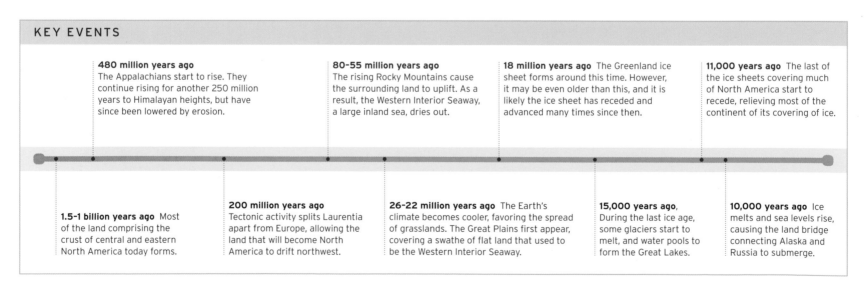

▽ ISOSTATIC REBOUND
As North America's ice sheets melted, the weighed-down land beneath slowly began to rise—a process called isostatic rebound. This effect can be seen, for example, in the exposure of old shorelines.

▽ CRACKS IN THE LAND
The San Andreas faultline visibly slices across the Mojave Desert, dividing the Pacific and North American tectonic plates.

△ DRY FALLS
The Palouse River flows through a giant gorge that was carved out by the Missoula floods.

Mountains rose as the submerged oceanic plate melted to create volcanoes. Islands slammed against the continent, and the collisions piled up layers of rock on the continent's coast. The Rocky Mountains are unusual in that they rose up farther inland than coastal mountain ranges, such as the Andes in South America, normally do. The reason remains a mystery for geologists—it may be that the ocean plate descended at a much flatter angle and so melted in the mantle at a point farther beyond the coast.

Arctic Ocean had permanent sea ice. The ice sheet's massive weight gouged deep scars into the land that would later become basins for the Great Lakes, and a huge lake called Lake Agassiz was ponded up against it. As the giant ice sheet finally melted, it unleashed cataclysmic torrents called the Missoula floods—in Washington and Oregon—carving out giant canyons and vast waterfalls that are now left as huge, bare cliffs, such as those seen at Dry Falls, Washington.

American icebox

There have been five major ice ages in Earth's history. The most recent began 1.8 million years ago and lasted until just 11,700 years ago. During that time, much of Canada and the northern U.S. were covered by a vast ice sheet, and most of the

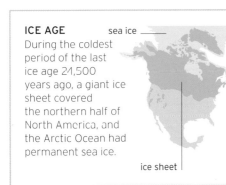

ICE AGE
During the coldest period of the last ice age 24,500 years ago, a giant ice sheet covered the northern half of North America, and the Arctic Ocean had permanent sea ice.

sea ice
ice sheet

Active faultline

The small Farallon Plate was once sandwiched between the Pacific and North American Plates. It was entirely swallowed up 30 million years ago as the two bigger plates collided and subducted. When the subduction stopped, the two gigantic plates began sliding past each other rather than smashing together. It is here that the world's most famous strike-slip fault formed—the San Andreas Fault. The fault is 810 miles (1,300 km) long and stretches from the Mendocino coast south to the San Bernardino Mountains and the Salton Sea. Since it formed, the land on either side has moved at least 340 miles (550 km). Every juddering movement unleashes devastating earthquakes.

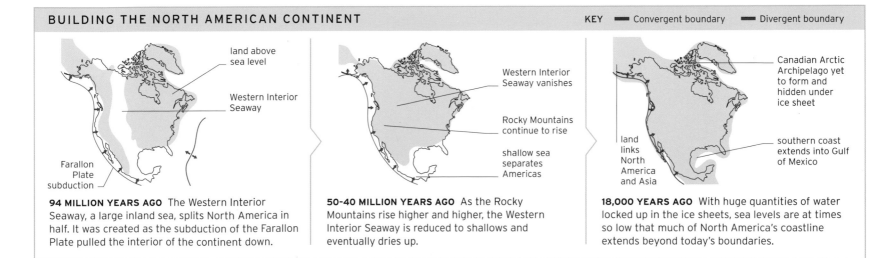

BUILDING THE NORTH AMERICAN CONTINENT

KEY ▬ Convergent boundary ▬ Divergent boundary

land above sea level

Western Interior Seaway

Farallon Plate subduction

Western Interior Seaway vanishes

Rocky Mountains continue to rise

shallow sea separates Americas

Canadian Arctic Archipelago yet to form and hidden under ice sheet

land links North America and Asia

southern coast extends into Gulf of Mexico

94 MILLION YEARS AGO The Western Interior Seaway, a large inland sea, splits North America in half. It was created as the subduction of the Farallon Plate pulled the interior of the continent down.

50-40 MILLION YEARS AGO As the Rocky Mountains rise higher and higher, the Western Interior Seaway is reduced to shallows and eventually dries up.

18,000 YEARS AGO With huge quantities of water locked up in the ice sheets, sea levels are at times so low that much of North America's coastline extends beyond today's boundaries.

NW. North
America

The Alaska Range

The world's highest mountain range outside of Asia and the South American Andes

Stretching for some 400 miles (650 km) like a crescent moon across south-central Alaska, this range appears as an awe-inspiring barrier of high, snow-capped peaks. The loftiest mountains are clustered around Denali, a name given by the local Native American Koyukon people to the highest summit in the Alaska Range. The mountains act as a formidable obstruction to the flow of moist air from the Gulf of Alaska, which lies farther to the south, and—as a result—they experience very high levels of snowfall. Due to this, the valleys in the range are filled with some large glaciers, such as the Black Rapids Glacier (see p.43). These, along with the dramatic Arctic scenery, have made the Alaska Range a magnet for hikers.

HIGHEST PEAKS IN THE ALASKA RANGE

1 2 3 4 5

1 Denali 20,308 ft (6,190 m)
2 Mt. Foraker 17,402 ft (5,304 m)
3 Mt. Hunter 14,574 ft (4,442 m)
4 Mt. Hayes 13,832 ft (4,216 m)
5 Mt. Silverthrone 13,219 ft (4,029 m)

▽ **HIGH POINT**
Snow-clad Denali, also known as Mount McKinley, is the highest mountain peak not just in the Alaska Range but in all of North America.

ON THE DIVIDE
Sunburst Peak overlooks Cerulean Lake in the Canadian Rockies. To the left, partially hidden by clouds, Mount Assiniboine lies exactly on the Continental Divide.

THRUST FAULTING

A thrust fault is a break in Earth's crust along which deeper (usually older) layers of rock are pushed up and over shallower layers. An overthrust occurs when they are pushed hundreds of miles forward, too. After erosion, ancient rocks may sit on top of younger layers.

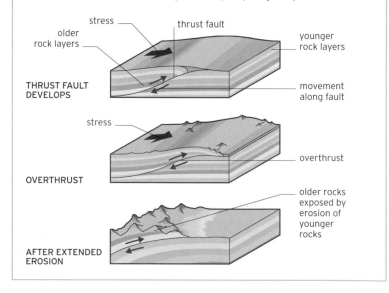

older rock layers · stress · thrust fault · younger rock layers

THRUST FAULT DEVELOPS · movement along fault

stress · overthrust

OVERTHRUST

older rocks exposed by erosion of younger rocks

AFTER EXTENDED EROSION

The Rockies boast **more than 100 summits** that are **over 10,000 ft (3,000 m)** high

The Rocky Mountains

A region of sublime natural beauty comprising the central part of one of the greatest mountain belts on Earth, the Western Cordillera of North America

W. North America

The Rocky Mountains, or Rockies, extend for almost 3,000 miles (4,800 km), running southward from the northern part of British Columbia in Canada and passing through parts of six U.S. states, from Montana to New Mexico. Located within the range is North America's Continental Divide: to the west of this divide, rivers flow into the Pacific, and to the east they flow into the Atlantic or Arctic Oceans. The highest peak in the Rockies is Mount Elbert in Colorado, at 14,440 ft (4,401 m).

Shaping the Rockies

The Rockies mainly formed in stages over the last 80 million years and have a geologically complex history. The range contains many examples of folding and thrust faulting (see panel, left), for example, in the Canadian Rockies and Montana. The Rockies also include areas, such as Yellowstone National Park (see pp.30–33), that have been extensively affected by volcanic activity, as well as regions impacted by glaciation during various ice ages.

A number of the peaks are covered in coniferous forests, with alpine tundra above the treeline. Many animals inhabit the Rocky Mountains, including elk,

moose, mountain goats and sheep, bears, wolverines, and bald eagles. With its stunning scenery, this range is a popular leisure-time destination, especially for activities such as camping, hiking, fishing, and mountain biking.

▽ **YOUNG AND RUGGED**
Pushed up in the past 13 million years, the Teton Range in Wyoming is one of the youngest parts of the Rockies. It features several peaks exceeding 12,100 ft (3,700 m) in height.

△ YOSEMITE SUNRISE
El Capitan, on the left in this image, is made of granite, while Half Dome, visible slightly right of center, is composed of granodiorite. This variation is a result of different magmas invading different areas of the existing rock.

El Capitan and Half Dome

Massive rock outcrops in Yosemite National Park, California, made almost entirely of granite

W. North America

EL CAPITAN ROCK TYPES

El Capitan is composed mainly of a 100-million-year-old igneous rock known as El Capitan granite. Taft granite, a slightly younger rock, makes up much of the rest.

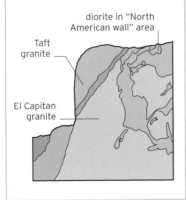

diorite in "North American wall" area

Taft granite

El Capitan granite

Tens of millions of years ago, numerous rounded masses of igneous rock, called plutons, formed underground in what is now central California as enormous globs of magma rose from deep in Earth's crust and then cooled. Following several periods of uplift and erosion, many of these plutons are now visible above the surface as massive rock cliffs and outcrops. Two of the most famous, both located in Yosemite National Park, are El Capitan and Half Dome.

Imposing cliffs

El Capitan ("the Chief") is a granitic behemoth that is around 1 mile (1.5 km) wide and some 2,950 ft (900 m) from base to summit along its tallest face. It was given its name in 1851 by a roving Spanish militia force. Equally impressive, overlooking Yosemite Valley about 5 miles (8 km) away, is Half Dome, named for its distinctive shape. One of its sides is a vertical face around 2,000 ft (600 m) high, while the other three sides are smooth and round, making it appear from some angles like a dome cut in half.

The two formations are famous for the challenge they pose to rock climbers. Once considered impossible to climb, El Capitan has now been conquered thousands of times over, and there are more than 70 different routes to the top, of varying difficulty. At Half Dome, an 8-mile (13-km) hiking route has been constructed to the top from the valley floor for non–rock climbers. This includes several hundred steps cut into the granite.

◁ COMEBACK RAPTOR
Peregrine Falcons have been nesting on El Capitan and other sites in Yosemite since 2009, following a 16-year absence.

W. North
America

The Cascade Range

A stunning array of volcanoes and other peaks in the North American West

In May 1980, a violent eruption tore apart the snow-capped peak of Mount St. Helens, the most active volcano in the Cascade Range. Before the eruption, a bulge had been growing for several weeks on the volcano's flank, due to magma being forced up inside. At 8:32 a.m. on May 18, an earthquake triggered a landslide and a sideways blast. This was soon followed by total disintegration of a part of the flank and the summit of the volcano.

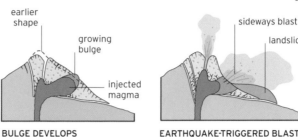

earlier shape

growing bulge

injected magma

BULGE DEVELOPS

sideways blast

landslide

EARTHQUAKE-TRIGGERED BLAST

cataclysmic blast of hot gases, ash, and pulverized rock

FLANK DISINTEGRATION

The Cascade Range is a part of the "Ring of Fire"—a horseshoe-shaped band of volcanoes and associated mountains that encloses the Pacific Ocean. Extending 680 miles (1,100 km), the Cascades formed over the last few million years as a result of the process of subduction as the Juan de Fuca oceanic plate slides underneath the North American Plate.

Volcanic hotbed

The Cascade Range includes many large and magnificent-looking volcanoes, such as Mounts Baker, Rainier, and St. Helens in Washington; Mount Hood in Oregon; and Mount Shasta in California. Several Cascade volcanoes have erupted at least once in the last 200 years, and all volcanic eruptions during this period in the contiguous United States (the whole country other than Alaska and Hawaii) have been from Cascade volcanoes.

▽ **BROODING GIANT**
Glacier-shrouded Mount Rainier is considered one of the world's most dangerous volcanoes, due to its eruption potential and its proximity to the city of Seattle.

▷ **MEGABLAST**
The 1980 eruption of Mount St. Helens is the deadliest to have occurred in the U.S. in recorded history. Most of the ash emitted settled over vast areas of the northwestern U.S.

The eruption of **Mount St. Helens** in 1980 produced an **ash plume** that **rose 15 miles (24 km)** into the air

◁ **DACITE CLIFFS**
The cliffs shown here, which rise vertically nearly 1,800 ft (550 m) above the lake and center around an edifice called Llao Rock, are made from a type of solidified lava called dacite.

Caldera rim
The rim is blanketed in a thick layer of pumice and ash from the eruption that formed the caldera.

Submerged platform
This broad mound is made of solidified lava that erupted onto the caldera floor soon after Mount Mazama's collapsed.

Underwater cinder cone
The summit of Merriam Cone is 505 ft (154 m) below the lake surface.

▶ **CRATER LAKE CALDERA**
This cross-section of Crater Lake cuts through Wizard Island. The caldera is about 6 miles (9 km) across, and the lake that fills it is famous for its exceptionally high-water clarity and deep-blue color, as well as for being one of the deepest lakes in the world, at almost 2,000 ft (600 m) deep.

Hillman peak
Rising 1,988 ft (606 m) above the lake surface, this is the highest point on the caldera rim.

Basalt layer
This is the uppermost of the rock layers on which Mount Mazama grew. It is solidified lava from an earlier volcano.

Collapsed block
Part of the floor under Mount Mazama fractured into blocks, which dropped downward.

Magma conduit
Rising magma produced volcanic features on the caldera floor. The conduit is now blocked off with solidified lava.

Breccia
The remains of the top part of Mount Mazama formed a thick layer of cemented sedimentary rock fragments called breccia.

sediments are a thousand feet thick

Magma chamber
Hydrothermal activity on the lake floor suggests that a magma chamber still exists under the caldera.

Ring fracture
This developed during the eruption that caused Mount Mazama's collapse.

◁ **WIZARD ISLAND**
This volcanic cinder cone grew up within a few hundred years of the collapse of Mount Mazama and currently reaches 750 ft (230 m) above the lake surface. It has not erupted in thousands of years.

△ **VOLCANIC REMNANT**
This oddly shaped island, protruding 165 ft (50 m) above the lake surface, is known as Phantom Ship. Made primarily from a rock called andesite, it is a remnant of a volcano that existed long before Mazama.

W. North America

Crater Lake

The collapsed crater of an ancient volcano in the Cascade Range now filled by a deep lake

water level is fairly constant at 6,178 ft (1,883 m) above sea level

Mazama rock layers
Buried around the sides of the lake are the undisturbed remains of Mount Mazama, consisting of lava and ash.

Crater Lake, in south central Oregon, is one of the world's finest examples of a caldera volcano: a stratovolcano that has partially collapsed (usually during the final phase of a catastrophic eruption), creating a gigantic, cauldronlike depression called a caldera. A caldera is sometimes described as a large crater, but it is actually a type of sinkhole, since it forms as a result of collapse rather than explosion. Over time, a caldera often fills with water, resulting in a deep lake like Crater Lake.

Collapse of a stratovolcano

The ancient stratovolcano from which Crater Lake caldera formed is called Mount Mazama. Before its top part collapsed, Mazama stood at around 12,000 ft (3,700 m), making it one of the highest peaks in the Cascade Range. During its eruption (see panel, right), a column of pumice and ash reached some 30 miles (50 km) high. In the final eruption phase, a circular system of cracks, called a ring fracture, opened around the volcano's lower slopes, causing its top to disintegrate.

Following this event, volcanic activity continued for some time, creating relatively small volcanic features on the caldera floor. Sediments and material from landslides covered the floor, and the caldera gradually filled with water. Although no significant volcanic activity has been seen at Crater Lake for thousands of years, it is possible that in the future it may erupt again.

◁ **PINNACLES**
These needlelike formations, a little distance from the lake, were once gas vents under a mass of hot volcanic pumice. The vents hardened into tubes and now stand alone because surrounding softer materials have been eroded away.

At **1,949 ft (594 m)** deep, **Crater Lake** is the **deepest lake** in the **United States**

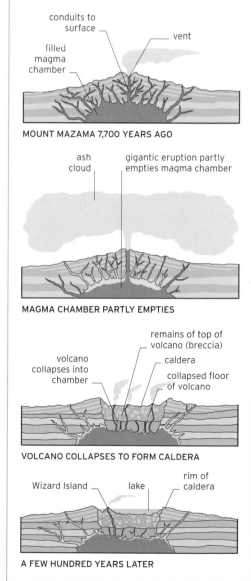

HOW CRATER LAKE FORMED

Around 7,700 years ago, a large stratovolcano called Mount Mazama underwent a cataclysmic eruption that left its magma chamber partly empty. Much of the volcano's floor and its top part collapsed downward. This left a huge, bowl-like depression—a caldera—with a raised rim. The caldera later filled with water, forming Crater Lake.

conduits to surface
vent
filled magma chamber

MOUNT MAZAMA 7,700 YEARS AGO

ash cloud
gigantic eruption partly empties magma chamber

MAGMA CHAMBER PARTLY EMPTIES

remains of top of volcano (breccia)
volcano collapses into chamber
caldera
collapsed floor of volcano

VOLCANO COLLAPSES TO FORM CALDERA

Wizard Island
lake
rim of caldera

A FEW HUNDRED YEARS LATER

W. North
America

The San Andreas Fault

Part of the boundary along which two tectonic plates grind past each other, producing frequent earthquakes

The San Andreas Fault is one of few places on Earth where the boundary between two tectonic plates can be seen running across land. Extending for approximately 810 miles (1,300 km) through California, the San Andreas is known as a continental transform fault. Along the fault, the Pacific Plate is slowly slipping past the North American Plate (see panel, below). Each slippage—which typically occurs only along a section of the fault—is accompanied by an earthquake. A large tremor that devastated San Francisco in 1906 is the most famous quake associated with movements on the fault.

MOVEMENT ALONG THE FAULT

The Pacific and North American plates are moving about 1³⁄₄ in (4.6 cm) per year in relation to each other. This motion results in jerky movement on the fault itself, with periods of little or no displacement interspersed with sudden big slippages during major earthquakes.

relative motion of Pacific Plate

San Andreas Fault

relative motion of North American Plate

△ **LINEAR SCAR**
In parts of California, the San Andreas Fault visibly bisects the landscape, as seen here in the Carrizo Plains, around 70 miles (120 km) northwest of Los Angeles.

△ **COLORPLAY**
Grand Prismatic Spring is the largest hot spring both in Yellowstone and in the U.S. Its colors come from pigments produced by heat-loving microbes in the water.

▷ **SCALDING HOT**
The temperatures of these hot springs can be close to boiling point, which means they continually give off water vapor. This turns into dramatic plumes of steam when the air temperature above the spring's surface cools.

NW. North
America

Yellowstone

*A national park in Wyoming, famous
for its spectacular geysers, hot springs,
and other geothermal features*

America's oldest national park, Yellowstone covers nearly
3,500 square miles (9,000 square km) of forest, grassland,
mountains, lakes, and canyons. It contains the greatest
concentration of geothermal features in the world,
including geysers, colorful hot springs, and mud pools.

Heat supply
Yellowstone's geothermal activity results from the fact
that part of the area rests above a huge underground
chamber of magma, which serves as a constant heat
source (see pp.32–33). The features are arranged in
distinct areas called geyser basins. Upper Geyser Basin
contains around 150 geysers, including the famous Old
Faithful, and several colorful springs. Many of the
geysers erupt more than 100 ft (30 m) into the air.
Plentiful rainfall seeps through Earth's crust and is
heated by the magma chamber, providing a continuous
hot water supply for the springs and geysers.

Geological diversity
Rugged mountain scenery, deep glacier-carved valleys,
waterfalls, canyons, and rich wildlife are among other
interesting features of Yellowstone. The Grand Canyon
of Yellowstone, as well as two waterfalls on the
Yellowstone River, constitute some of its most dramatic
attractions. Considered the world's largest intact
ecosystem in the northern temperate region, Yellowstone
is home to many large mammals such as American
bison, wolves, black and grizzly bears, and moose.

Yellowstone has more than
300 geysers, almost **one-third**
of all the geysers in the world

◁ **BLOWHOLE**
Every 10-12 hours, Castle
Geyser throws up a
fountain of hot water
nearly 90 ft (27 m) high
for roughly 20 minutes.
It is thought to have
been doing this for about
1,000 years.

Underneath a SUPERVOLCANO

There are a handful of volcanic sites in the world known as supervolcanoes: sites that have seen at least one eruption rating an 8 on a scale called the Volcanic Explosivity Index. These are the largest eruptions that geologists know about; intervals between them tend to be measured in tens or hundreds of thousands of years. These sites are also capable of future eruptions that could radically alter landscapes and severely impact the world's climate. All supervolcanoes are calderas that sit above large, active magma chambers. Yellowstone National Park lies over a supervolcano that last erupted about 640,000 years ago. There are two large, interconnected magma chambers under Yellowstone caldera. The heat from the upper chamber drives geothermal features: hot springs, geysers, fumaroles, and boiling mud pools.

△ COLORFUL POOLS
The colors in hot spring pools and around their edges come from pigments produced by heat-loving microorganisms or from minerals.

Grand Prismatic spring

▷ HOT SPRING TERRACES
Hot springs can create terraces of travertine (a form of limestone) as seen at Mammoth Hot Springs, just outside the caldera boundary.

Old Faithful geyser

center of caldera is about 6 miles (10 km) east of Old Faithful geyser

rim of Yellowstone caldera

Teton Range

▽ MUDPOT
Acid hot springs, such as Sulfur Caldron, dissolve nearby ground into a pool of boiling, sulfurous-smelling mud, called a mudpot.

Yellowstone Lake

the Absaroka Range is one of several mountain ranges of the Rocky Mountains that surround the caldera

Upper crustal magma reservoir
This contains hot, mostly spongelike solid rock and about 9 percent magma at a temperature of around 1,400°F (760°C).

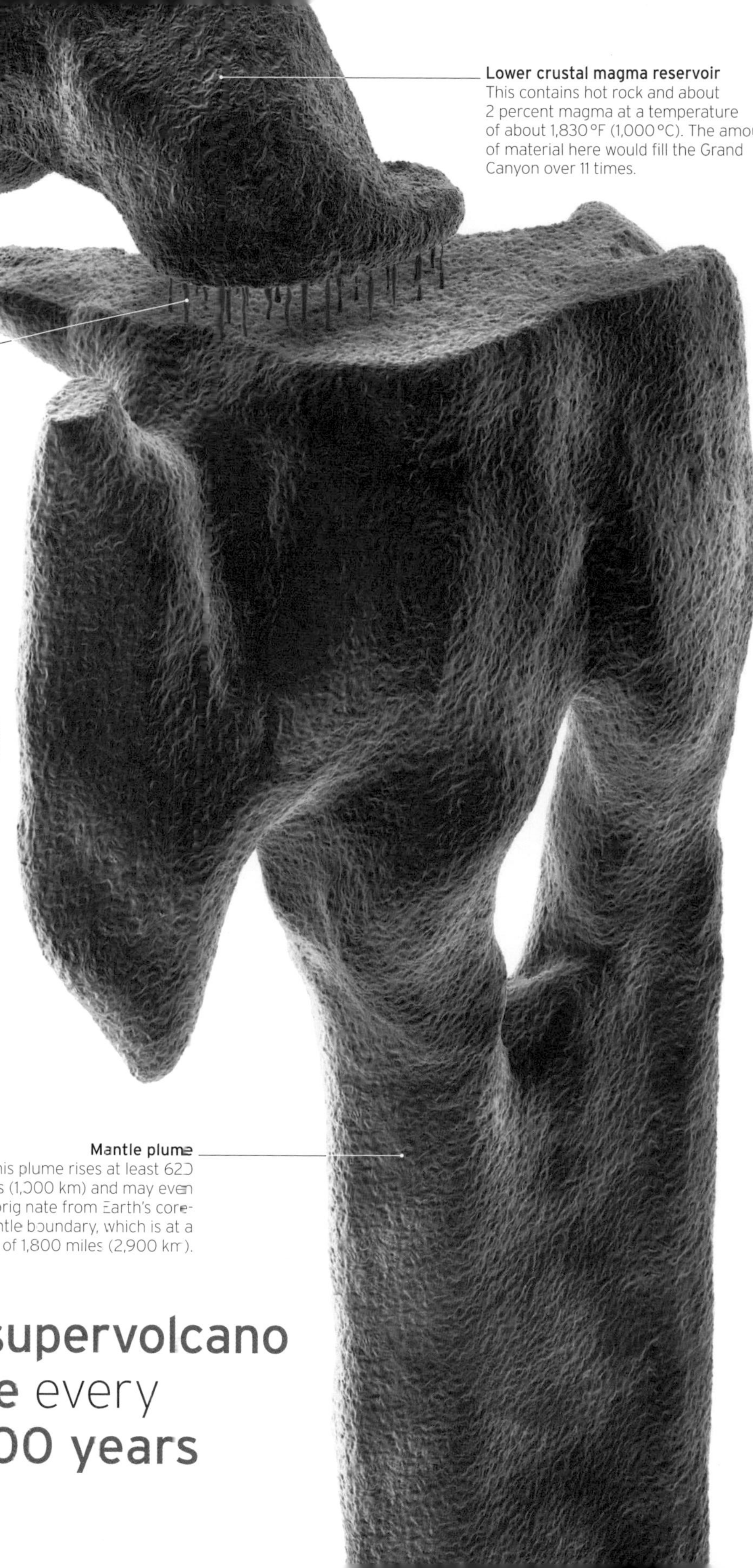

Lower crustal magma reservoir
This contains hot rock and about 2 percent magma at a temperature of about 1,830°F (1,000°C). The amount of material here would fill the Grand Canyon over 11 times.

▶ **YELLOWSTONE CALDERA**
The heat source for the Yellowstone Caldera is a mantle plume—a mass of hot, semisolid rock moving slowly upward in the mantle. This supplies heat and material to the magma reservoirs in the crust under Yellowstone.

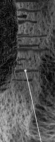

Dikes or vertical cracks
The upward movement of hot rock or magma occurs through dikes.

Mantle plume
This plume rises at least 620 miles (1,000 km) and may even originate from Earth's core-mantle boundary, which is at a depth of 1,800 miles (2,900 km).

△ **FUMAROLES**
Around 4,000 of these holes in the ground occur within the boundaries of the Yellowstone Caldera. They release steam and other gases—some of them foul-smelling, such as hydrogen sulfide.

The Yellowstone supervolcano erupts about once every 600,000–800,000 years

EVIDENCE OF PAST VOLCANISM

The North American Plate is moving to the southwest at the rate of about 14 miles (22 km) per million years, whereas the mantle plume currently under Yellowstone stays in approximately the same place. As a result, traces of past eruptions and calderas stretch for hundreds of kilometres to the southwest of Yellowstone.

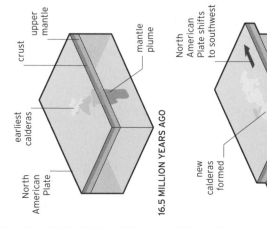

crust
upper mantle
mantle plume
earliest calderas
North American Plate

16.5 MILLION YEARS AGO

North American Plate shifts to southwest
new calderas formed

8 MILLION YEARS AGO

North American Plate moves farther to southwest
Henry's Fork Caldera (close to Yellowstone Caldera)

1.3 MILLION YEARS AGO

Ship Rock

A twin-peaked volcanic plug with spectacular radiating dikes, which stands high above the deserts of New Mexico

SW. North America

The most famous volcanic plug in the American Southwest, Ship Rock towers 1,585 ft (483 m) above a high-desert plain in the Navajo Nation—a Native American territory in Arizona and New Mexico. Also called a volcanic neck, this type of feature originates when a mass of erosion-resistant rock forms inside a volcano after it has stopped erupting (see panel, right).

Rock ribs

Regarded as a sacred monument by the Navajo people, Ship Rock's Navajo name is *Tsé Bit'a'í*, which means "winged rock." This probably alludes to the presence of prominent walls of volcanic rock, called dikes, that radiate from the monolith. The three most prominent of these rock ribs run at roughly 120 degrees to each other. Seen from certain perspectives, the whole monument can have a birdlike appearance.

HOW SHIP ROCK FORMED

Ship Rock is composed of a hard igneous rock called lamprophyre that solidified in the throat of an ancient volcano about 30 million years ago. The accompanying radiating dikes formed around the same time. The rest of the volcano, which consisted of softer rock, has since eroded away.

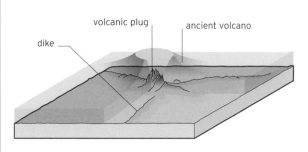

volcanic plug   ancient volcano

dike

△ **KNIFE EDGE**
Parts of Ship Rock's dikes have a knifelike form and tower at a height of 65 ft (20 m) above the high-desert plain.

▽ **NAVAJO MONUMENT**
Ship Rock and three of its dikes can be seen here. According to a Navajo legend, birdlike monsters once nested on the rock's high peaks, feasting on human flesh.

The **longest** of **Ship Rock's dikes** stretches for nearly **2 miles (3 km)**

W. North
America

Devil's Tower

A 40-million-year-old massive domed monolith in Wyoming with columnar fluting on its sides

Devil's Tower is one of the most iconic U.S. landmarks. Often referred to as a butte (an isolated hill with vertical sides and a small, flat top), its exact origins are debated—in particular, whether or not it formed inside a volcano. However, experts agree that it started off as a mass of magma that cooled and solidified underground. In the process, remarkably uniform polygonal columns formed in the cooling body, and this fluting is still visible today.

Regarded as a sacred spot by Native Americans, in 1906, it became the first site in the U.S. to be declared a National Monument. It is also famous as the location for a climactic meeting between humans and visitors from space in the 1977 film *Close Encounters of the Third Kind*.

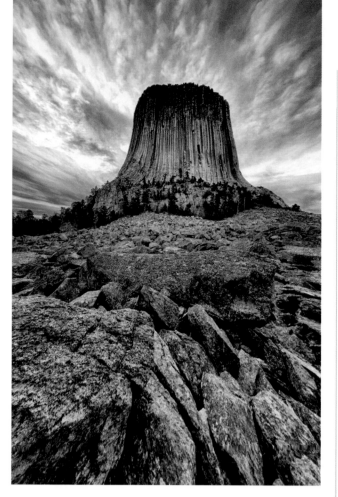

△ **STANDING TALL**
Devil's Tower rises 866 ft (264 m) from its rocky base to its relatively flat top, which is about the size of a football field.

HOW DEVIL'S TOWER FORMED

One theory about the formation of Devil's Tower suggests that a volcano formed, and later, some magma within the volcano solidified into a hard mass. This was exposed as the tower when surrounding rock layers were eroded away.

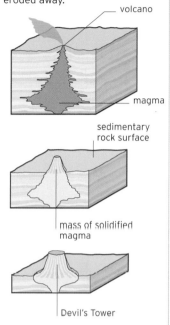

volcano

magma

sedimentary rock surface

mass of solidified magma

Devil's Tower

The Appalachians

A complex and highly scenic series of mountain ranges, stretching from Alabama to southeastern Canada

E. North
America

The Appalachians are the eroded remnants of an ancient mountain belt that formed about 500 million years ago through collisions between ancient continents. Over the last 65 million years, the whole belt has been uplifted and eroded to what is seen today—a vast area of gently rounded, mostly forested hills and mountains. It consists of many parallel provinces (see panel, below), including the Blue Ridge Mountains, noted for a bluish haze when seen from a distance. The color is in part due to a chemical released into the air by trees in the region.

▽ **FALL PALETTE**
In fall, foliage creates an explosion of color across the Appalachians, as seen here in the White Mountains range in the northern Appalachians.

INSIDE THE APPALACHIANS

This west-to-east cross-section through the central part of the Appalachians shows folding, faults, and ridges. It also shows the provinces into which the whole belt is split.

Appalachian plateaus | Valley and Ridge province | Blue Ridge Mountains | Piedmont | Atlantic Ocean

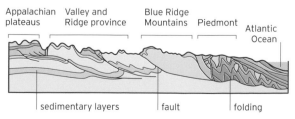

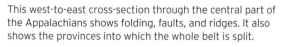

sedimentary layers | fault | folding

The Cave of Crystals

A horseshoe-shaped underground cavern in Mexico containing some of the most enormous natural crystals ever discovered

S. North America

The Cave of Crystals lies 980 ft (300 m) below a mountain at Naica in northern Mexico. In April 2000, workers at the Naica Mine were excavating a new tunnel when they discovered the Cave of Crystals. The chamber is filled with gigantic, beamlike crystals of the mineral selenite.

Birth of giants

The cave, which has been eroded out of the limestone that forms much of Naica Mountain, has existed for several thousand years. At some point after its formation, when it was flooded with water, mineral-rich groundwater rising from below came into contact with cooler, oxygen-rich surface water in the vicinity of the cave. The oxygen diffused into the groundwater, causing

The cave's **largest crystal** is **39 ft (12 m) long** and weighs **55 tons (50 tonnes)**

the selenite to be deposited in the form of crystals. This continued at an extremely slow rate for at least 500,000 years, forming the gigantic crystals that exist today.

Terra incognita

The cave is relatively unexplored because of its depth and the continued presence of heated water below the surface. Air temperatures can reach 136°F (58°C), and at present, the cave lacks public access.

△ **GYPSUM TROVE**
The crystal reserves found within the Cave of Crystals are made of selenite. A variety of gypsum, selenite is a soft, light-colored mineral containing calcium sulfate and water.

◁ **CRYSTAL FOREST**
The Cave of Crystals in Mexico houses some of the largest natural crystals ever found. Here, massive beams of selenite dwarf explorers in the cavern.

LOCATION OF THE CAVE OF CRYSTALS

Naica Mountain has several caverns, including the Cave of Crystals. Some of these can be explored only because the Naica Mine Company pumps water out of the ground to keep the water level down.

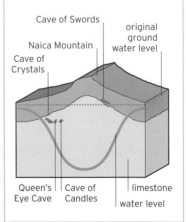

Cave of Swords

original ground water level

Naica Mountain

Cave of Crystals

Queen's Eye Cave

Cave of Candles

limestone

water level

The Sierra Madre

A majestic system of mountain ranges extending diagonally across Mexico and forming a part of the North American Cordillera

S. North America

HIGHEST PEAKS IN THE SIERRA MADRE

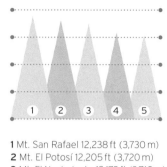

1 2 3 4 5

1 Mt. San Rafael 12,238 ft (3,730 m)
2 Mt. El Potosí 12,205 ft (3,720 m)
3 Mt. El Nacimiento 12,172 ft (3,710 m)
4 Sierra de La Marta 12,156 ft (3,705 m)
5 Mt. Teotepec 11,647 ft (3,550 m)

Mexico's Sierra Madre has three main parts—the Sierra Madre Occidental (in northwest Mexico), the Sierra Madre Oriental (in the northeast), and the Sierra Madre del Sur (in the south). Between the Occidental and Oriental ranges lies the Mexican Plateau. This consists of layers of sedimentary rock, laid down in shallow seas 250–65 million years ago, that have subsequently been eroded into an angular, gouged landscape in the Sierra Madre Occidental and a more rounded terrain in the Oriental. Parts of the range also show signs of past volcanism in the form of igneous rock intrusions and layers of hardened lava poured out by long-disappeared volcanoes.

Labyrinthine landscape

The western side of the Sierra Madre Occidental is a series of high escarpments (steep slopes that form due to faulting or erosion) whose sides drop into deep canyons called *barrancas*. The dimensions of some of these, notably the Copper Canyon, or Barrancas del Cobre, are comparable to those of the Grand Canyon (see pp.54–57) in the U.S.

The Sierra Madre Oriental is cut through by many steep-walled narrow valleys. Many of these are aligned in a north–south direction, but there are also several passages through these mountains that run from the lowlands of the Gulf of Mexico to the east. The Sierra Madre del Sur consists of a labyrinth of narrow ridges and steep-flanked valleys.

◁ **GREAT DESCENT**
A part of the Copper Canyon of the Sierra Madre Occidental, Urique Canyon is the deepest canyon in North America, descending 6,135 ft (1,870 m) from its highest point.

Popocatépetl

North America's second highest volcano and the second highest peak in Mexico, with a history of violent eruptions

S. North America

Towering to a height of 18,491 ft (5,636 m), Popocatépetl, which means "smoking mountain" in the ancient Aztec language, is famous because of the ferocity of its eruptions, its tongue-twisting name, and its proximity to Greater Mexico City, the largest metropolitan area in the Americas.

Violent nature

A stratovolcano, Popocatépetl, or "El Popo," can be seen erupting from most parts of Mexico City. Over the past few thousand years, it has had three Plinian eruptions—extremely violent blasts that produce huge plumes of gas and ash. Such explosions can cause large loss of life in populated areas. After a massive eruption in 1947, the volcano lay dormant until it came back to life in 1994. Ever since, smoke has been seen emanating almost daily from its huge summit crater. The year 2000 saw the most dramatic eruption in more than a thousand years. In April 2016, it erupted again, spewing out lava, ash, and rock, and it has continued to produce powerful explosions at irregular intervals.

▽ **HIGH CONE**
Popocatépetl's height, symmetrical shape, and snow-covered summit give it an imposing presence.

△ **MELTWATER STREAMS**
In summer, meltwater streams appear all over
the ice sheet. An estimated 60 cubic miles
(250 cubic km) of ice is currently being
lost from the ice sheet each year

N. North
America

The Greenland Ice Sheet

The largest glacier in Earth's Northern Hemisphere, covering 80 percent of the island of Greenland and holding a significant fraction of the world's freshwater

The Greenland ice sheet is the world's second largest ice sheet (and second largest glacier) after the Antarctic Ice sheet (see pp. 306–307), covering 660,000 square miles (1.71 million square km) of Greenland. It has an estimated volume of 680,000 cubic miles (2.85 million cubic km), with an average thickness of about 5,480 ft (1,670 m) and a maximum thickness of 10,515 ft (3,205 m). Some of its ice is thought to be up to 110,000 years old.

Shape and movement

The ice sheet's surface is slightly domed, reaching a height of about 10,800 ft (3,290 m) above sea level at the top of the dome. From here, ice moves slowly down toward the ice sheet's edges, where it is mostly constrained by coastal mountains; it only reaches the sea along a broad front in a few places. As a result, Greenland (unlike Antarctica) has no ice-shelves. In many

places, the ice flows through gaps in the mountains in the form of outlet glaciers. When these reach the coast, they discharge enormous numbers of icebergs into the sea. One outlet glacier, the Jakobshavn Glacier in western Greenland, is the world's fastest flowing glacier. At its sea end, the ice flows at about 3 ft (1 m) per hour.

Threats to sea level and freshwater

The Greenland ice sheet is melting due to global warming. If it melted completely, the world's oceans would rise by about 24 ft (7.2 m), drowning many of the world's major cities and leading to loss of about 6.5 percent of the world's freshwater reserves.

◁ **ARCTIC PREDATOR**
Although rarely venturing onto the ice sheet itself, polar bears are active around its edges on the Greenland coast, where they hunt and feed on seals.

△ **ICEBERG MAKER**
A helicopter flies past the end of Graah Glacier, an outlet glacier that flows out of the Greenland ice sheet. From this wall of ice, huge icebergs break off, or calve, from time to time.

◁ **WASHED OUT**
From the snout of Russell Glacier in central western Greenland emerges a river of meltwater, skirting rock debris that has been carried there by the ice.

ICE MOVEMENT AND ICE DIVIDES

Although ice covers most of Greenland, there is an ice-free zone of variable width around much of the island's coast. Snow falling on the ice sheet compacts to ice, which then slowly flows down in the direction of the coast along the flow lines shown here. Imaginary lines called ice divides separate regions of ice moving toward different coasts.

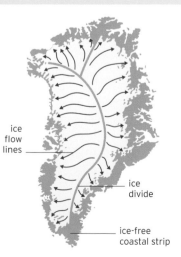

ice flow lines

ice divide

ice-free coastal strip

The **Jakobshavn Glacier** produces about **99 million tons (90 million tonnes)** of **icebergs** every day

The Columbia Glacier

An enormous Alaskan glacier that moves quickly but is also rapidly shortening in length

NW. North America

The Columbia Glacier is one of Alaska's largest glaciers, with an area of about 360 square miles (920 square km) and a length of 25 miles (40 km). It descends from the Chugach Mountains in southern Alaska to Prince William Sound (an inlet of the Gulf of Alaska), where it calves large numbers of icebergs into the sea.

Shrinking record

The Columbia Glacier is one of the fastest moving in North America, but its fast forward movement is offset by extremely rapid calving of icebergs at its terminus, at the rate of some 14 million tons (13 million tonnes) every day. As a result, since the 1960s, the glacier's terminus has moved backward, and since 2011 it has become detached from some of its largest tributary glaciers. The retreat cannot be attributed solely to global warming, as other nearby glaciers are not shrinking at the same pace. Instead, experts believe it is connected to the shape of the bedrock channel under the glacier.

THE RETREAT OF THE COLUMBIA GLACIER

Since the 1960s, the iceberg-calving front of the Columbia Glacier has retreated about 12 miles (20 km). Marked here are its positions at roughly 15-year intervals since 1969.

Chugach Mountains   Columbia Glacier

2016
2001
1985
1969

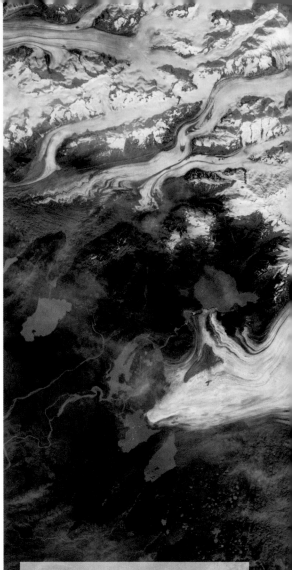

SPRAWLING GIANT
This satellite view shows the Bagley ice field with its linked arms at top right, the glacier's main descending trunk at center right, and the terminal lobe at the bottom.

◁ **RETREATING TERMINUS**
About two-thirds of the area of the glacier visible in this photograph, which was taken in the mid-2000s, is now open water.

▽ **MILITARY FORMATION**
The higher part of the glacier is criss-crossed by crevasses, producing an army of ice pinnacles on its surface.

△ **ICE SHARDS**
Massive icebergs regularly break off the end of the glacier and may then float around for months in Vitus Lake.

The Bering Glacier

The longest glacier in North America—and the largest by surface area—which terminates in a large lake and has thick vegetation growing on part of its surface

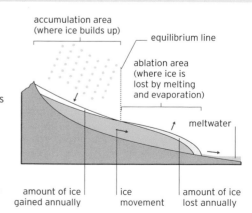

NW. North America

The Bering Glacier is around 111 miles (178 km) long and flows down from Alaska's Chugach Mountains toward the Gulf of Alaska. In parts, the ice is about 2,600 feet (800 m) thick. The glacier has two very different main parts. The upper part, known as the Bagley ice field, lies at an altitude of $^3/_4$–$1^1/_4$ miles (1.1–2 km) and is a vast basin of clean ice formed from compaction of fallen snow. This ice field is about 56 miles (90 km) long and completely fills several interconnecting valleys. The lower part is a broad trunk of ice ending in a 26-mile- (42-km-) wide, roughly oval terminal lobe, which is partly covered in rock debris and vegetation. A large lake of meltwater, Vitus Lake, fringes the glacier's terminus. From this lake and the edge of the terminal lobe, streams flow a short distance to the coast, which is on the Gulf of Alaska.

△ **GLACIER HERB**
Some areas on the lower part of the glacier, consisting of stagnating ice, have a variety of plants growing on them, including this herb called mistmaiden.

ICE GAINS AND LOSSES

Ice from the glacier descends to the ablation area where it is lost by melting and evaporation. The equilibrium line marks the ice gain and ice loss balance. The Bering Glacier's accumulation area is a vast ice field at an altitude of $^3/_4$ mile (1.1 km), while the ablation area comprises much of the rest of the glacier.

accumulation area (where ice builds up)

equilibrium line

ablation area (where ice is lost by melting and evaporation)

meltwater

amount of ice gained annually

ice movement

amount of ice lost annually

The Kaskawulsh Glacier

A vast glacier in Canada that is notable for its sheer size and impressive medial moraines

NW. North America

Around 400 square miles (1,000 square km) in area, the Kaskawulsh Glacier snakes for some 45 miles (70 km) through the St. Elias Mountains of Canada's Yukon Territory. At its terminus, or snout, it produces meltwater that maintains the level of one of the Yukon's largest lakes, Kluane Lake.

Bold stripes

All glaciers constrained by valleys have dark bands of material running along their sides, called lateral moraines (see panel, p.150). These contain rock fragments that have been plucked off the walls of the valley and are being carried along by the glacier. When two glaciers merge to form a larger glacier, the lateral moraines of the smaller glaciers typically combine to form a bold dark stripe running down the merged glacier. This stripe is called a medial moraine. The lower half of the Kaskawulsh Glacier has several bold medial moraines, formed through a series of minimergers of its different arms, or tributaries, upstream. The glacier has three main arms, each with its own tributaries.

The Kaskawulsh's **broadest medial moraine** is over **1,300 ft (400 m) wide**

▷ **MERGER OF EQUALS**
In this view, different arms of the glacier can be seen to merge in the distance. The merger produces the widest medial moraine (dark band) on the glacier's downstream surface.

The Malaspina Glacier

The world's largest and finest example of a piedmont glacier— a huge lobe of ice spread out over a broad, flat plain

NW. North America

PIEDMONT GLACIERS

A piedmont glacier occurs where the flow of ice within a valley spills out through a narrow opening onto a broad plain, often near a coast. The ice spreads out to form a wide lobe.

valley glacier / narrow outlet / broad, flat plain

wide lobe

The enormous Malaspina Glacier is up to 40 miles (65 km) wide and extends up to 28 miles (45 km) forward from where its ice "root" emerges from between a gap in some mountains near the coast of Alaska. Studies of earthquake waves passing through it have shown that its ice is up to 2,000 ft (600 m) thick.

Patterned lobe

The glacier has complex patterns of light and dark bands on its surface—the dark areas are moraines, containing rock debris. The bands are arranged in zigzags and swirls, and are thought to have been caused by alternating, perhaps seasonal, patterns of fast and slow ice flow. Also on the surface are colonies of tiny creatures called ice worms.

△ **ICE SWIRL**
The glacier, with its vast swirling ice patterns, is named after an Italian explorer, Alessandro Malaspina, who visited it in 1791.

The Black Rapids Glacier

An unusual glacier, the largest in the Alaska Range, with a history of phases of exceptionally rapid advance

NW. North America

The Black Rapids Glacier flows down a valley in south central Alaska and is one of the world's more accessible glaciers, because its snout lies only a few miles from a road. It is about 25 miles (40 km) long and has several tributaries. In 2002, an earthquake caused a huge part of it to be buried in rock debris.

The Black Rapids is a surge-type glacier, in which periodic rapid transfers of ice occur from the upper to the lower parts, which increase in thickness and then surge forward. The Black Rapids Glacier made national news in the U.S. for three months in 1937, when its ice suddenly began moving at 100 ft (30 m) a day—more than 20 times its normal rate—and its snout began to rapidly advance down its valley. During this period, it became known as the Galloping Glacier. Since this surge, however, the Black Rapids has steadily retreated to its current length, which is at a historic minimum.

▽ **RARE FIN**
An ice fin—caused by pressure squeezing ice upward as a glacier grinds against hard bedrock—is a rare glacial feature. This fin was sighted on the Black Rapids in 2013.

◁ **SURFACE MELT**
Surface pools, and even streams, can develop on the higher parts of the Black Rapids when meltwater forms faster than it can be absorbed by the glacier.

NW. North America

The Kennicott Glacier

A spectacular and accessible glacier in the Wrangell– St. Elias National Park of south central Alaska

The Kennicott Glacier flows down from the southern Wrangell Mountains, its principal source of ice being the flanks of Mount Blackburn—a snow-clad, eroded, extinct volcano and the fifth highest peak in the U.S. The glacier is about 26 miles (42 km) long.

Arms and tributaries

The upper part of the Kennicott Glacier has two arms, west and east, that flow around a ridge of rock and a nunatak (a small rocky peak poking through the ice) known as Packsaddle Island. After these two arms have joined, the impressively striped main trunk of the glacier flows south, where it merges with two large tributaries—the Gates and Root Glaciers—and later passes an abandoned mining community called Kennecott. It ends in a broad, rock-covered snout, surrounded by shallow meltwater pools and lakes. Several streams emerge from underneath the glacier and merge to form the Kennicott River.

The Kennicott Glacier and Root Glacier are accessible from a road built to supply the old mining camp. Guided hikes onto the glaciers provide a chance to view the ice formations, as well as vivid blue pools and streams on the glaciers' surfaces, or to explore the ice caves underneath them. Crossing the Root Glacier gives access to Donoho Peak. A strenuous scramble up this can give brilliant views of the whole area.

▷ **SUBGLACIAL CAVE**
Under the glacier's edge is a world of glowing blue ice, dripping water, and light seeping through cracks. These caves can be safely visited only at certain times of year due to seasonal flood risks.

▶ **KENNICOTT GLACIER**
Kennicott is a typical valley glacier— one hemmed in by surrounding mountains. It covers an area of about 100 square miles (250 square km). The full length of the glacier is shown here, flowing down from Mount Blackburn to its terminus at the head of the Kennicott River.

The glacier has **lost** about **245 ft (75 m)** of its **thickness** since the 1960s

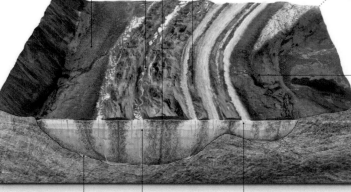

surface stream of flowing meltwater

surface debris made up of rocks that have fallen on to the glacier

Englacial debris
This consists of rocks and dust carried inside the glacier.

Medial moraine
A line of debris is formed by the union of two lateral moraines when glaciers merge.

Lateral moraine
Rock debris is dragged along and deposited at a glacier's sides.

Water channel
Channels carry meltwater along the glacier's base.

moulin (vertical water channel)

Crevasses
These are deep, wedge-shaped clefts in the ice surface.

△ **ANATOMY OF A VALLEY GLACIER**
A typical valley glacier like the Kennicott is hundreds of feet deep at its center and contains rock debris on its surface, inside it, and dragged along at its sides and base. The ice is not solid but fractured by crevasses and carved by meltwater channels.

Mount Blackburn
The highest peak in the Wrangell Mountains, Mount Blackburn reaches 16,390 ft (4,996 m).

Kennicott Glacier West Upper Arm

Kennicott Glacier East Upper Arm

Packsaddle Island (nunatak)

The Gates Glacier
A tributary of the Kennicott Glacier, the Gates Glacier is 8 miles (13 km) long.

△ **WINDING TRAILS**
This view of a middle section of the Kennicott Glacier shows its well-developed medial moraines (dark stripes containing rock debris). Some of these are more than 1,150 ft (350 m) wide.

Donoho Peak provides excellent views from its summit at 6,696 ft (2,041 m)

The Root Glacier
This 14-mile- (22-km-) long tributary consists of relatively clean, rock-free ice.

◁ **MELTWATER CHANNEL**
The surface of both the Kennicott Glacier and its tributary, the Root Glacier, feature numerous turquoise meltwater channels that snake across the surface.

Kennecott (old mining camp)

Meltwater pools
The glacier's terminus, or snout, which consists of a gray-colored slush, is surrounded by meltwater pools.

Snout edge
Over the past several decades, the snout edge has retreated markedly, leaving mounds of moraine behind.

Kennicott River

▷ **BLUEST BLUE**
Some meltwater pools on the glacier's surface are more than 20 ft (6 m) deep and an exceptionally vivid blue. The blueness is due to the depth and purity of the water, which absorbs all other light wavelengths.

NW. North
America

The Mendenhall Glacier

A partially hollow glacier close to the Alaskan city of Juneau, best known for its extraordinary blue ice caves

The Mendenhall is one of 38 large glaciers that flow down off the Juneau Ice-field. The glacier is about 13 miles (21 km) long and ends in an iceberg-filled lake. It has been retreating and shrinking since at least the 1700s, with some phases of rapid retreat in recent years, and it is expected that it will recede out of the lake by about 2020.

Exploring the glacier

By paddling a canoe across the meltwater lake, it is possible to reach the side of the glacier's terminus, clamber across its surface, and then explore its ice caves, with their deep blue ceilings and streams of water running over the rocks inside. Wildlife in the area includes beavers, bears, spawning salmon in season, and the occasional wolverine (a carnivore resembling a small bear).

▷ **BLUE DESCENT**
An ice climber descends through a moulin into one of the large ice caves in the glacier.

◁ **ICE PROWLER**
Occasionally, a wolverine is spotted in the area. It is thought that wolverines use glaciers for temporarily storing killed prey.

MELTWATER DRAINAGE

The lower parts of many glaciers have extensive systems for draining meltwater. These include vertical channels, called moulins, and streams at the base. Changes in the drainage pattern can cause outlet channels to dry up, leaving them as internal ice caves, as in the case of Mendenhall.

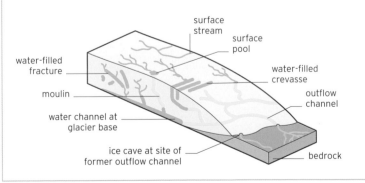

surface stream
surface pool
water-filled fracture
water-filled crevasse
moulin
outflow channel
water channel at glacier base
ice cave at site of former outflow channel
bedrock

NW. North America

The Margerie Glacier

A glacier with a spectacular terminal wall that drops blue icebergs into Glacier Bay

The Margerie Glacier originates from the eastern end of the St. Elias Mountains, in southeastern Alaska, and discharges icebergs into Glacier Bay (an inlet of the Gulf of Alaska). Glaciers that flow down to the sea are called tidewater glaciers. The Margerie Glacier's terminal wall is in Tarr Inlet, at the northern end of the Bay, and is 260 ft (80 m) high. Unlike most other Alaskan glaciers, it has not retreated in recent years.

History of the bay

Glacier Bay was entirely filled with a single massive glacier when the English naval officer Captain George Vancouver discovered it in 1794. But due to higher temperatures and lower average snowfall over the last few centuries, this has been transformed into a 60-mile- (100-km-) long fiord with many smaller glaciers feeding ice and meltwater into it, including the Margerie.

◁ **FALLING APART**
A huge quantity of ice drops from the 1-mile- (1.5-km-) wide terminus of Margerie Glacier into Glacier Bay. This occurs several times each day, accompanied by thunderous cracking noises.

ICEBERG CALVING

Glaciers that flow to the coast extend only a short distance over the sea. The stress created by the glacier's weight causes chunks to break off and float away as icebergs. This is called iceberg calving.

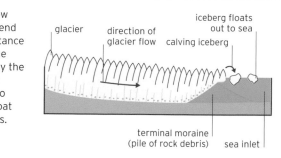

glacier

direction of glacier flow

calving iceberg

iceberg floats out to sea

terminal moraine (pile of rock debris)

sea inlet

W. North
America

The Yukon

One of North America's longest watercourses, meandering through some of the continent's last wild frontiers

The Yukon River is 1,980 miles (3,185 km) long, making it the third longest river in North America after the Missouri and Mississippi. Eight major rivers flow into the Yukon, and its tributaries drain an area of about 328,000 square miles (850,000 square km)—the fourth largest drainage basin in North America.

Epic journey

From its mountain source in British Columbia, the Yukon flows slowly northwest through the lowland forests of the Yukon Territory before entering Alaska, where it widens into a vast wetland area called the Yukon Flats. From here, its lower course turns southwest to wind across the entire breadth of the relatively flat central Alaska landscape to its mouth in the Bering Sea.

The vast scale of the Yukon and the remoteness of the landscapes that it crosses have led to the romantic mythology of frontier life associated with its name. Once the principal means of transportation for the pioneers of the Klondike Gold Rush at the turn of the 19th century, the river remains an epic watercourse for those with an adventurous spirit.

▷ **MIGRATION ROUTE**
In early May, large flocks of sandhill cranes can be seen at various points along the Yukon, migrating to nesting grounds in western Alaska and northwestern Siberia.

Despite its great length, **only four bridges** cross the Yukon River

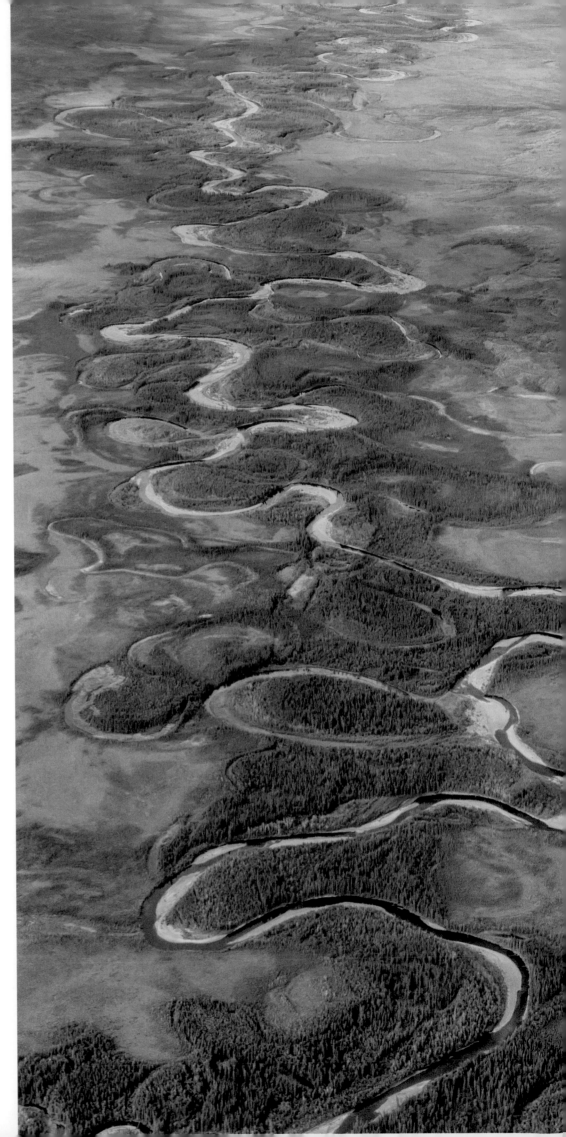

△ HAZARDOUS PASSAGE
One of the best-known landmarks on the Yukon River, the Five Finger Rapids is a point at which the river splits into channels around four large rock towers. It presented a hazardous obstacle to early gold prospectors.

◁ WINDING COURSE
For much of its lower course, the Yukon is a slow, winding river with many large meanders and oxbow lakes scattered throughout its floodplain.

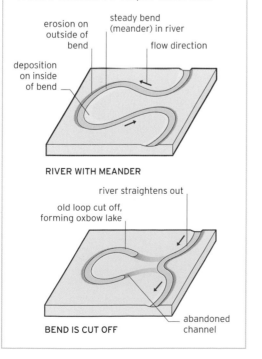

HOW AN OXBOW LAKE FORMS

Erosion and deposition make the shape of a river meander more and more accentuated over time. Eventually, the river cuts its course across the neck of the loop and follows a more direct route. The abandoned section forms a distinctive U-shaped oxbow lake.

erosion on outside of bend
steady bend (meander) in river
flow direction
deposition on inside of bend

RIVER WITH MEANDER

river straightens out
old loop cut off, forming oxbow lake
abandoned channel

BEND IS CUT OFF

W. North America

Lake Abraham

A glacial lake with unusual natural sculptures captured within its frozen surface during winter

Lake Abraham is an artificial lake created by the damming of the North Saskatchewan River in northern Alberta, Canada, in 1972. It stretches 20 miles (32 km), oriented north to south along the river valley, and is 2 miles (3.3 km) across at its widest.

What lies beneath

For much of the year, Lake Abraham displays the turquoise waters seen in most glacial lakes of the Canadian Rockies. However, during winter, beneath its clear, icy surface, bizarre frozen stacks of cotton-wool-like bubbles appear. These amazing formations actually contain highly flammable methane gas produced as bacteria at the bottom of the lake decompose dead organic material. In summer, the methane simply rises to the surface and escapes, but in winter the gas is trapped—and held on dazzling display—until the spring thaw.

△ BUBBLE STACKS
Bubbles of methane gas rising from the bottom of the lake are trapped in ice as they reach the colder surface, with successive bubbles freezing in stacks.

MAKING WAVES
Although unaffected by tides, the Great Lakes can take on sealike features. Sustained winds may produce strong currents and huge waves, like these in Lake Erie.

ELEVATIONS AND DEPTHS

This profile shows the water surface elevations and depths of the Great Lakes. Four of the lakes are at varied elevations, producing the effect of a stepped waterway flowing from west to east. Lake Superior, the deepest lake, holds more water than the other four lakes combined.

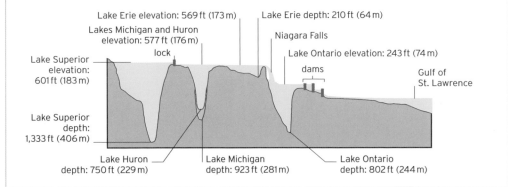

Lake Erie elevation: 569 ft (173 m)

Lakes Michigan and Huron elevation: 577 ft (176 m)

lock

Lake Superior elevation: 601 ft (183 m)

Lake Superior depth: 1,333 ft (406 m)

Lake Erie depth: 210 ft (64 m)

Niagara Falls

Lake Ontario elevation: 243 ft (74 m)

dams

Gulf of St. Lawrence

Lake Huron depth: 750 ft (229 m)

Lake Michigan depth: 923 ft (281 m)

Lake Ontario depth: 802 ft (244 m)

△ **WINTER FREEZE**
During winter, large areas of the Great Lakes freeze over, causing much disruption to shipping. Here, on Lake Huron, distinctively shaped pancake ice has formed.

E. North
America

The Great Lakes

*Earth's largest system of freshwater
lakes, often referred to as inland seas*

North America's Great Lakes have a combined surface
area of 94,000 square miles (244,000 square km)—
exceeding that of the UK—and contain one-fifth of
the world's fresh surface water. The system comprises
five interconnected lakes—Superior, Michigan, Huron,
Erie, and Ontario—draining from west to east and
emptying into the Atlantic Ocean via the St. Lawrence
River. Although the gradient between the lakes is
generally shallow, the descent between Lakes Erie and
Ontario features the dramatic Niagara Falls (see p.52).

Diverse shorelines

The Great Lakes began to form at the end of the last
ice age, when moving ice sheets gouged huge basins
that subsequently filled with water as the ice melted.
The current shapes of the lakes started to become
established about 10,000 years ago. Their shorelines
vary considerably, from sandy beaches to rocky
cliffs, and even wetlands. Also located on their
shores are several major cities including
Toronto, Chicago, Cleveland, Buffalo,
and Detroit.

◁ **ONE OF
THOUSANDS**
The lakes contain
35,000 islands, ranging
in size from 10 feet
wide—as shown here
in Lake Superior—to
several thousand square
miles in area.

▷ **LARGEST NATIVE**
The lake sturgeon is the
largest of more than
150 native fish species
in the Great Lakes,
growing to about 6 ft
(2 m) in length and
weighing 200 lb
(90 kg) or more.

rows of
protective
bony
plates
(scutes)

barbels used to sense
bottom-dwelling prey

shovel-like nose

The **inland** U.S. state of **Michigan**
has **more coastline** than
California or **Florida**

Niagara Falls

A powerful and world-famous waterfall system with one of the highest flow rates of any waterfall on Earth

E. North America

△ HORSESHOE FALLS
Visitors can experience the incredible power of Niagara Falls by taking a boat trip to Horseshoe Falls—one of the world's most accessible natural wonders.

Niagara Falls is a system of three waterfalls located on the Niagara River, through which water flows from Lake Erie into Lake Ontario. The falls lie on the Canada–U.S. border and form the southern end of the Niagara Gorge.

The Horseshoe Falls, the largest of the three, has a dramatic crest line that is 2,200 ft (670 m) wide and a vertical drop of more than 188 ft (57 m) into a plunge pool 115 ft (35 m) deep. Separated by Goat Island, the American Falls and the diminutive Bridal Veil Falls crash down 66–115 ft (20–35 m) onto a talus— a slope of rocks—formed by a huge rock slide in 1954. At peak times, about 45 million gallons (170 million liters) of water pass over the three falls each minute.

Cutting back

Niagara Falls was formed as the Niagara River plunged over a huge escarpment on its way to the Atlantic Ocean. In a process that continues today, the river eroded layers of soft rock sitting below a harder dolomite caprock. As the soft layers eroded, undercutting the dolomite above, great chunks of the caprock gave way, leaving a vertical cliff face.

▽ DISTINCTIVE GOOSE
The Canada goose is one of the most conspicuous birds in the Niagara Falls area. Some are resident, but large flocks of these geese also travel through during their spring and fall migrations.

THE RETREATING FALLS

The great erosive power of Niagara Falls is demonstrated by its fast rate of movement, tracked since the end of the 17th century. It is estimated that, at the present rate of erosion, the falls will reach Lake Erie about 50,000 years from now.

direction of flow of river

2005
1886
1842
1819
1764
1678

edge of waterfall

Niagara Falls has **retreated 7 miles (11 km)** over the last **12,500 years**

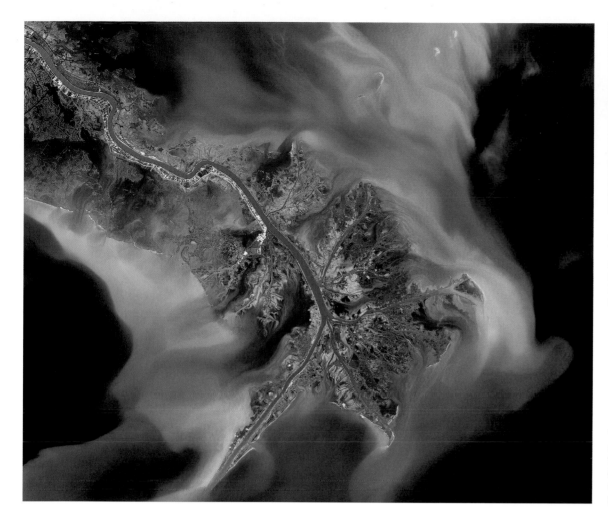

△ **MEANDERING WATERWAY**
The lower Mississippi River and its tributaries are slow-moving waterways. They wind through a gently contoured landscape that includes lowland forest habitats, as shown here, and open floodplains.

◁ **BIRD'S FOOT DELTA**
Also known as a bird's foot delta, due to its shape, the Mississippi River Delta deposits more than 606 million tons (550 million tonnes) of nutrient-rich sediment into the Gulf of Mexico each year, leading to the formation of algal blooms.

The Mississippi-Missouri

A vast river system that extends over much of North America, from the Rocky Mountains in the west to the Appalachians in the east

C. North America

The Mississippi–Missouri river system is the largest watershed in North America, draining more than 40 percent of the contiguous U.S. (all of the country except for Alaska and Hawaii). The Mississippi River's source is Lake Itasca in northern Minnesota, which it leaves as a stream just 10 ft (3 m) wide. The Mississippi's western tributaries, including those of the Missouri, drain the Great Plains, while those to the east drain the Appalachian Plateau.

Great River

In combination, the Mississippi–Missouri stretches 3,710 miles (5,970 km), making it the fourth longest river in the world after the Nile, Amazon, and Yangtze. Also known as the Big Muddy, the silt-laden Missouri discharges into the clearer Mississippi just north of St Louis in Missouri. However, it is only after its confluence with the Ohio River—the Mississippi's largest tributary by volume—at Cairo, Illinois, that the river takes on the full grandeur that led Algonquian-speaking inhabitants to name it *Misi-ziibi*—"Great River."

▷ **FALSE MAP TURTLE**
The endemic false map turtle is an inhabitant of large streams and rivers of the Mississippi-Missouri. It is often found basking on logs and rocks.

SHIFTING DELTAS

The modern Mississippi Delta evolved after the last ice age, when water levels began to stabilize. Over the last 9,000 years, it has shifted across a distance of about 200 miles (320 km), in a sequence of at least seven episodes. The seven major delta lobes and changing river courses are shown here.

present course of Atchafalaya River

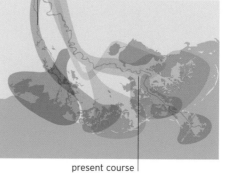

present course of Mississippi

KEY
- 4,600 years before present (BP)
- 4,600–3,500 years BP
- 3,500–2,800 years BP
- 2,800–1,000 years BP
- 1,000–300 years BP
- 750–500 years BP
- 550 years BP

SW. North America

The Grand Canyon

A vast gorge cut by the Colorado River, the Grand Canyon holds a record of 1.7 billion years of Earth history in its walls

One of North America's most iconic landscapes, the Grand Canyon is a spectacular valley carved into the Colorado Plateau by the Colorado River over millions of years. Renowned for its dramatic rock formations, precipitous cliffs, and striking coloration, the canyon is named for its impressive scale. Averaging 4,000 ft (1,220 m) deep for its entire length of 277 miles (446 km), it drops 6,000 ft (1,830 m) below the rim at its deepest and reaches 18 miles (29 km) across at its widest.

A record in the walls

An extraordinary demonstration of the powerful forces of erosion, the Grand Canyon is one of Earth's best-studied geological features. The layers of rock exposed on its walls present a remarkable record of geological history reaching back almost 2 billion years, although the canyon itself is much younger.

▷ **CLIFFS AT TUWEEP**
Sunlight pierces the early morning fog, illuminating the sheer face of the canyon wall and revealing the Colorado River far below.

◁ **GAINING TRACTION**
Forked hooves that spread apart to improve grip enable the desert bighorn sheep to move freely on steep slopes. It can survive for long periods without drinking water.

In some places, **the canyon** is more than **1 mile (1.8 km) deep**

THE SHAPE OF THE CANYON

The Grand Canyon extends from Lees Ferry in the east to Grand Wash Cliffs in the west. As well as the main river channel, there are hundreds of miles of tributaries contained within side canyons. The canyon's North Rim is about 980 ft (300 m) higher than the South Rim as the plateau slopes downward from north to south.

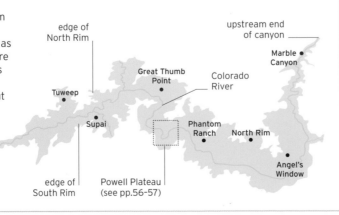

edge of North Rim
upstream end of canyon
Marble Canyon
Great Thumb Point
Colorado River
Tuweep
Supai
Phantom Ranch
North Rim
downstream end of canyon
edge of South Rim
Powell Plateau (see pp.56–57)
Angel's Window

The formation of a
RIVER CANYON

The Grand Canyon is a spectacular example of a gorge—a steep-sided mountain valley eroded by a river. Weathering processes have both shaped the inside of the canyon—producing features such as mesas, buttes, and spires that are often seen in arid and semiarid places—and helped make it wider (see panel, below).

A river requires a lot of erosive power to carve such a deep gorge. The Colorado owes its power to its steep gradient, which increases the speed at which the water moves along its channel, and to a relatively high sediment load, which enables the water to scour and deepen the channel.

The Grand Canyon is a relatively young canyon carved into old rocks. It seems likely that the Colorado gained the power to carve the canyon when the Colorado Plateau was pushed upward by tectonic forces, steepening the slope of its bed. This sudden increase in a river's power is called rejuvenation.

HOW THE GRAND CANYON WAS FORMED

The Colorado Plateau is made up of layers of sedimentary rock (or strata), including limestone, sandstone, and shale. The shape of the canyon is partly determined by how resistant these rocks are to processes of weathering and erosion.

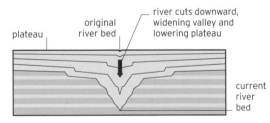

plateau — original river bed — river cuts downward, widening valley and lowering plateau — current river bed

DOWNCUTTING
The Colorado River has cut down progressively into the plateau. The canyon widens significantly whenever the river encounters a layer of relatively soft rock.

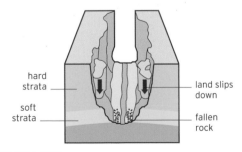

hard strata — land slips down — soft strata — fallen rock

CLIFF COLLAPSE
The river is able to cut relatively easily through soft strata, undercutting cliffs above made of harder rock and causing them to collapse.

▶ **THE GRAND CANYON**
This cross section of the Grand Canyon, at a point near the Powell Plateau, displays an array of geological formations created by the cutting power of the Colorado River. Each layer of rock has put up a different level of resistance to the river.

The South Rim
The South Rim tilts away from the canyon, so water falling on its surface does not flow over the edge into the Colorado.

Shallow slopes
Shallow slopes are formed when a river encounters a layer of soft rock, such as shale.

◁ **COLORADO CONFLUENCE**
The Little Colorado River is one of the Colorado River's largest tributaries. Tributaries add both water and sediment to the main channel, often altering the color of the water. The point where two rivers meet is called the confluence.

The **oldest rocks** exposed in the walls at the **bottom of the canyon** are almost **2 billion years old**

Side canyon
Side canyons are carved by tributaries of the main river channel. Those of the Grand Canyon have contributed to its width.

◁ **INCISED MEANDER**
Meanders are winding bends usually found in the flat lower courses of rivers. Those occurring on an entrenched river—one that is confined to a canyon—are known as incised meanders. These cut deep gorges, like those seen in the Grand Canyon, along preexisting routes.

△ **KAIBAB PLATEAU**
The Grand Canyon cuts through the southern end of the Kaibab Plateau, which is capped with limestone.

The North Rim
More elevated than the South Rim, the North Rim experiences much colder temperatures and snowfall during winter.

Steep cliffs
Cliffs are often made of more durable rocks, usually strongly jointed sandstone and limestone.

Rock layers
Sedimentary rock strata are usually laid down in horizontal beds, with younger layers above older layers. This creates a geological record that is revealed through erosion.

Colorado River
The Colorado flows along the floor of the canyon and continues to cut through layers of rock.

Oldest rock layer
Unlike the sedimentary rocks above, the oldest rock layers are made up of igneous and metamorphic rocks. These hard rocks resist weathering and erosion, so the bottom of the canyon is narrow with steep sides.

▷ **VISHNU BASEMENT ROCKS**
These very hard, crystalline rocks—a combination of schist, granite, and gneiss—exposed near the bottom of the Grand Canyon are among its oldest rocks, having formed more than 1.7 billion years ago.

▷ **NIGHT HUNTER**
Living in the aptly named Bat Cave at the start of the cavern complex are thousands of Mexican free-tailed bats, which roost there by day and leave each evening to feed.

tail extends beyond tail membranes

large external ears

S. North America

Carlsbad Caverns

A vast underground cave system containing a stunning array of calcite formations

Beneath the Chihuahuan Desert in the Guadalupe Mountains, New Mexico, lies a series of massive underground chambers and passages decorated with stalactites and stalagmites. The caverns formed through an unusual process in which groundwater, rich in sulfuric acid, invaded and dissolved the limestone rock on a huge scale. Only later did percolating rainwater enter the resulting caves and furnish their interiors with the elaborate calcite formations seen today. The famous Big Room, a rambling chamber up to 625 ft (191 m) wide in places, contains many remarkable features, including the Hall of Giants—a series of gigantic stalagmites rising almost 65 ft (20 m) from the cave floor—and the Painted Grotto, colored by iron oxide and hydroxide pigments.

STALACTITES, STALAGMITES, AND PILLARS

As mineral-saturated water seeps through the roof of a cave and drips down, a residue starts to build up and a stalactite forms. As drops of water hit the ground, they gradually create a mound of residue that grows into a stalagmite. Over time, the two may join to create a pillar.

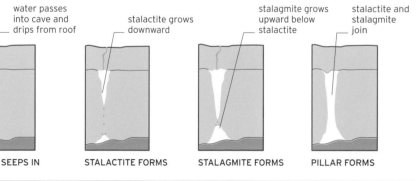

water passes into cave and drips from roof

stalactite grows downward

stalagmite grows upward below stalactite

stalactite and stalagmite join

WATER SEEPS IN   **STALACTITE FORMS**   **STALAGMITE FORMS**   **PILLAR FORMS**

◁△ **ELABORATE DECORATIONS**
Many of the caverns' roofs are covered with delicate tubular stalactites called straws, formed by water moving slowly through cracks in the rock. Curtainlike deposits of calcite, called flowstones (left), form when water runs down the cavern walls.

The Okefenokee Swamp

One of the largest swamps in North America, a diverse wetland ecosystem displaying remarkable biodiversity

SE. North America

Straddling the border between Georgia and Florida, the Okefenokee Swamp is a large, shallow wetland area lying in a basin that was once beneath the Atlantic Ocean. The swamp is dominated by peat-filled mires but also contains lakes, flooded prairies, and mixed woodlands. With an area of 685 square miles (1,770 square km), it is considered the largest "blackwater" swamp in North America. Although its waters are free of sediment, they are stained a dark color by tannins released by peat and decomposing vegetation.

Over the 7,000 years that the swamp has been forming, peat has collected to depths of more than 15 ft (4.5 m). In some areas, floating sections shudder when trodden on; translated from the Hitchiti language, Okefenokee means "water-shaking." The swamp is drained by two rivers: the Suwannee River flows into the Gulf of Mexico, while the St. Marys River empties into the Atlantic.

Flourishing wildlife

A great variety of animals and plants thrive in the Okefenokee Swamp's mosaic of habitats. A haven for more than 230 bird species, including many wading birds and waterfowl, the swamp is well known for its abundance of amphibians and reptiles including alligators, snakes, and the alligator snapping turtle, the world's largest freshwater turtle. It is also a key habitat for a population of Florida black bears and home to other mammals, including bobcats and otters. More than 600 plant species include the carnivorous sundews, bladderworts, and pitcher plants, which have developed their unusual diets to compensate for the nutrient-poor soils.

◁△ **ACIDIC WATERS**
The acidic waters and soils of the Okefenokee Swamp have led to unusual plant adaptations. The hooded pitcher plant is one of several varieties that capture and digest insects in their modified leaves.

HOW A MIRE FORMS

Mires (or bogs) form as flooded areas are filled in by sediment. Fen peat forms as mineral-rich groundwater prevents dead plant matter from decomposing. As the surface rises above the groundwater, mosses grow and their remains become bog peat.

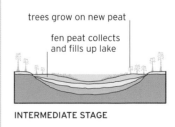

accumulating mud
aquatic plants take root in shallows

lake clay | impermeable bedrock
WETLAND SUCCESSION

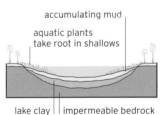

trees grow on new peat
fen peat collects and fills up lake

INTERMEDIATE STAGE

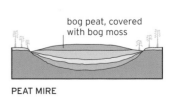

bog peat, covered with bog moss

PEAT MIRE

The Everglades

North America's only subtropical wetland system, dominated by iconic sawgrass prairies, and known locally as the "river of grass"

SE. North America

Located at the southern tip of the Florida peninsula, the Everglades is a vast wetland wilderness containing a variety of lowland habitats. Much of the inland area is covered by the famous sawgrass prairies that are permanently inundated with water. This so-called river of grass, which also contains lakes, waterways (sloughs), and cypress marshes at its edges, is dependent on fires—usually started by lightning strikes—for maintenance and regeneration. Islands of hardwood forests, known as hammocks, contain a mixture of trees including tropical mahogany and temperate oaks and maples, while some of the highest areas, including dry ridges, are covered in South Florida slash pine. In coastal areas, where freshwater and saltwater merge, salt marshes and tidal mangroves dominate. The combined mangrove system of the Everglades lining the coast, creeks, and estuaries is the largest in the Western Hemisphere.

Critical habitats

A major migration stopover, the Everglades is home to more than 400 species of birds. Some 50 species of reptiles live in the area, and it is the only place in the world where alligators and crocodiles coexist. The wetland provides critical habitats for many threatened and endangered species.

▷ **LOCAL WADER**
Roseate spoonbills roost in trees, then fly to their feeding grounds at sunrise, where they search for fish and other small prey.

One in three people living in Florida **rely** on the Everglades for their **water supply**

△ **UNDERWATER GRAZERS**
A West Indian manatee mother and nursing calf graze in shallow waters on seagrasses and other aquatic plants.

◁ **WINDING SLOUGHS**
A series of slow-moving sloughs snake a route through sawgrass before emptying their freshwater into the brackish waters of an Everglades estuary.

▷ **CYPRESS SWAMP**
Found throughout the Everglades, cypress swamps are populated by trees that have adapted to flooded conditions. They are home to many species, including crocodilians.

ECOLOGICAL SUCCESSION

This cross-section of the Everglades shows the relationship between different vegetation types and water inundation levels. The forested areas develop slowly as layers of organic material build up and elevate the land. The change from one type to the next, known as succession, may be affected by fire or environmental factors and may even be reversed.

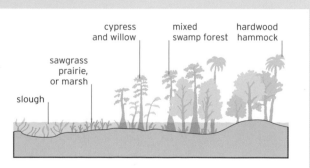

cypress and willow · mixed swamp forest · hardwood hammock · sawgrass prairie, or marsh · slough

W. North
America

Thor's Well

A gaping hole in a rocky stretch of Oregon's coastline that acts like a giant drainpipe

Situated on an area of coast called Cape Perpetua, Thor's Well is circular in shape, about 20 ft (6 m) deep, and has a hole at the bottom that connects with the open sea. As the waves crash in at high tide, the well fills up, overflows, and spouts seawater into the air. The water is then sucked back for several seconds, making it appear as if the sea is being drained.

Churning chasm

Thor's Well churns most spectacularly around the time of high tide. The stirring action depends on the height of the high tide and the direction and size of the waves. The wind can also be a factor. During storms, water can burst out of the well extremely violently, making it too dangerous to approach closely.

△ **NATURAL DRAIN**
Every 10 seconds or so, during high tide, it seems as if the ocean is being drained away down a bottomless hole. This can be a dramatic sight.

A large number of **mussels** gamely **cling on** to the well's **internal walls** as the water churns

W. North
America

Monterey Bay

A diverse ecosystem, with an underwater canyon and kelp forests on the sea bottom

Monterey Bay occupies a large stretch of the coast of central California. Underwater is a thick forest of kelp and the shore end of one of the world's longest undersea canyons, which runs for 60 miles (100 km) to the southwest. Around the bay are some pristine beaches and jewel-like tide pools.

Marine sanctuary

Offshore, the Monterey Bay National Marine Sanctuary covers a much larger area of around 6,100 square miles (15,700 square km). It is home to a fantastic diversity of sea life, including more than 30 species of mammals, over 300 species of fish, and nearly 100 bird species.

◁ **ALGAL FOREST**
Several species of kelp, a form of seaweed, grow in the bay's nutrient-rich waters. Kelp can grow up to 20 in (50 cm) a day, with some specimens becoming as tall as trees.

Big Sur

A long, rugged stretch of the central California coastline, with steep mountains and stunning views

W. North
America

Widely regarded as one of the most spectacular meetings of sea and land in the world, Big Sur is a roughly 100-mile- (160-km-) long section of undeveloped, wild coastline. Some sections of the hillsides just above the shoreline have been shaped into flat areas called marine terraces. These are former beaches lifted up by tectonic activity. A winding, narrow road extends down its length, affording spectacular views of the Pacific. The road and Big Sur in general, however, are susceptible to landslides as a result of wave action, weakening of the cliffs by faulting and fracturing, the destruction of vegetation by summer fires, and heavy winter rainfall.

BIG SUR'S COASTAL SLOPE

One area of the Big Sur coastline, to the west of Cone Peak (in the Santa Lucia Range), has the steepest coastal slope in the U.S. outside Alaska and Hawaii. Here, the land rises sharply in a series of steps to Cone Peak, which is just 3 miles (5 km) from the Pacific Ocean.

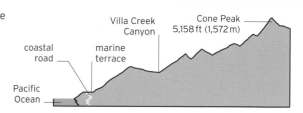

Villa Creek
Canyon

Cone Peak
5,158 ft (1,572 m)

coastal
road

marine
terrace

Pacific
Ocean

quartz
crystal

diopside
crystal

▷ **DEEP-GREEN JEWEL**
The mineral diopside is common in the Big Sur region. Here, typical dull green diopside crystals are embedded with quartz crystals in a rock groundmass.

◁ **MISTY HEADLANDS**
Big Sur, seen here at sunset, winds in and out for much of its length, forming a near-endless array of small bays and headlands, often shrouded in mist.

▽ **WAVE-BATTERED ROCKS**
Around Big Sur's headlands, small rocky islets typically extend up to 330 ft (100 m) offshore. These are continuously pounded by waves surging in from the Pacific.

The Bay of Fundy

A deep inlet into North America's eastern seaboard, best known for its huge tides but also for its dinosaur fossils and semiprecious minerals

E. North America

Located on Canada's east coast, the Bay of Fundy is a unique coastal environment with a landscape of breathtaking cliffs, sea caves, and unusual rock formations. The rock layers and outcrops around the bay display an enormous geological diversity and range of ages. They tell a story of hundreds of millions of years of natural history with, for example, dinosaur discoveries from the Triassic Period and fossils of marine invertebrates from the Carboniferous Period.

Massive tides

High cliffs channel the tidal waters flowing into the bay until they separate into two narrow arms (called Chignecto Bay and the Minas Basin) at its northeastern end. This channeling effect and the bay's overall shape give rise to some

of the world's fastest-running tides and highest tidal ranges (the difference between high and low tides)—the highest ever being a range of 71 ft (21.6 m) recorded one night in October 1869 at Burntcoat Head in the Minas Basin. The tidal surge in the northern arm, Chignecto Bay, produces a bore up to 6 ft (1.8 m). The twice-daily tides transform the shorelines, tidal flats, and exposed sea bottom as they flood into the Bay and its harbors. They also create unique conditions for whales and seabirds.

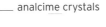 analcime crystals

◁ **MINERAL CLUSTER**
Cape Blomidon, which borders the Minas Basin, and other sites in the bay are important source areas for the white mineral analcime.

176 billion tons (160 billion tonnes) of seawater flow in and out of the bay twice a day

TIDAL RANGES

At a typical Fundy location like Cape Blomidon (see graph below), the average tidal range is about 39 ft (12 m). But at some locations, it can be as high as 65 ft (20 m). As well as depending on location, tidal ranges are affected by the monthly lunar cycle.

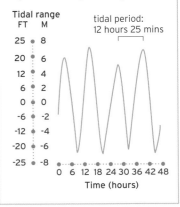

Tidal range
FT   M

tidal period:
12 hours 25 mins

25 • 8
20 • 6
12 • 4
6 • 2
0 • 0
-6 • -2
-12 • -4
-20 • -6
-25 • -8

0 6 12 18 24 30 36 42 48
Time (hours)

E. North
America

Acadia
Coastline

*An area of rugged coastal scenery, interspersed
with tranquil inlets and harbors*

Forming part of the coast of Maine, Acadia National Park
consists principally of one large island, Mount Desert Island,
and several smaller islands nearby. It is best known for its
distinctive rocky coastline—the result of a complex series of
geological processes over hundreds of millions of years. These
include the intrusion of magma into existing sedimentary rock
layers about 360 million years ago to form granite masses. From
about 2 million years ago, a huge ice-sheet blanketed the area
and sculpted out a series of mountains separated by U-shaped
valleys. Today, waves and tides are major agents of change at
Acadia, gradually eroding cliffs and depositing the debris as
sediments around the coastline.

The **base rocks** of the Acadia coastline are up to **500 million years old**

△ **MEANDERING CHANNELS**
The bay's vast intertidal zone (the area
covered at high tide and exposed at
low tide) has a web of deep channels
scoured out by tidal currents.

▽ **LOW TIDE**
At St. Martins, on Fundy's northern
shore, it is possible at low tide to "walk
the ocean floor" out to sea caves. At
high tide, the sea fills the caves.

△ **SHELTERED INLET**
In some areas, wave-pounded cliffs give way to tranquil
inlets. Sea-level rise since the last ice age has resulted
in the flooding of these glacier-sculpted valleys.

North American Boreal Forest

The largest intact forest on Earth, one of only five that contain expanses of untouched wilderness, and a breeding ground for billions of birds

N. North America

Boreal forest, or taiga, is the world's largest terrestrial biome, consisting of vast expanses of mostly conifer species. In North America alone, 2 million square miles (6 million square km) of boreal forest stretches from Alaska to Newfoundland, and from the Great Lakes as far as the Arctic tundra.

Cold-climate forest

The boreal forest is also known as a snow forest due to its climate. Short, humid summers of 50–100 frost-free days are followed by long, cold winters with up to six months of temperatures below freezing. Precipitation falls mainly as snow. Conifers, such as spruce, are well adapted to survive such rigorous conditions (see panel, below), and they dominate the landscape. Lakes, rivers, and wetlands are numerous, making the North American boreal forest Earth's largest source of unfrozen freshwater. Because it is still relatively intact, this is a vital habitat for wildlife such as moose and brown bears, and a breeding ground for up to 3 billion North American birds.

▷ **BOREAL SPECIALIST**
The mainly solitary Canadian lynx roams the boreal forest in all seasons, following its primary prey species, the snowshoe hare.

◁ **DENSE COVER**
Conifers such as spruce, pine, and fir grow close together, forming an almost unbroken canopy that allows little light to reach the forest floor.

SHEDDING SNOW

A conifer's conical shape and flexible, downward-pointing branches prevent heavy snow accumulation. Narrow, needlelike leaves also prevent snow buildup and have a waxy outer coat that prevents water loss by evaporation in freezing conditions.

snow slides off branches when too much builds up

△ **CANOPY OF MOSSES**
The continually wet conditions of the coastal temperate rain forest provide an ideal habitat for moss. Over 100 varieties can be found draped over the branches of the conifer trees that dominate this environment.

Pacific Northwest Rain Forest

The world's largest undeveloped coastal temperate rain forest, with the greatest amount of wood—living and dead—of any forest on Earth

W. North America

△ **FOREST FERNS**
The rich forest floor provides nutrients for many fern species, including lady, deer, and sword ferns.

Coastal temperate rain forests are a very rare forest type. Today, they are believed to cover just 116,700 square miles (302,200 square km)—less than 0.2 percent of Earth's land area. Oceans, mountains, and high rainfall are features of these forests, and the largest intact areas stretch from northern California to Canada and the Gulf of Alaska. The Pacific Northwest measures annual rainfall in feet, and 8–14 ft (2.5–4.2 m) is not unusual. Constantly decaying cedar, Sitka spruce, Douglas fir, and coast redwood needles and deciduous leaves create rich soil that supports hundreds of moss, fern, and invertebrate species, while rivers, lakes, and streams are home to salmon and trout, which in turn attract bears and bald eagles.

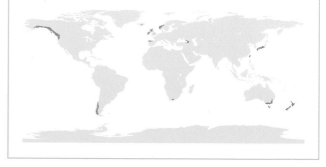

▷ **WASTE MANAGEMENT**
A vital part of the ecosystem, the Pacific banana slug breaks down rotting plants, spreading seeds and spores in the process.

THE WORLD'S COASTAL TEMPERATE RAIN FORESTS

Apart from North America's Pacific Northwest, coastal temperate rain forests occur in southwest South America; the northeastern Atlantic region (including Iceland, Ireland, Scotland, Norway, and northern Spain); the eastern Black Sea; southwest Japan; New Zealand, Tasmania, and Australia's New South Wales; and the tip of South Africa.

W. North
America

The Giant Forest

An ancient, sky-scraping forest of more than 8,000 giant sequoias, home to half of the largest and longest-lived trees on Earth

Growing at the heart of central California's Sequoia National Park, in the southern Sierra Nevada mountains, the Giant Forest is home to more than 8,000 giant sequoias (*Sequoiadendron giganteum*). While they may not be the world's tallest trees, they are easily the largest by volume.

A tale of two redwoods

Giant sequoias are often called redwoods and vice versa, but while both species are native to California and share similar cinnamon-colored bark and heady heights, they are decidedly different. The trees of the Giant Forest generally grow at elevations of 4,900–6,900 ft (1,500–2,100 m) on western Sierra Nevada slopes, making them much more cold-hardy than coast redwoods (*Sequoia sempervirens*), which

grow at lower coastal elevations. Giant sequoias live for around 3,000 years (as opposed to the redwoods' 2,000), and their bark is up to 2 ft (60 cm) thick. Sequoias do not grow as tall as redwoods—the maximum is around 310 ft (95 m) compared to the redwoods' 380 ft (115 m)—but what they lack in height is made up for in girth. The largest tree in the world, the Giant Forest's General Sherman, stands just 275 ft (84 m) high but is more than 100 ft (30 m) wide and weighs an estimated 1,350 tons (1,225 tonnes). At 52,513 cubic feet (1,487 cubic meters), it is the single largest living organism by volume in the world.

◁ **OPPORTUNISTIC SCAVENGER**
The Steller's jay's raucous call often breaks the silence of the Giant Forest. Omnivorous, jays eat anything from seeds and nuts to hikers' leftovers.

The **high tannin** content in giant sequoia **bark prevents rotting**

▷ **MAMMOTH PROPORTIONS**
The President is 245 ft (75 m) tall, making the scientists shown with it antlike by comparison.

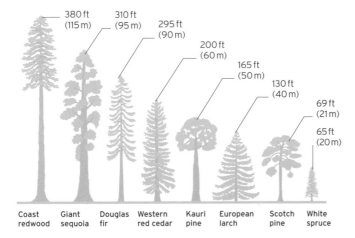

CONIFER HEIGHTS

Because they lack the spread of deciduous trees, conifers grow much taller than the latter, but their heights vary greatly according to species. Coast redwoods are the tallest trees on Earth.

- 380 ft (115 m)
- 310 ft (95 m)
- 295 ft (90 m)
- 200 ft (60 m)
- 165 ft (50 m)
- 130 ft (40 m)
- 69 ft (21 m)
- 65 ft (20 m)

Coast redwood	Giant sequoia	Douglas fir	Western red cedar	Kauri pine	European larch	Scotch pine	White spruce

△ **SHADE TOLERANT**
Although the trees of the Giant Forest cast giant shadows, small trees such as the Pacific dogwood survive well in their shade.

Pando Aspen Grove

W. North America

One of the oldest organisms in the world, a quivering grove with a surprising genetic heritage

△ **SHIVERING TREES**
Quaking or trembling aspens are so named because their leaves, which are attached via long, flattened stalks, flutter in even a slight breeze, making the trees appear to shiver.

It contains what seem to be around 47,000 individual trees, but the Pando grove of central Utah is in fact one giant living organism—a clonal colony of quaking aspen. All of its trees are genetically identical, and all began as ramets, or suckers, growing from and sharing a single root system.

Ancient roots

Aspen clonal colonies are common in North America. What makes Pando—Latin for "I spread"—special is its size: it sprawls over 108 acres (44 hectares), with an estimated weight of 6,615 tons (6,000 tonnes). Pando's accepted age is also unusual. Although individual trunks live around 100–150 years, the clonal root system is thought to be at least 80,000 years old. Some scientists, however, believe that the figure could be closer to 1 million. Recently, Pando's survival has been threatened because deer and moose eat its ramets. Parts of the forest are now fenced off so that new suckers grow undisturbed.

SHARED ROOT SYSTEM

Aspen clonal groves begin with a single seed. As a "parent" tree grows, its lateral root system spreads. Suckers sprout from these roots, forming genetically identical trees that send out their own lateral roots, and the process continues.

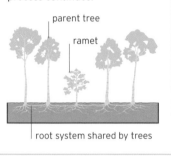
parent tree
ramet
root system shared by trees

The Cherokee National Forest

E. North America

At the heart of an ancient mountain range, a mist-shrouded forest that is one of the most biologically diverse in North America

The Cherokee National Forest follows Tennessee's state border with North Carolina, covering around 1,015 square miles (2,630 square km) of the Appalachian Mountains' eastern slopes. Because the southern Appalachians were not covered by glaciers during the last ice age, diverse plant and animal species flourished, and the forests that covered the mountains provided a relatively isolated habitat until the mid-19th century. The Cherokee forest is a natural sanctuary, with more than 20,000 plant and animal species. Beech, hickory, sassafras, birch, maple, and numerous types of oaks are just some of the forest's common trees, attracting more than 120 bird species, including the threatened red-cockaded woodpecker.

▷ **MORNING FOG**
The Cherokee forest is bisected by the Great Smoky Mountains National Park, and the thick, blue-tinged fogs are common to both areas.

▽ **ON THE PROWL**
Around 1,500 black bears roam the forest, relying on the dense cover provided by its trees.

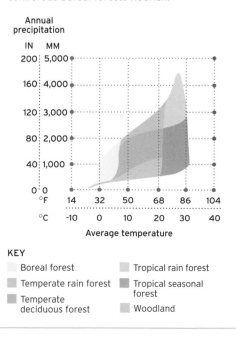

△ **BLAZE OF RED**
Sugar maples make up 20-50 percent of the trees in the northern Green Mountains and are noted for their fiery-red leaves in fall.

The Green Mountain National Forest

One of two national forests in northeastern North America, covering the Green Mountains with many colors in fall

E. North America

Vermont's Green Mountain National Forest consists of more than 625 square miles (1,620 square km), spread over and around the mountains that share its name. Many of the peaks in the forest are over 2,950 ft (900 m), and the spruce, maple, beech, and paper birch that cover them make the region a popular tourist destination, particularly in fall, when the forest becomes a blaze of color. Softwood as well as hardwood tree species flourish in the region, and these, coupled with a dense understory fed by mountain streams and winter snowfall, provide food and shelter for a wide variety of bird and other animal species. Beavers, moose, deer, and wild turkeys are common, as well as gray wolves and black bears.

△ **FALL COLORS**
The wide variety of Green Mountain trees produces a multihued landscape in the fall, ranging from light gold to deep red.

The **Green Mountain Bear Corridor** serves as a vital **north-south passage** for bears

FOREST TYPES AND CLIMATE

Forest types vary with latitude. Closest to the equator are tropical forests, which include evergreen rain forests or those with two seasons—dry and wet. At higher latitudes lie temperate forests, which consist of broadleaf rain forests as well as deciduous forests with defined seasons. At latitudes of 50° to 60° North, where summers are warm and short and winters cold and long, coniferous boreal forests flourish.

Annual precipitation

IN	MM
200	5,000
160	4,000
120	3,000
80	2,000
40	1,000
0	0

°F	14	32	50	68	86	104
°C	-10	0	10	20	30	40

Average temperature

KEY

- Boreal forest
- Temperate rain forest
- Temperate deciduous forest
- Tropical rain forest
- Tropical seasonal forest
- Woodland

North American Tundra

Stretching across the continent's northern reaches, a cold, treeless landscape where plants huddle to the ground to survive

N. North America

△ **POLYGONAL PATTERN**
The continual Arctic freeze-thaw cycle creates ice wedges that eventually shape tundra soil into polygons. These become freshwater ponds during the summer thaw.

In North America, tundra covers northern Alaska in the U.S., extends eastward through northern Canada, and forms a narrow band around the coast of Greenland. North of the Arctic Circle, this ecoregion is known as Arctic tundra, while in the mountains above the timberline, it is classified as alpine tundra.

Challenging conditions
Arctic tundra conditions are the coldest: summer high temperatures are just 40°F (4°C), plunging to -25°F (-32°C) or below in winter. The alpine tundra climate is less extreme: summers may reach 54°F (12°C) while winters rarely fall below 0°F (-18°C). However, in both cases, this is mainly a treeless landscape, where vegetation hugs the ground to survive freezing winds that contribute to blizzard conditions in winter, especially on alpine tundra. Elsewhere, the constantly frozen underlying ground

layer called permafrost (see p.175) blocks the movement of water through the soil during the summer thaw, turning some low-lying Arctic tundra areas into bogs. Despite these challenges, almost 2,000 plant species grow in the tundra, primarily grasses, sedges, mosses, and shrubs. All are shallow-rooted to cope with the tundra's thin active soil layer and to survive the region's freeze-thaw cycle.

Flowering plants such as saxifrage bloom vigorously during the brief summer, while mosses and lichens provide vital nourishment in lean months for mammals such as reindeer and Arctic hares. In areas where the soil is slightly deeper and richer, mat-forming shrubs such as bilberry and crowberry produce fruit in summer that may last beneath winter snows until the following spring.

▽ **SUMMER COAT**
The Arctic fox is well-adapted to its environment. In summer, its white winter coat thins, changing to brownish-grey to match the tundra's rocks and low-growing vegetation.

The Great Plains

The largest expanse of temperate grassland in North America, a fragile landscape shared between two nations and increasingly under threat

C. North America

At around 3,000 miles (4,800km) long and 310–680 miles (500–1,100km) wide, the vast grassland known as the Great Plains covers the heart of the North American continent. It slopes gently downward from the Rocky Mountains east to the Mississippi River, and stretches from southern Canada as far south as Texas.

Types of prairies

Once immense seas of grass known as prairies, much of the Great Plains today consists of seemingly endless fields of grain—primarily corn and wheat—planted on the most fertile soil. Other sections have been converted to cattle ranching. The natural grassland that remains consists of shortgrass and bunchgrass in the drier west, giving way to medium-grass and tallgrass prairie in central and eastern regions that receive more rainfall. Tallgrass prairie is among the world's rarest ecosystems. It is dominated by up to 60 different grass species, some of which reach heights of up to 8 ft (2.4 m), towering over an understory of flowers, lichens, and liverworts. These support a wide range of wildlife, from tiny western harvest mice and shrews to coyotes. All prairies are fragile habitats, and various programs are underway to preserve them.

Less than 1 percent of the **original tallgrass prairies** that existed prior to European settlement **remains**

NORTH AMERICA'S PRAIRIE TODAY

Land surveys estimate that up to 70 percent of the original grasslands that once constituted the Great Plains have disappeared, due to conversion of land for agriculture or for oil and gas exploration. Losses have been greatest in tallgrass prairies, while around 52 percent of shortgrass prairie remains intact.

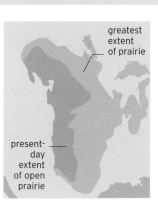

greatest extent of prairie

present-day extent of open prairie

△ **SUPER STORM**
Huge, violent supercell thunderstorms are common summer occurrences in the Great Plains. Their rotating updrafts often spin off tornadoes.

▷ **RETURN OF THE HERDS**
American bison were hunted almost to extinction by the late 19th century. Today, around 500,000 roam the Great Plains.

The Great Basin Desert

North America's largest desert, where moisture arrives as snow—if it arrives at all

W. North America

With an area of about 200,000 square miles (518,000 square km), the Great Basin is the largest desert in North America. It lies mainly in Nevada and western Utah, but stretches into California and Idaho.

Classified as a cold desert, the Great Basin's elevations range from 3,900 to 9,800 ft (1,200 to 3,000 m). The Sierra Nevada and the Rocky Mountain ranges block most moisture from the Pacific Ocean and Gulf of Mexico, forming a rain-shadow area. So the average annual precipitation is only 12 in (30 cm), and usually arrives as snow. A single "wet year" can often be followed by many years of drought.

Habitats here vary greatly, with mountaintop "islands" of bristlecone or pinyon and juniper forests separated by sagebrush-filled valleys or arid salt lakes—in some places, the soil is so salty that no plants can survive.

Bristlecones can live for more than 5,000 years

▷ BONNEVILLE SALT FLATS
The flats are covered with water in winter, which evaporates in summer, leaving behind crystallized salt.

▽ BRISTLECONE LANDSCAPE
Bristlecones are found at high altitudes and in harsh conditions. They are more disease-resistant and at lesser risk of wildfires than many other desert plants.

▷ MOJAVE ICON
The Joshua is actually a tree-sized succulent that grows up to 40 ft (12 m) in height. It takes about 50 years to mature and can live for 150 years or more.

HOW STONES MOVE

Shallow water covers the playa (dry lakebed) surface. It freezes overnight but begins to melt in sunlight, separating the ice into panels. Light winds move these panels over the water beneath, pushing the stones along. Once the water evaporates, the "push trails" of the stones are left behind in the mud.

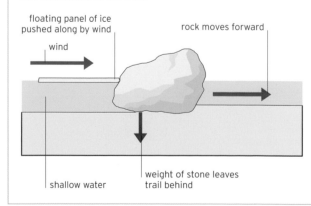

floating panel of ice pushed along by wind

rock moves forward

wind

shallow water

weight of stone leaves trail behind

sharp, thin, spiky leaves
minimize water loss

SW. North
America

The Mojave Desert

Home of the lowest, hottest point in North America, giant yuccas, and moving stones

North America's smallest, driest desert is sandwiched between the cold Great Basin Desert to the north and the hot Sonoran Desert to the south. Due to its proximity to two contrasting desert areas, experts rarely agree on the Mojave's boundaries, size, or type.

Levels of extremes

Badwater Basin in the Mojave's Death Valley lies 282 ft (86 m) below sea level—the continent's lowest point—yet Telescope Peak in the Panamint Range soars 11,043 ft (3,366 m) above it. North America's highest air temperature of 134°F (57°C) was recorded here, at Furnace Creek in 1913, yet winter temperatures regularly drop below freezing. Winds above 50 mph (80 kph) often buffet the Mojave's western edge but are rare in the east. In the north, low-growing scrubland resembles the Great Basin Desert, but to the south, creosote bush takes hold.

The Joshua tree, the largest North American yucca, is a symbol of the Mojave and is often used by ecologists as an indicator of the desert's southern boundary. Death Valley is also home to "sailing stones"—boulders that appear to move unaided (see panel, left).

branching stem
covered in fire-resistant
fibrous bark

◁ **DEVIL'S GOLF COURSE**
In Death Valley, growing crystals push against one another, forming an inhospitable, jagged landscape called the Devil's Golf Course.

△ **MYSTERIOUS MOVERS**
Stones, some as large as 700 lb (320 kg), move across a dry lakebed in Death Valley. They leave behind them trails up to 1,500 ft (460 m) long.

SW. North America

The Sonoran Desert

One of the wettest deserts on Earth, with an abundance of plant species

A place of transition between temperate and tropical climates, the Sonoran Desert merges into the Chihuahuan Desert to the southeast and the Mojave to the northwest. Calderas bear witness to the prehistoric volcanic eruptions that created the desert's mountain ranges, valleys, and alluvial fans known as *bajadas*. The Sonoran's annual rainfall of 3–20 in (7–50 cm)—the most in any North American desert—supports over 2,000 plant species. These must survive average summer temperatures of 104°F (40°C), but winters are relatively mild at 50°F (10°C), with few frosts.

▽ DESERT IN BLOOM

Winter rains transform the Sonoran with patches of color, as plants such as dune evening primrose and pink sand verbena bloom in profusion.

ROOT NETWORKS

Sonoran plants use varied strategies to survive drought conditions. Within hours of rainfall, saguaro cacti send out new roots close to the surface to store water in their expanding stems. Velvet mesquite roots burrow 165 ft (50 m) downward to tap into the water table.

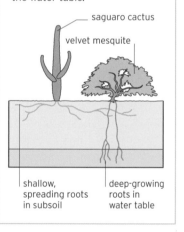

saguaro cactus

velvet mesquite

shallow, spreading roots in subsoil

deep-growing roots in water table

▽ HANDS UP

The valley's distinctive West and East Mitten Buttes have intrigued visitors for decades. Viewed from the south, they resemble two giant mittens, held thumbs-upward.

Monument Valley

A remote valley, where thousands of acres of wide, red plain are broken only by isolated stone monoliths, formed naturally over millennia

W. North America

Part of the Colorado Plateau, Monument Valley is a high-elevation desert, averaging about 5,600 ft (1,700 m) above sea level. It is located in one of the Southwest's driest and least-populated areas, across Utah and Arizona.

During the last 50 million years, wind and water have eroded what was once an even higher plateau, peeling away layers of shale, sandstone, and conglomerate. What remains are incredible stone formations, 400–980 ft (120–300 m) high. Settlers in the 19th century considered the desert terrain hostile, and this, along with its harsh climate and isolated location on Navajo Nation-owned land, saved it from exploitation. Today, all of Monument Valley is owned by the Navajo Nation, which restricts visitors to a 17-mile (27-km) road tour and one hiking trail.

△ **RESIDENT RATTLER**
The prairie rattlesnake is one of several species that inhabit the desert. It often takes over burrows abandoned by prairie dogs or burrowing owls.

The valley's majestic **rock formations** began as **sand** on a seafloor **270 million years ago**

◁ **ICONIC LANDSCAPE**
Monument Valley's buttes, mesas, and spires are familiar to millions via Hollywood's classic westerns. Each formation has a descriptive name, such as (from left to right) Brigham's Tomb, Castle Rock, Bear and Rabbit, and Stagecoach.

MESAS AND BUTTES

Once part of a plateau, both mesas and buttes have flat tops and sheer sides. Their top layers are hardened rock, more resistant to erosion, but the layers beneath are made of less-resistant sedimentary rocks. Over time, these sedimentary layers are eroded by the action of wind and water, separating mesas from the parent plateau and forming canyons between them. Eventually, the larger mesas are worn down into smaller buttes.

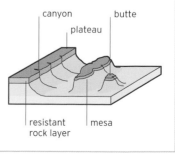

canyon   butte
plateau
resistant rock layer   mesa

Bryce Canyon

A natural amphitheater filled with thousands of strange stone spires, columns, and towers—more than anywhere else on Earth

W. North America

△ BRYCE AMPHITHEATER
The rim of Bryce Canyon is 8,000–9,000 ft (2,400–2,700 m) high–perfect for overlooking the national park's "amphitheater" area.

Bryce is not a conventional canyon because it was not created by flowing water cutting through rock. However, an ancient lake that once covered much of what is now southwestern Utah did play a role in its formation. The process began 55–35 million years ago.

A sea of stone totems

Mineral-rich sediment from surrounding mountains was washed down by rainfall and deposited on the lake's base, eventually turning into the pink limestone that geologists today know as the Claron Formation. Geological upheaval forced the lakebed upward by nearly 1 mile (1.6 km) to create a plateau. Millennia of weathering and erosion subsequently carved the Claron Formation first into cliffs, then into tall columns of rock, some a mere 5 ft (1.5 m) tall, others rising as high as 150 ft (45 m) (see panel, right). Known as hoodoos, these multicolored towers contain sections of varying thicknesses and are composed mainly of limestone but with silt and mudstone as well. Minerals such as iron oxide and manganese oxide add pink, purple, and blue tints. The different hues and lumpy, unequal shapes of the stone spires remind some visitors of wooden totem poles.

Freezing and thawing

While hoodoos occur on every continent, they are more common in the northern part of Bryce Canyon than anywhere else on Earth. This is partly due to the region's climate. On average, Bryce Canyon undergoes 200 freeze-thaw cycles a year; temperatures rise above freezing point during the day but drop below it at night, causing any meltwater that has seeped into the rock to freeze and expand. Known as frost-wedging, this process shatters stone, reshaping it into the hoodoos that the Paiute Indians, who settled in the area around 1200 CE, named the Legend People. As frost-wedging continues, many Legend People collapse each year.

◁ THOR'S HAMMER
Top-heavy hoodoos, such as Thor's Hammer, are fragile and at greatest risk of collapse, especially during spring and fall, when frost-wedging is at its peak.

HOW A HOODOO FORMS

Rainwater washes debris away from the plateau's edges, forming fins. Frost-wedging creates ice windows in the fins, separating them into hoodoos. Rain then rounds off the edges, lending a lumpy appearance.

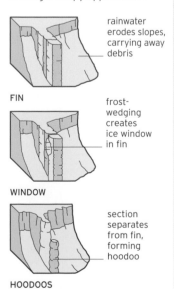

FIN — rainwater erodes slopes, carrying away debris

WINDOW — frost-wedging creates ice window in fin

HOODOOS — section separates from fin, forming hoodoo

Meteor Crater

The best-preserved meteor impact site on Earth and evidence of an explosive 50,000-year-old collision whose effects can still be seen several miles away

W. North America

A massive, bowl-shaped cavity carved into the Arizona Desert, Meteor Crater is one of the most studied meteorite craters on Earth. Its shape has been remarkably well preserved, and it is comparatively young for an impact site. Nearly 1 mile (1.6 km) across, more than 550 ft (168 m) deep, and with a circumference of 2½ miles (3.9 km), the depression known as Meteor Crater was initially believed to be an extinct volcano. It was not until the early 20th century that it was identified as having been created by an extraterrestrial impact, and details about its origins are still being discovered.

Around 50,000 years ago, a meteorite estimated at 130 ft (40 m) across and weighing 300,000 tons (272,000 tonnes) hurtled into Earth's atmosphere at 26,800 mph (43,120 kph)—10 times faster than gunshot. Half of it slammed intact into what is now the northern Arizona Desert, forming the crater; the rest fragmented before the collision, falling to the surface as a cloud of debris. Iron-rich meteorites weighing 1–1,000 lb (0.5–454 kg) have been found in a 6-mile (10-km) radius around the main impact site.

△ **EXPLOSIVE IMPACT**
Scientists estimate that the force released by the meteorite impact was greater than that generated by 20 million tons (18 million tonnes) of TNT.

CRATER FORMATION

The impact created when a single meteorite hits Earth often destroys the meteorite itself in a shock-wave-inducing explosion that fractures bedrock and melts surface rock. It leaves a bowl-shaped depression with a raised outer rim of ejected debris and a base lower than the surrounding landscape.

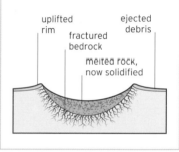

uplifted rim   ejected debris
fractured bedrock
melted rock, now solidified

△ **CLIFFTOP ARCH**
Mesa Arch stretches across the top of the Island in the Sky mesa, a headland that rises 2,000 ft (600 m) above the base of Canyonlands National Park. Below it, the mesa wall drops vertically for more than 490 ft (150 m).

W. North America

Mesa Arch

A dramatic natural archway formed on the edge of a high mesa

It looks like a bridge, but the 88-ft (27-m) long Mesa Arch, in southeastern Utah's Canyonlands National Park, is a pothole arch, created not by running water but by weathering and erosion. Water accumulated in a small depression in a sandstone mesa known as the Island in the Sky. Over time, the depression was enlarged by weathering and erosion into a larger pothole, undercutting more resistant rock at the cliff edge and eventually forming what is now thought to be the most-photographed natural arch in the world. Although pothole arches survive for thousands of years, they do eventually collapse. However, seismic research in 2015 indicated that Mesa Arch is stable—for now.

SW. North
America

Antelope Canyon

Beautiful chambers sculpted by flash floods, hidden beneath the Arizona desert

In northern Arizona lies one of the American Southwest's most-visited canyons, but few passing drivers know that it exists. This is because, unlike most other canyons, whose broad gorges open to the sky, Antelope Canyon is a slot canyon. Also called narrows, slot canyons have nearly vertical, twisting walls divided by fissures that can be just a handspan's width at the surface yet plunge downward for thousands of feet.

A tale of two canyons

Antelope Canyon was discovered in 1931 on land owned by the Navajo Nation. It twists and turns beneath the desert for 5 miles (8 km); some sections are only a few feet wide, but can be 120 ft (36 m) or more deep. The canyon has two parts: Lower Antelope Canyon, known in Navajo as *Hazdistazí* ("spiral rock arches"), and Upper Antelope Canyon, or *Tse' bighanilini* ("the place where water runs through rocks").

The spiral-patterned chambers have been created by running water, but the process has been intermittent. Over millennia, the local red Navajo sandstone has been subjected not to constant scouring from a river but to rare yet violent flash floods. Whenever the nearby Antelope Creek overflows, water forces its way through the desert surface via tiny cracks; rocks and other debris carried by the floods help carve out the rainbow-colored corridors. Where the canyon widens enough to allow in sunlight, winds blow in sand, abrading the walls and changing the patterns and the floor levels.

> Were it not for **floods** periodically **washing the sand away**, the canyon would have **disappeared long ago**

△ **SANDSTONE SPIRALS**
Shafts of light, created briefly when the Sun is directly overhead, illuminate the chambers of Upper Antelope Canyon. The lighting effect transforms the sandstone into a multicolored spectacle.

◁ **RIBBON IN THE SAND**
Antelope Canyon's snakelike shape is best seen from above. Antelope Creek periodically forces its way through the canyon, its waters cleansing and altering the winding course before draining into nearby Lake Powell.

The Chihuahuan Desert

One of North America's largest deserts and the most species-rich in the Western Hemisphere

SW. North America

Bisected by the Rio Grande, the Chihuahuan Desert is bordered to the west and east by the Sierra Madre Occidental and Oriental (see p.37). It is a rain-shadow desert, formed because the mountains block the flow of moisture-laden air from coastal waters.

High life

Most of the desert lies at an altitude of about 4,500 ft (1,370 m), making the summers slightly milder than in other deserts, although daytime temperatures still exceed 100°F (38°C). The nights can be cold, with snow at higher elevations. Despite such extremes, more than 3,500 plant species flourish here, including more cacti species than in any other desert. While higher slopes are characterized by thick-leaved, frost-hardy yuccas, shrubs such as mesquite and creosote bush dominate the lower region. Such varied habitats support a wide range of wildlife, including bats, roadrunners, and tarantulas. A surprising number of fish are able to survive in pools and oases.

△ WORLD'S LARGEST GYPSUM DUNEFIELD
White gypsum deposits punctuate the Chihuahuan Desert, with the largest concentration in northern New Mexico's Tularosa Basin. Its 275 square miles (712 square km) include the White Sands National Monument.

STORING WATER

Cacti such as the golden barrel are adept at surviving on a limited water supply. The plant's pleated "barrel" is actually a fleshy stem that stores water, expanding and contracting as the supply waxes and wanes. Instead of having leaves that lose moisture through transpiration, it is covered in thorns.

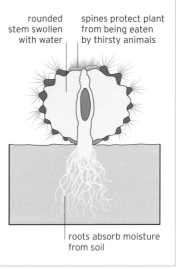

rounded stem swollen with water

spines protect plant from being eaten by thirsty animals

roots absorb moisture from soil

◁ SPIKY MEAL
The Chihuahuan Desert contains over one-fifth of the world's estimated 1,500 cacti species, including this prickly pear. Many animals, such as desert tortoises, eat its pads and fruit—barbs and all.

Mist rising
Water vapor collects above the canopy of the Amazon
Rain Forest on the eastern slopes of the Peruvian Andes.
Hot, wet, and humid all year round, rain forests reach their
greatest extent in the tropical zones of South America.

Central & South America

ANDES TO AMAZON

Central and South America

South America is roughly triangular in shape, with a curved hook of land at either end: the narrow neck known as the Isthmus of Panama in the north; and the aptly named Cape Horn and the islands of Tierra del Fuego in the south.

The soaring Andes mountains, South America's sturdy spine, extend 4,500 miles (7,200 km) down

the continent's western edge, making them the world's longest mountain range by far. They are also the second highest, after the Himalayas, and are crowned by the 22,838 ft (6,961 m) peak of Aconcagua in Argentina. To the northeast, high plateaus rise above the vast, dense tropical rain forests, which are watered by the mighty Amazon River and its tributaries—together, they carry huge quantities of water east into the Atlantic. Southward, the continent narrows to a point, and wedged between the Andes and the sea is flat pampas grassland.

KEY DATA

▲ **Highest point** Aconcagua, Argentina: 22,838 ft (6,961 m)

▼ **Lowest point** Laguna del Carbón, Argentina: -344 ft (-105 m)

● **Hottest record** Rivadavia, Argentina: 120°F (49°C)

● **Coldest record** Sarmiento, Argentina: -27°F (-33°C)

CLIMATE

Much of Central and South America lies within the tropics, but the south extends close to icy Antarctica. The Amazon forests are warm and moist, but plateaus in the mountains are dry.

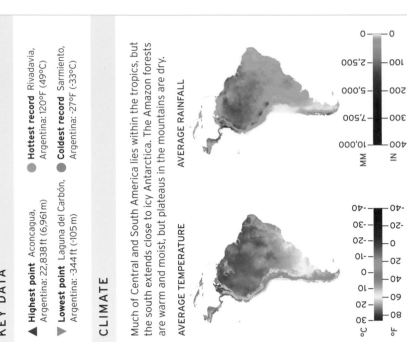

AVERAGE TEMPERATURE

AVERAGE RAINFALL

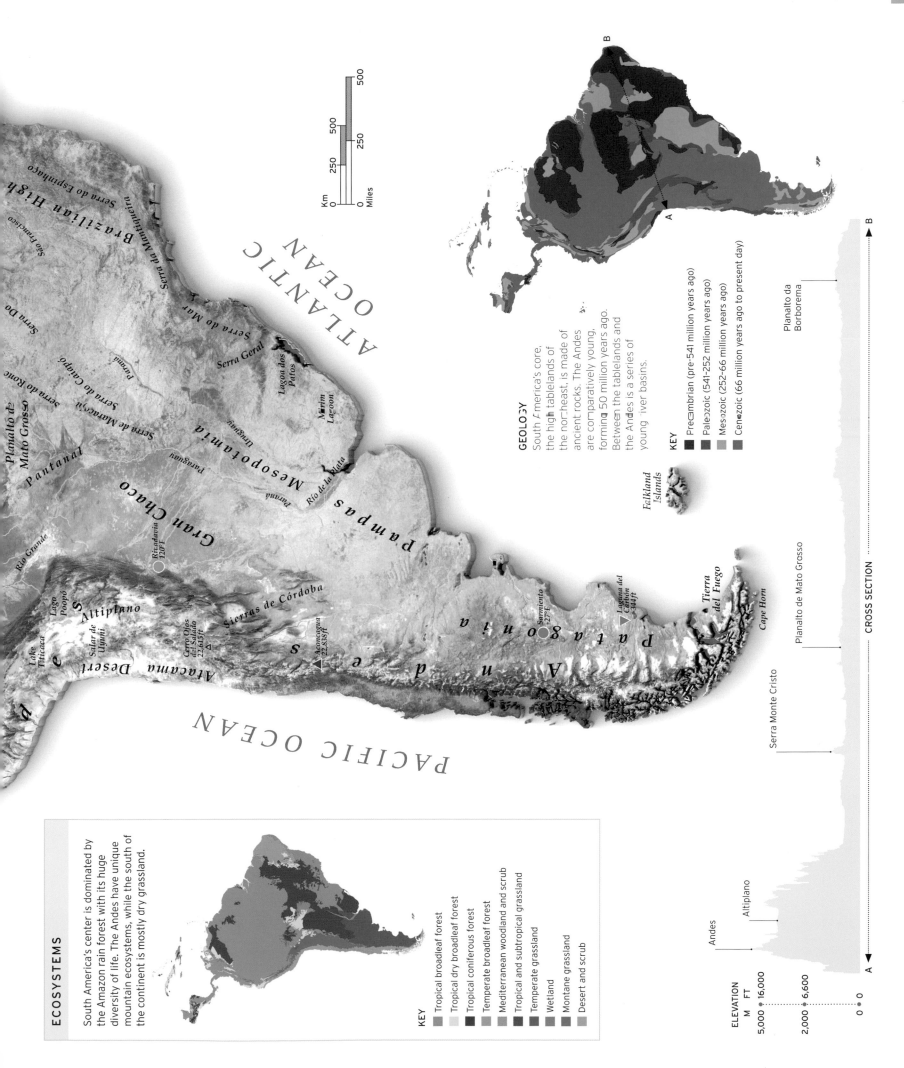

GEOLOGY

South America's core, the high tablelands of the northeast, is made of ancient rocks. The Andes are comparatively young, forming 50 million years ago. Between the tablelands and the Andes is a series of young river basins.

KEY

- Precambrian (pre-541 million years ago)
- Paleozoic (541–252 million years ago)
- Mesozoic (252–66 million years ago)
- Cenozoic (66 million years ago to present day)

ECOSYSTEMS

South America's center is dominated by the Amazon rain forest with its huge diversity of life. The Andes have unique mountain ecosystems, while the south of the continent is mostly dry grassland.

KEY

- Tropical broadleaf forest
- Tropical dry broadleaf forest
- Tropical coniferous forest
- Temperate broadleaf forest
- Mediterranean woodland and scrub
- Tropical and subtropical grassland
- Temperate grassland
- Wetland
- Montane grassland
- Desert and scrub

CROSS SECTION

ELEVATION
M   FT
5,000 16,000
2,000   6,600
0   0

Andes
Altiplano
Serra Monte Cristo
Planalto de Mato Grosso
Planalto da Borborema

ATLANTIC OCEAN

PACIFIC OCEAN

Brazilian High
Serra do Espinhaço
São Francisco
Serra da Mantiqueira
Serra do Mar
Serra Geral
Serra Do
Planalto de Mato Grosso
Serra do Roncador
Serra de Maracaju
Serra do Caiapó
Pantanal
Paraná
Mesopotamia
Lagoa dos Patos
Mirim Lagoon
Uruguay
Río de la Plata
Paraná
Paraguay
Pampas
Gran Chaco
Rivadavia 120°F
Sierras de Córdoba
Rio Grande
Altiplano
Lago Poopó
Lake Titicaca
Salar de Uyuni
Atacama Desert
Cerro Ojos del Salado 22,615ft
Aconcagua 22,838ft
Patagonia
Andes
Sarmiento -27°F
Laguna del Carbón -344ft
Tierra del Fuego
Cape Horn
Falkland Islands

△ ANCIENT PLATEAU
This plateau in Venezuela is one of hundreds of dramatic tabletop mountains, or tepuis, formed from the remains of a sandstone plateau that built up on top of the ancient Guiana Shield.

THE SHAPING OF CENTRAL AND SOUTH AMERICA

Attached to Central America, South America is a huge continent – with high plateaus of old rocks in the east and the towering Andes to the west. It became an independent continent about 90 million years ago, when it split away from Africa.

Shields and tablelands

To the east, a large portion of South America's land is dominated by three giant slabs of ancient rock, typically made of gneiss and granite, forming the Patagonian Shield, the Brazilian Shield, and the Guiana Shield. These slabs survived and moved together as a group for billions of years, even when South America was part of the southern supercontinent Gondwana. South America, along with Africa, split from Gondwana about 180–170 million years ago. Then, 140–90 million years ago, South America slowly split from Africa and the shields eventually became the bedrock around which the rest of South America has been gradually built.

Today, the shields form a base under some of the world's largest high plateaus. In places, these old plateaus, or tablelands, end so abruptly in high cliffs that people once believed they were lost worlds where ancient creatures might survive.

The world's highest waterfall, Angel Falls, plunges over the edge of the Guiana Shield, dropping 3,212 ft (979 m). The slabs are by no means intact. In places, they have cracked and slipped to open up huge valleys—for example, the Parana, Sergipe, and São Francisco rivers now flow through such valleys in the Brazilian Shield.

The rise of the Andes

After it separated from Africa and became a separate continent, South America drifted slowly west, allowing the South Atlantic to expand. Its relentless movement brought it into direct collision with the Nazca Plate under the Pacific Ocean. The collision forced the eastward-moving, smaller but heavier Nazca Plate to subduct under the oncoming, larger but lighter South American Plate. As the

Central and South America have only been linked by land for 3 million years

KEY EVENTS

80-70 million years ago Complex tectonic activity in a shallow ocean north of South America causes uplifting magma to cool in an arc of new islands—forming the Caribbean.

50 million years ago The Nazca Plate under the Pacific Ocean subducts beneath the South America Plate to the west of the continent, and this causes the Andes to start rising. Most of the mountains have grown over the last 10 million years.

15-10 million years ago The Andes rise even higher and block the Amazon River from flowing into the Pacific. It flows eastward and pools into a lake. About 10 million years ago, the Amazon starts to flow into the Atlantic, lowering the lake's water level, and the lush Amazon Rain Forest grows in its place (see opposite page).

180 million years ago The supercontinent Gondwana comprises South America, Africa, India, Australia, and Antarctica. South America and Africa separate from the rest about 180-170 million years ago, and South America starts its own journey when it splits from Africa 140-90 million years ago.

16-13 million years ago The area that will become the Andean Altiplano undergoes its first bout of elevation, with two more instances occuring 13-9 and 10-6 million years ago, which bring the area closer to its present-day height.

14 million years ago Desertification starts to occur in the area that will form the Atacama Desert.

3 million years ago A land bridge called the Isthmus of Panama forms, connecting South America with North America.

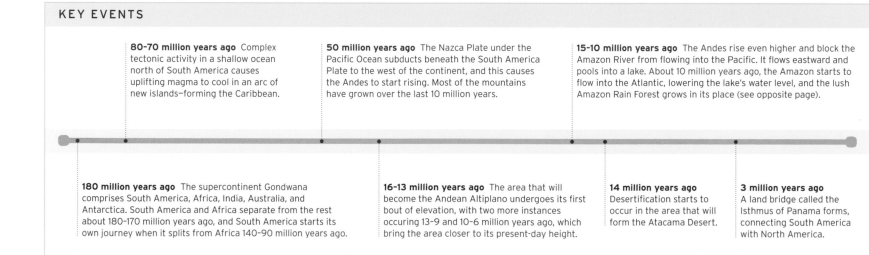

◁ **HIDDEN CAVE**
This underground lake in Chapada Diamantina National Park, Brazil, contains some of the continent's oldest rocks.

◁ **CONTINENTAL SPINE**
The Andes are the world's longest mountain range, stretching 4,500 miles (7,200 km) – almost along the entire western edge of South America.

◁ **CHANGED COURSE**
The Amazon River is the world's largest river, carrying huge amounts of water through its enormous, lush lowland basin. Until 10 million years ago, it used to be a huge inland lake.

◁ **RUPTURING VOLCANO**
The 16,480-ft- (5,023-m-) high Tungurahua volcano in Ecuador is one of many towering volcanoes that have burst through the Andes.

Nazca Plate was driven down into the mantle, a deep trench opened up in the ocean floor along the west coast of South America. The descending Nazca Plate dragged against the South American Plate, fracturing and deforming it.

The Andes started to form 50 million years ago as rocks were crumpled and piled on top of each other. Meanwhile, the Nazca Plate, melting as it was pushed into the mantle, caused magma to punch through the young mountain range as volcanoes, adding a series of towering volcanic peaks such as Cotopaxi in Venezuela, and Chimborazo and Tungurahua in Ecuador.

River basins

Between the three ancient shields and the newer fold mountains of the Andes are lowland river basins, filled over time with layer upon layer of sediment. There are three main basins, forming a linked lowland belt running north to south down the middle of the continent. The Orinoco in the north flows from the Venezuelan Andes into the Atlantic south of Trinidad. The sprawling Amazon River in the center flows right across South America from the Andes to emerge in the Atlantic at the equator. Until about 10 million years

ago, the Amazon pooled into a giant, inland lake and drained westward into the Pacific Ocean. The Chaco-Pampa in the south is a large basin through which the Parana, Paraguay, and other rivers flow into the Rio de la Plata.

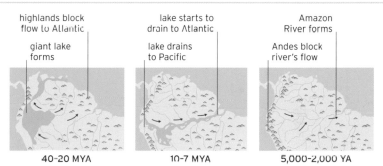

highlands block flow to Atlantic

giant lake forms

lake starts to drain to Atlantic

lake drains to Pacific

Amazon River forms

Andes block river's flow

40-20 MYA   **10-7 MYA**   **5,000-2,000 YA**

HOW THE AMAZON BASIN FORMED
The Amazon River flowed from the Guiana Highlands into a huge lake, which drained into the Gulf of Mexico and the Pacific Ocean. Eventually, the rising Andes forced the lake to drain into the Atlantic Ocean instead.

BUILDING THE CENTRAL AND SOUTH AMERICAN CONTINENT

KEY ━ Convergent boundary ━ Divergent boundary

Nazca Plate subducts under South American Plate

South America starts to split from Africa

Gondwana supercontinent

Central American Seaway

three shields form basis of continent

Mid-Atlantic Ridge spreads, South Atlantic opens

Africa drifts east

Central and South America linked by land

94 MILLION YEARS AGO South America begins to break away from Africa and move west. The South American Plate abuts the Nazca Plate, which subducts beneath it.

50-40 MILLION YEARS AGO The Mid-Atlantic Ridge spreads, forcing South America and Africa farther apart and creating the Atlantic Ocean. The Central American Seaway separates the Americas.

18,000 YEARS AGO The Central American Seaway, which once separated the continent from Central America, has now closed and South and North America have become linked by land.

W. Central America

Santa María

A dangerous Central American volcano that produced the third largest eruption of the 20th century

Santa María is one of the most prominent of a chain of large volcanoes that rise high above the Pacific coastal plain of Guatemala. This chain is the result of the process of subduction as the Cocos oceanic plate slides beneath the Caribbean Plate. For much of recorded history, forest-covered Santa María has been quiet, but in October 1902, it erupted, tearing a huge crater in its southwestern flank and killing more than 5,000 people.

Activity today

In 1922, a lava dome—a huge, mound-shaped mass of gray lava—began growing in the massive flank crater formed in the 1902 eruption. This has since developed into a complex of four overlapping domes. Right up to the present day, frequent eruptions have occurred from these domes, sometimes generating dangerous pyroclastic flows (see panel, p.89) or blasting plumes of ash to heights of a couple of miles. Smaller explosions occur almost daily and can be viewed from the summit of Santa María's quieter main cone.

△ **BURSTING FORTH**
Most current activity at Santa María consists of regular, often spectacular eruptions from the lava domes on its southern flank.

◁ **DAWN VIEW**
In this view at sunrise, Santa María's main volcanic cone, standing high in the background to the left, overlooks its group of dangerous lava domes.

SANTA MARÍA'S LAVA DOMES

Santa María's four lava domes, collectively known as the Santiaguito Complex, are called El Caliente ("the hot one"), La Mitad ("middle one"), El Monje ("the monk"), and El Brujo ("the sorcerer").

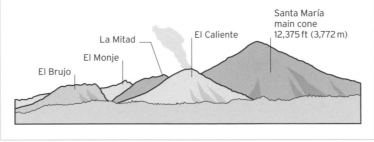

El Brujo

El Monje

La Mitad

El Caliente

Santa María main cone 12,375 ft (3,772 m)

Santa María's **1902 eruption** produced about **1.3 cubic miles (5.4 cubic km)** of **ash and pumice**

Masaya Volcano

A huge volcano in Nicaragua that sometimes features a rare, spectacular lake of fiery lava

Masaya is an example of a complex volcano: it has more than one volcanic cone, as well as multiple vents and craters. Its two main cones lie within a larger, roughly oval structure called the Masaya Caldera, which formed during a cataclysmic eruption some 2,500 years ago. The caldera is about 7 miles (11 km) long, 3 miles (5 km) wide, and has a rim that in places is 980 ft (300 m) high.

At the summit of Masaya's main cones are several deep holes called pit craters. One of these, the Santiago Crater, has been the main site of volcanic activity at Masaya for many years and sometimes contains a red-hot pool known as a lava lake.

THE MASAYA CALDERA

Masaya Volcano lies within the much larger Masaya Caldera. The volcano itself consists of two main overlapping volcanic cones. One contains at its summit two pit craters—the Santiago and San Pedro craters—while the other is indented by the larger San Fernando Crater.

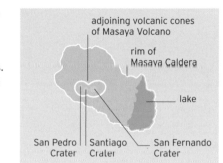

adjoining volcanic cones of Masaya Volcano

rim of Masaya Caldera

lake

San Pedro Crater | Santiago Crater | San Fernando Crater

△ **HELLISH PIT**
The Santiago Crater sometimes fills with a lava lake that produces spectacular bursting bubbles as it boils off noxious gases. It is one of just a handful of lava lakes in the world.

PYROCLASTIC FLOWS

Also known as a nuée ardente (glowing cloud), a pyroclastic flow is an avalanche of volcanic gases, ash, and lava fragments. It often forms from the collapse of a volcanic eruption cloud or a lava dome. Most pyroclastic flows travel for about 3-6 miles (5-10 km), typically moving downhill at around 60 mph (100 kph) or faster, burning, flattening, and burying everything they encounter along the way.

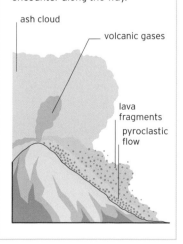

ash cloud

volcanic gases

lava fragments

pyroclastic flow

Soufrière Hills

A Caribbean volcano that has devastated over half the island of Montserrat since 1995

Caribbean

There are several volcanoes named Soufrière in the Caribbean—*soufrière* is French for "sulfur outlet"—but the most famous is on the island of Montserrat, called Soufrière Hills. Since 1995, several destructive events called pyroclastic flows (see panel, left) have caused it to be the most studied volcano in the world.

Collapsing lava domes

Soufrière Hills produces a viscous lava that, instead of flowing away, piles up to form a lava dome. This emits a lot of steam and sometimes glows at night. The danger with lava domes, though, is that they can disintegrate to produce pyroclastic flows. At Soufrière Hills, a series of dome collapses and flows between 1995 and 2000 destroyed Montserrat's capital city, Plymouth, obliterated many settlements, and caused several deaths. Two-thirds of the population had to leave the island. However, since 2010, the volcano has quieted, and Montserrat remains a compelling place to visit because of its scenic beauty and volcanic history.

△ **DANGEROUS DOME**
The gray conical object in the background here is a lava dome. Part of its flank has begun to break up, triggering a pyroclastic flow, which resembles an avalanche.

N. South
America

The Guiana Highlands

A spectacular area of South America with astonishing flat-topped mountains and some of the world's most impressive waterfalls

The Guiana Highlands extend across the northern part of South America, occupying much of southern Venezuela, as well as parts of Guyana and northern Brazil. The Highlands are famous for their ancient and magnificent flat-topped mountains called tepuis. The word *tepui* means "house of the gods" in the language of the local Pemon people. Many of these landforms appear otherworldly, with their sheer cliffs and flat, vegetated, seemingly inaccessible tops. Some of the most famous include Auyan Tepui, Mount Roraima, and Kukenan Tepui. Mount Roraima is said to have inspired the Scottish author Sir Arthur Conan Doyle to write his novel *The Lost World*, about the discovery of a remote plateau where prehistoric creatures still exist.

Auyan Tepui and Angel Falls

Auyan Tepui is the largest of the tepuis, with a surface area of 270 square miles (700 square km), partly covered by cloud forest. The world's highest uninterrupted waterfall, Angel Falls, drops from a cleft in its summit. The Churun River, a tributary of the Orinoco, flows over the falls, descending a total of 3,212 ft (979 m) to the valley floor. The main plunge, of 2,648 ft (807 m), is followed by a series of sloping cascades and rapids.

◁ **ENDEMIC FLORA**
A long period of isolation means that many animal and plant species, such as the example from the genus *Orectanthe* shown here, have evolved on tepui summits and are found only on tepuis.

The **rock** from which **Mount Roraima** was **carved out** was laid down about **1.8 billion years ago**

△ **LOST WORLD**
One of the highest tepuis, Mount Roraima is surrounded by stepped cliffs up to 1,300 ft (400 m) high and is the source of several rivers.

◁ **LONG DROP**
In Angel Falls' uppermost thousand feet, the water barely makes contact with the sheer sandstone cliffs behind.

HOW TEPUIS FORM

The Guiana Highlands were once an extensive sandstone plateau. Over a period of some 1.5 billion years, weak sections of the plateau were eroded away by rivers and streams, leaving behind the tepuis.

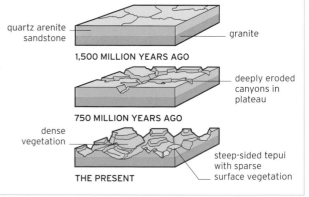

quartz arenite sandstone — granite

1,500 MILLION YEARS AGO

deeply eroded canyons in plateau

750 MILLION YEARS AGO

dense vegetation

steep-sided tepui with sparse surface vegetation

THE PRESENT

The Andes

The world's longest continental mountain range, comprising a series of impressive snow-capped peaks, volcanoes, and high plateaus

W. South America

The Andes extend for 4,500 miles (7,200 km) north to south along the western edge of South America, passing through Venezuela, Colombia, Ecuador, Peru, Bolivia, Argentina, and Chile. The range comprises several distinct subranges that formed in the last 50 million years. Some of these sit on either side of high-plateau regions, one of which, the Altiplano (see pp.96–97), is the world's second highest plateau after the Tibetan Plateau.

Deserts and glaciers

The Andes are a formidable physical and biological barrier between the Pacific coast and the rest of South America, ranging in width from 60 miles (100km) wide at the southern tip of the continent to 440 miles (700km) in the central region. A variety of physical features can be found across the range, including one of the most arid places on Earth—the Atacama Desert (see pp.124–25)—

and some of the world's deepest canyons. In the Altiplano, there are vast salt flats and lakes. The southern sector of the Andes has been heavily glaciated in the past, and glaciers still extend down to sea level in southern Chile, where there is an extensive network of fjords (see p.117).

Volcanoes and earthquakes

Interspersed within the Andes are about 180 active and numerous extinct volcanoes. They are the result of tectonic activity (see pp.94–95) that results in the production of magma under the Andes, where it rises to create substantial volcanic activity. This tectonic activity also generates earthquakes, including the most powerful ever recorded, which had its epicenter off the coast of Chile in 1960.

▽ **MOUNTAIN PACK ANIMAL**
Along with alpacas, vicuñas, and guanacos, llamas are one of four relatives of camels found in the Andes. While llamas and alpacas are domesticated, the others are wild.

△ **DIRTY THUNDERSTORM**
Lightning in and around a volcanic ash plume—as in this 2015 eruption of the Calbuco volcano in Chile—is referred to as a dirty thunderstorm. The flashes of lightning result from friction between ash particles and the atmosphere.

THE SOUTHERN ANDES RAIN SHADOW

In the southern Andes, the prevailing winds blow from the west and push moisture-laden air from the Pacific Ocean toward the mountains. As this air meets the Andes, it rises, expands, and cools, dropping rain on the mountains' western flanks. However, it then continues over and down the eastern side as a dry wind, creating arid conditions (a rain shadow) in the southeastern part of South America.

rising air cools and condenses

rain falls on western slopes

rain shadow on eastern slope

prevailing wind

evaporation

ocean

△ **ROOF OF THE AMERICAS**
With its highest point at 22,838 ft (6,961 m), Mount Aconcagua, seen here at the center, is the highest peak not just in the Andes but in the entire world outside of Asia. It lies in Argentina, just east of the country's border with Chile.

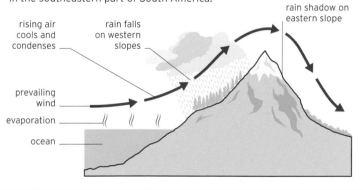

Cotopaxi, in Ecuador, is one of the **highest active volcanoes** in the world. It has **erupted 50 times** since **1738**

Building a
MOUNTAIN CHAIN

Most mountain chains form where two tectonic plates, at least one of which carries continental crust, move toward each other, or converge, driven by convection in Earth's mantle. If both carry landmasses, these collide and mountains are pushed up; the Himalayas formed in this way. But a mountain chain can also form if there is a landmass on just one of the plates—the formation of the Andes provides a prime example.

The creation of the Andes

The process that built the Andes unfolded over tens of millions of years and began with the development of a new convergent plate boundary to the west of South America. The ocean-dominated plates to the west of the boundary, which include the precursor of today's Nazca and Cocos Plates, as well as the Antarctic Plate, began to subduct, or move under, the South American Plate to their east. This movement squeezed the lithosphere (the crust and top layer of mantle) of the western part of South America, causing major fracturing, called thrust faulting, and uplift of blocks of lithosphere. In parts of the region, a single mountain chain was created; in others, parallel subranges were pushed up in different phases. The subduction process led to the production of magma at depth, which rose up to create volcanism in many parts of the chain.

ANDEAN VOLCANIC ZONES

The Andean volcanoes are clustered in four regions—the northern, central, southern, and austral volcanic zones. Between these belts lie large gaps containing no volcanoes. For example, there are around 30 volcanoes in southern Peru, but none in the rest of the Peruvian Andes. The reasons for these gaps are not fully known. One theory is that the tectonic plates that subduct under South America do so at a narrower angle of dip in these "gap" regions.

northern volcanic zone

central volcanic zone

southern volcanic zone

austral volcanic zone

▶ **THE CENTRAL ANDES**
This is a section through the central Andes of Bolivia and northern Chile, which also extends in the north into Peru. In this region, two roughly parallel subranges–the western and eastern cordilleras–are separated by a high, wide plateau called the Altiplano.

▷ **WESTERN SLOPES**
This region, the Atacama Desert, is extremely dry. Winds in these latitudes blow mainly from the east. As they push moist air up the eastern side of the Andes, rain falls on that side, leaving the western slopes and coastal settlements dry.

Ocean trench
The Peru-Chile Trench marks the place where one plate moves under the other. It is 4-5 miles (7-8 km) deep.

Nazca Plate
This plate is about 30-40 miles (50-60 km) thick.

the Nazca Plate is slowly moving toward and under the South American Plate at a rate of around 3 in (7.5 cm) a year

Lithospheric mantle
This is the uppermost layer of the mantle. With the overlying crust, it comprises a relatively strong layer called the lithosphere, which is divided into plates.

Asthenosphere
This relatively deformable layer of Earth's upper mantle is warmer than the overlying lithosphere.

Oceanic crust
The top part of the descending plate is oceanic crust. Made of rocks such as basalt and gabbro, it holds substantial amounts of water.

As the Andes formed, South America shortened from west to east

the Western Cordillera is a subrange of the central Andes

Lake Titicaca

◁ SNOW-CAPPED VOLCANO
Recent volcanic activity in the Western Cordillera is confined to an arc of over 40 active volcanoes with many adjacent extinct volcanoes, such as Nevado Sajama (pictured here).

◁ SALINE LAKES
There are numerous salty lakes on the Altiplano (such as Lake Poopo, seen here). Their salinity is due to the fact that all rivers in the region drain inland, which means that salts produced by weathering of the nearby mountains are trapped here.

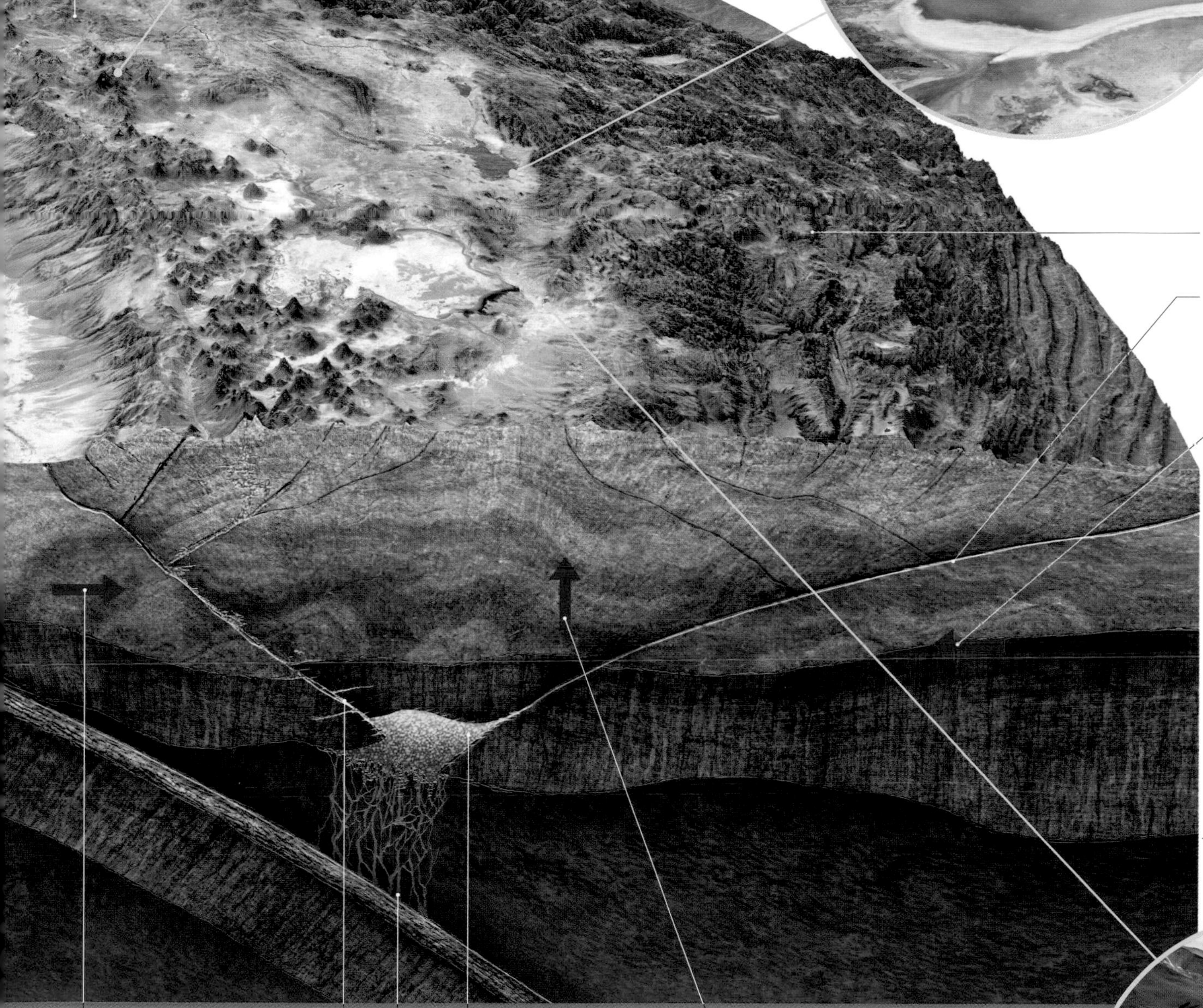

the Eastern Cordillera is broad in parts, with topography resulting from multiple thrust faults and the deformation of crustal layers

Thrust fault
Faults or fractures along which movement and uplift have occurred are called thrust faults.

South American Plate moving west relative to subducting Nazca Plate

South American Plate
In the Andean region, the South American Plate is around 90 miles (150 km) thick. The upper 25 miles (40 km) are continental crust; the rest is lithospheric mantle.

coastal block of continental crust being pushed east relative to stable South American Plate

Magma
Pockets of molten magma are generated by lower melting temperature.

continental crust squeezed and pushed upwards

the magma flow often follows fractures and other lines of weakness in the crust on its way to the surface

water is released from descending oceanic crust and moves into the mantle, where it lowers the melting temperature

▷ HIGH PLATEAU
Overlying a crustal block that has risen about 2 miles (3 km) in the past 25 million years, the Altiplano is a windswept region around 90–120 miles (150–200 km) wide of plains, lakes, and salt flats.

W. South
America

The Altiplano

A vast, high plateau of the central Andes with stunning, otherworldly scenery and harsh living conditions

Set against a backdrop of volcanoes, the Altiplano is a windswept landscape of sparsely vegetated high plains, lakes, and salt flats, occupying about 77,000 square miles (200,000 square km) of the central Andes. Much of it lies in Bolivia, but parts of it extend into Peru, Chile, and Argentina in the south. It contains one large urban area—the adjacent cities of La Paz and El Alto in Bolivia.

Land of fire, ice, salt, and wind

The Altiplano is bordered on the west by a subrange of the Andes, the Western Cordillera, which includes some massive volcanoes. The Altiplano is mostly arid, especially in the south, where the annual rainfall is only about 8 in (200 mm). It also has some large salt flats, such as the Salar de Uyuni (see p.126).

Because of the high altitude, the air is thin and the climate is quite harsh. Although temperatures during the day can reach 50–77°F (10–25°C), at night they plunge to as low as -4°F (-20°C) in the southern Altiplano. The local wildlife, which is adapted to these conditions, includes vicuñas (llama relatives), viscachas (rabbitlike rodents), and the Andean fox.

◁ PROTECTIVE COAT
The viscacha's thick, fine fur allows it to withstand year-round freezing nighttime temperatures.

The **average height** of the Altiplano is 12,300 ft (3,750 m)

BENEATH THE ALTIPLANO

In the central Andes, two or more phases of mountain building pushed up the Eastern and Western Cordilleras, which are separated by the Altiplano. The details of how the Altiplano formed are fiercely debated, although recent research suggests that the crust under it was lifted about 2 miles (3 km) in a series of spurts in the last 25 million years.

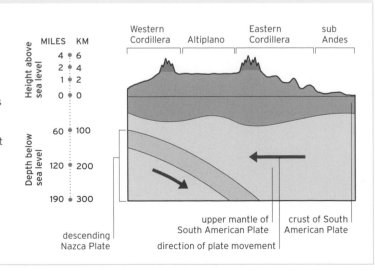

	MILES	KM
Height above sea level	4	6
	2	4
	1	2
	0	0
Depth below sea level	60	100
	120	200
	190	300

Western Cordillera · Altiplano · Eastern Cordillera · sub Andes

descending Nazca Plate

direction of plate movement

upper mantle of South American Plate

crust of South American Plate

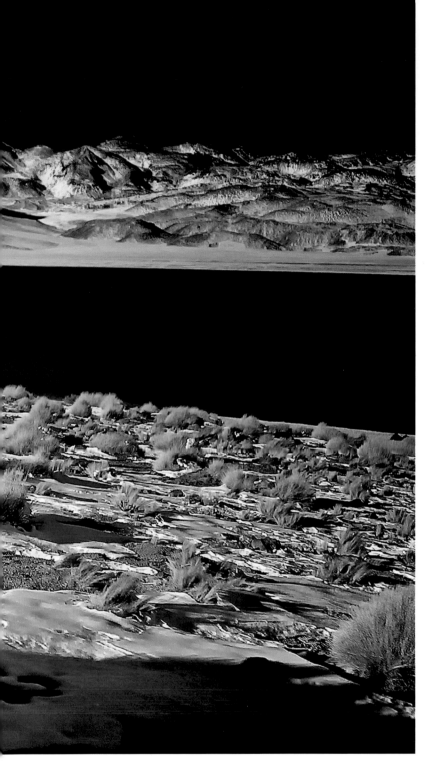

W. South
America

El Tatio Geysers

*One of the highest concentrations of geysers in the
world, high in the Chilean Andes*

El Tatio is the third largest geyser
field in the world and by far the
highest-altitude large field, at
14,170 ft (4,320 m) above sea
level. It lies in a volcanic region
on the edge of Chile's Atacama
Desert (see pp.124–25) and can
be reached from San Pedro de
Atacama—one of the nearest
towns—via a minor road.

The geysers here do not shoot
up to enormous heights—the
tallest eruptions observed are
to around 20 ft (6 m). However,
the sheer density of geysers
makes up for this: more than
20 of them are within a few
minutes' walk. Interspersed
between the geysers are hot
springs and fumaroles (gas-
emitting volcanic vents).
Visitors have to be warned
not to approach the springs
and geyser pools too closely
because fatalities have occurred
when people have fallen
into the scalding water.

LAYOUT OF THE GEYSER FIELD

El Tatio contains around
80 geysers spread over an
area of about 1.9 square miles
(5 square km). Most are
concentrated within three
main areas: the upper, middle,
and lower geyser basins.

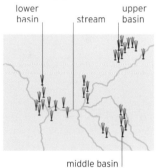

lower
basin   stream   upper
basin

middle basin

▽ SCALDING WATER SPOUT
The water spouted out by this and
other geysers at El Tatio is extremely
hot—up to 185°F (85°C), which is
close to the boiling point of water
at this altitude.

△ CHILEAN
ALTIPLANO
This part of the Altiplano
in Chile features a dark
blue lake, Laguna
Miscanti, with brackish
water. Behind it towers
an extinct volcano, Cerro
Miscanti. It is surrounded
by yellow bunch grass.

◁ SCARLET LAKE
Situated near the
Chile-Bolivia border, and
known locally as Laguna
Roja, this is actually a hot
spring with a temperature
of 104-122°F (40-50°C).
A colony of bright red,
heat-loving algae
thrives in the water.

The Sugarloaf Mountain

A monolith made from granite and quartz that rises vertically from the water's edge in Rio de Janeiro

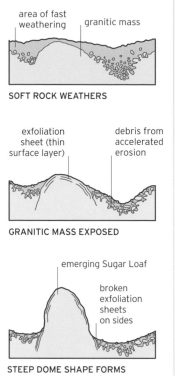

E. South America

Rising to a height of 1,299 ft (396 m) above Rio de Janeiro, the Sugar Loaf is situated on a peninsula that juts out into the mouth of Guanabara Bay—a large sea inlet that includes Rio's harbor.

The rock from which the Sugar Loaf and other nearby dome-shaped mountains are composed was formed millions of years ago from magma rising from deep underground and then solidifying beneath the surface. Starting around 100 million years ago, South America separated from Africa to form the southern Atlantic Ocean, resulting in fractures in Earth's crust around the eastern coast of South America. The fractures created weaknesses in the rocks surrounding the granitic masses, allowing them to be broken down by agents such as rain (see panel, right).

▷ **RISING STRAIGHT UP**
From the top of the Sugarloaf (in the center in this image), there is a bird's-eye view of both Rio's harbor, seen in the foreground here, and Guanabara Bay, visible on the mountain's left.

HOW THE SUGAR LOAF FORMED

The Sugar Loaf began as a mass of granite underground. It was exposed at the surface as the softer rock above and around it was weathered away. Further erosion created its shape.

area of fast weathering   granitic mass

SOFT ROCK WEATHERS

exfoliation sheet (thin surface layer)   debris from accelerated erosion

GRANITIC MASS EXPOSED

emerging Sugar Loaf

broken exfoliation sheets on sides

STEEP DOME SHAPE FORMS

The Hornocal Mountains

A small mountain range in northwest Argentina with a spectacular zigzag structure and incredible colors

E. South America

▽ **DELUGE OF COLOR**
The hues of the Hornocal Mountains' rock layers range from vivid reds to creams and gray-greens.

The Hornocal Mountains are part of a formation of sedimentary rocks that extends for some 310 miles (500 km) southward from Peru. Known as the Yacoraite formation, it is composed mainly of limestones, as well as some sandstones and siltstones, and was formed in a shallow sea between 72 and 61 million years ago (when this part of South America was underwater).

Variations in the mineral composition of the sediments deposited into the sea, the chemistry of the seawater, and the life-forms present at the time account for the striking variation in the rock colors. Over the years, these rock layers were lifted up and tilted as a result of tectonic plate movements, then weathered and eroded, resulting in the breathtaking sight that is visible today. The riot of colors lends a slightly surreal appearance to the location.

The Torres del Paine

An area of astonishing scenic beauty in Patagonia with snow-capped mountains, glaciers, waterfalls, lakes, and rivers

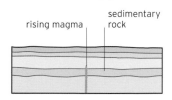

S. South America

Some 1,220 miles (1,960 km) south of Santiago, the capital of Chile, lies one of South America's most spectacular national parks, the Torres del Paine (Towers of Paine). *Paine,* pronounced "pie-nay," means "blue" in the native Tehuelche language—a reference to the fact that the park's mountains appear slightly azure from a distance.

Granite spires and sharp peaks

One of the most iconic features of the Torres del Paine is a group of three huge granite spires—the torres, or towers—rising 4,900 ft (1,500 m) above the Patagonian steppe. Close by, and part of the same massif, are the Cuernos del Paine (Horns of Paine)—three

enormous mountains surmounted at their tops by sharp black "tusks" of rock. For many visitors, these are the outstanding attraction because of their spectacular forms, sharp edges, and different rock colors. Other high peaks in the park include Cerro Paine Grande at 9,460 ft (2,884 m) and Mount Almirante Nieto at 8,660 ft (2,640 m).

One of the world's last remaining wildernesses, the Torres del Paine has a unique range of wildlife, including guanacos, foxes, rheas, and Andean condors. The park is adorned with beautiful vegetation, including evergreen shrubs, such as Chilean Fire Bush.

▽ TOWERING HIGH

The black-topped Cuernos del Paine, which soar to more than 8,200 ft (2,500 m), are visible on the left here, while Mount Almirante Nieto can be seen on the right.

◁ THE THREE TOWERS

Torre Sur on the left, Torre Centrale, rising to 9,200 ft (2,800 m), and Torre Norte on the right form the three tower granite centerpiece of the park.

HOW THE TOWERS FORMED

The Torres del Paine was formed partly from granite that originated from magma rising into the subsurface 13 million years ago. The magma cooled to form a granite block, which was later exposed when the area was eroded by glaciers.

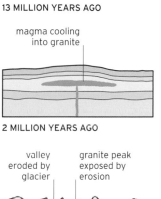

rising magma

sedimentary rock

13 MILLION YEARS AGO

magma cooling into granite

2 MILLION YEARS AGO

valley eroded by glacier

granite peak exposed by erosion

TODAY

The Quelccaya Ice Cap

The largest glaciated area in the world's tropical regions, located in the Cordillera Oriental of the Peruvian Andes

W. South America

The Quelccaya Ice Cap covers an area of 17 square miles (44 square km) of the High Andes and—reflecting how high the freezing zone is in the tropics—all of it lies above an altitude of 16,000 ft (5,000 m). The ice cap is of great importance to the surrounding region because its meltwater irrigates fields in the valleys below and provides drinking water for some large Peruvian cities, like Cusco. However, due to global warming, the ice cap has lost about one-fifth of its area over the past 30 years, and the rate at which it is shrinking is accelerating. The recent appearance of 5,700-year-old plant remains from under the ice cap indicate that more than 50 centuries have passed since it was smaller than it is today.

▷ **KEYHOLE LAGOON**
This unusual-looking keyhole-shaped feature is a meltwater lake—temporarily iced over—that developed on one edge of the ice cap in the 2000s.

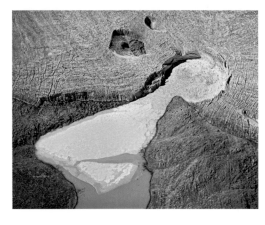

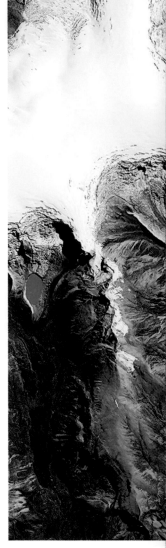

The **ice cap** lies close to the **Amazon rain forest**, where **temperatures** are about **80°F (27°C) higher**

The Pastoruri Glacier

A small Peruvian glacier that is shrinking but still a remarkable sight and a magnet for enthusiasts of various snow- and ice-related sports

W. South America

The Pastoruri Glacier lies in the southern part of the Cordillera Blanca in the Andes of north central Peru, occupying one flank of a 17,200-ft- (5,240-m-) high Andean peak called Nevado Pastoruri. It is a cirque glacier—a type of glacier that occupies a bowl-shaped depression high up on a mountain.

Steep sides

The Pastoruri Glacier has steep, clifflike edges, making it a popular destination for ice climbers. Although parts of its surface are heavily crevassed, others are often covered in deep, soft snow, which attracts skiers and snowboarders. The glacier has lost over 15 percent of its ice over the past 35 years, and it is expected to continue to shrink.

▷ **BLUE ICE CAVE**
Inside the Pastoruri Glacier, at a height of around 17,000 ft (5,200 m), there is an ice cave. Visitors can hike around the glacier and visit the glowing blue interior.

◁ **HIGH-ALTITUDE CAP**
In this false-color satellite image, part of the ice cap and some small lakes surrounding it are visible. Vegetation in nearby valleys appears red.

▽ **MOUNTAIN CACTI**
Various species of cacti grow in the high Peruvian Andes near the ice cap, including this variety in the *Tephrocactus* genus, often described as a mounding cactus.

The Southern Patagonian Icefield

The world's second largest continuous icefield, stretching almost 230 miles (370 km) down the southern Andes

SW. South America

An icefield is a single, extensive mass of ice at high altitude that is partly hemmed in by mountains (in contrast, ice caps and ice sheets almost completely submerge a region). The Southern Patagonian Icefield is a remnant of the last glacial period, around 18,000 years ago, when all of southern Chile and Argentina was covered in a thick sheet of ice, over an area estimated to be about 185,000 square miles (480,000 square km). Today, it occupies an area less than 3 percent of that size. Mount Fitz Roy at 11,020 ft (3,359 m) and the volcano Lautaro at 11,886 ft (3,623 m) are two of the main peaks surrounded by the icefield.

▷ **OUTLET GLACIER**
The Gray Glacier is one of dozens of outlet glaciers that carry ice down from the icefield. Some end in glacial lakes, while others discharge icebergs into Chilean fjords (see p.116).

THE ICEFIELD AND ITS OUTLET GLACIERS

The Southern Patagonian Icefield is long and thin, with an average width of only about 22 miles (35 km). Some of the largest of its numerous outlet glaciers are marked here.

Occidental Glacier

Bruggen Glacier

Jorge Montt Glacier

O'Higgins Glacier

Viedma Glacier

Lago Argentino

Perito Moreno Glacier

Gray Glacier

The Perito Moreno Glacier

An impressive Patagonian glacier that can be seen discharging huge quantities of ice into an Argentinian lake

S. South America

The Perito Moreno is an outlet glacier from the Southern Patagonian Icefield (see p.101), descending to the east of the icefield, where it discharges icebergs into Argentina's largest lake, Lago Argentino. The glacier is about 15 miles (24 km) long, has an area of about 80 square miles (200 square km), and terminates in a vast wall of ice that is some 3 miles (5 km) wide. Unlike most of the world's glaciers, the Perito Moreno is not retreating, although glaciologists are uncertain why that is.

Calving face

Situated in Los Glaciares National Park, the glacier is one of Patagonia's most visited sites. It is possible to go on a guided trek over the glacier's surface. Many sightseers enjoy witnessing huge shards of ice falling off the glacier's calving face. The best time to view this is in the afternoon when, due to the Sun's warmth, disintegration events occur about every half an hour.

THE BRAZO RICO ICE DAM

The glacier periodically advances and blocks off Brazo Rico, an arm of Lago Argentino. An ice dam forms, and Brazo Rico may then rise as much as 200 ft (60 m) before its waters break through the dam, discharging about a billion tons of water in 24 hours.

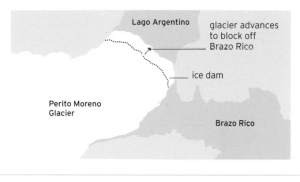

Lago Argentino

glacier advances to block off Brazo Rico

ice dam

Perito Moreno Glacier

Brazo Rico

The Perito Moreno's **iceberg-calving** wall of ice towers **240 ft (74 m)** above Lago Argentino

▷ **CLOSING IN**
Here, the glacier is forming a dam between Brazo Rico (at the bottom) and Lago Argentino (at the top). Icebergs can be seen calving into Brazo Rico.

◁ **SAILING BY**
Massive blocks and pinnacles of ice slowly move downhill. In the foreground are boulders and other rock debris deposited at the glacier's side.

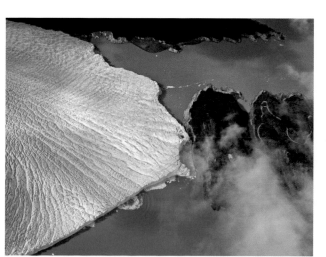

△ **CRASHING DOWN**
The ice dam eventually breaks down as water pressure forces a tunnel through it. Soon afterward, the ice bridge over the tunnel dramatically collapses.

The Llanos

A huge expanse of savanna plains, transformed by seasonal flooding and supporting a unique collection of animals

N. South America

△ **WATCHFUL EYE**
There are more than 100 species of reptiles living in the Llanos, including many different turtles, snakes, and crocodilians. One of the most abundant is the spectacled caiman.

The Llanos, or "Plains," is a sprawling tropical grassland region, stretching across Colombia and Venezuela from the foothills of the Andes to the Orinoco Delta, where it meets the Caribbean. Although the Llanos contains a mosaic of different habitats, including woodland and swamps, most is treeless savanna, which is subject to seasonal flooding by the Orinoco River's tributaries.

Unique wildlife

The Llanos is home to many species not typically found in savanna ecosystems. These include the critically endangered Orinoco crocodile, which is only found in the Llanos, and the largest rodent in the world, the semiaquatic capybara. The region supports several hundred species of birds—both residents and migrants—with a large proportion of waterfowl and waders.

long, gently curving bill

partly webbed feet

△ **DISTINCTIVE BILL**
The Llanos contains the world's largest population of scarlet ibis. The bird uses its long, thin bill to probe for food such as insects in soft mud and under plants.

SEASONAL FLOODING

The Llanos is a region of climatic extremes with seasonal flooding and droughts. More than 90 percent of its annual rainfall occurs between April and October. By May, the tributaries of the Orinoco have started to overflow, and at the height of the season much of the Llanos is wetland.

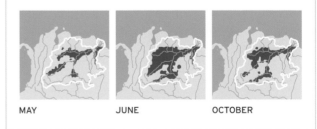

MAY   JUNE   OCTOBER

The Llanos is home to the **world's most powerful snake**, the green anaconda

N. South
America

Caño Cristales

Colorful aquatic plants and clear water combine to create a liquid rainbow

On the Serrania de la Macarena tableland in central Colombia runs a 60-mile- (100-km-) long river that for five months of the year turns a combination of red, green, yellow, and blue. The Caño Cristales' colorful display is due to the blooming of an endemic plant called *Macarenia clavigera*, which requires a precise combination of water and sunlight levels to take on its bright red hues. Other green plants, the sandy riverbed, and the clear water reflecting the sky provide the other colors. Despite being rich in aquatic plants, the river is thought to contain no fish.

△ **CRYSTAL CLEAR WATERS**
The Caño Cristales is a relatively fast-flowing river with waterfalls, rapids, and deep, swirling pools. The water lacks sediment and nutrients, and as a result is extremely clear.

W. South
America

Lake Titicaca

The world's highest navigable lake, and South America's largest

The largest freshwater lake in South America, Lake Titicaca covers 3,230 square miles (8,370 square km), straddling the border between Peru and Bolivia. It is situated in the Altiplano basin (see p.96), high in the Andes, and at 12,507 ft (3,812 m) above sea level is the world's highest commercially navigable lake.

The lake's two sub-basins are connected by the Strait of Tiquina. The larger section to the north, Lago Grande, is much deeper than the smaller Wiñaymarka. More than 25 rivers from five major systems drain into the lake, but it has just one outflow, the Desaguadero River, at its southern end. The river accounts for only about 5 percent of Lake Titicaca's water loss; the rest is lost by evaporation in its arid, windy environment.

◁ **FALLING LEVELS**
In the last few years, the water level of Lake Titicaca has been slowly falling. Lack of rainfall has been compounded by reduced levels of summer meltwater as glaciers shrink due to climate change.

ALTIPLANO LAKES

This cross-section of the northern and central drainage basins of the Altiplano shows the current levels of Lake Titicaca and Lake Poopó alongside levels of ancient lakes that have dried up, leaving salt flats behind. By the end of 2015, Lake Poopó had also completely dried up, partly due to low levels of water in Lake Titicaca, which ultimately feeds it.

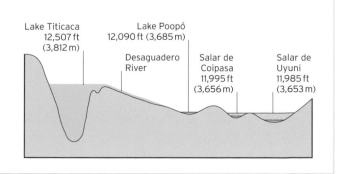

Lake Titicaca 12,507 ft (3,812 m)

Lake Poopó 12,090 ft (3,685 m)

Desaguadero River

Salar de Coipasa 11,995 ft (3,656 m)

Salar de Uyuni 11,985 ft (3,653 m)

The Amazon

The world's largest river, snaking through the most extensive rain forest on Earth and supporting a unique ecosystem

N. South America

The greatest river on Earth, measured by both the size of its catchment area and the volume of water flowing through its channels, the Amazon carries one-fifth of all the freshwater entering Earth's oceans. Its vast basin drains more than 40 percent of South America. About two-thirds of its main stream and by far the largest portion of its catchment is in Brazil. The Amazon and more than 1,000 tributaries support the great biodiversity of the world's largest rain forest, with which it shares its name (see pp.120–21). Although the source of the Amazon (see p.108) and its total length have been the subject of much debate, it is widely regarded the world's second longest river, after the Nile (see pp.190–91).

The Amazon rises high in the Andes of southern Peru, less than 125 miles (200km) from the Pacific Coast, and flows more than 4,000 miles (6,400km) to the northeast coast of Brazil, where it discharges an average of 53 billion gallons (200 billion liters) every second into the Atlantic Ocean. During flooding, some parts of the Amazon can reach a width of more than 30 miles (50km). The river's estuary spans an area more than 185 miles (300km) wide.

◁ **FIERCE REPUTATION**
Despite its reputation as a ferocious predator, the red-bellied piranha mainly forages and scavenges for food and travels in schools for protection against larger predators.

The Amazon carries **more water** than the **next seven largest rivers** combined

EARTH'S LARGEST RIVER BASINS

A river's drainage basin, also known as its catchment area, is the total area of land that is drained of water by the river and all its tributaries. The Amazon has by far the largest drainage basin of any river on Earth. Its area is about twice the size of the next largest, that of the Congo in Central Africa (see p.192).

Amazon:
2.7 million square miles
(7 million square km)

Congo:
1.4 million square miles
(3.5 million square km)

Nile:
1.3 million square miles
(3.4 million square km)

Mississippi-Missouri:
1.2 million square miles
(3.2 million square km)

Ganges: 0.4 million square miles
(1 million square km)

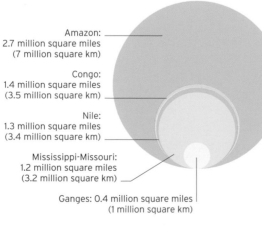

△ **VIGOROUS GROWER**
The water hyacinth is fast-growing, and outside its native range of the Amazon Basin it can be highly invasive, blocking up waterways and inhibiting other wildlife.

▷ **WIDE LOOPS**
A large part of the Amazon Basin is flat, and for much of its course the river and many of its tributaries meander slowly through vast areas of pristine rain forest.

The features of a
RIVER BASIN

A river basin is a geographic area in which all of the water that enters, usually by precipitation, drains to a common outlet through a river and its network of tributaries. River basins vary greatly in shape, complexity, and size—with the Amazon Basin being by far the largest on Earth. However, all river basins share a set of common features.

Source to mouth

A basin's watershed marks its perimeter, separating it from the basins of adjacent river systems. Naturally, watersheds run through areas of high ground, such as mountain ranges and plateaus. The source of a river system—which may be a spring, glacial meltwater, or a marsh—is the farthest point of a river's course from its eventual outlet, measured along the river's course itself. The elevated sections of a river basin feature streams that are usually fast flowing, with rapids and waterfalls, that descend onto lower, flatter terrain and become slower and wider, and often meander.

Most channels in a basin are tributaries that empty into the main river. Tributaries, such as the Amazon's Rio Negro, may themselves be substantial rivers, and waters from disparate sources, of different content and color, may come together at their meeting points, or confluences. A mature river in its lower course usually creates a floodplain by dropping sediment on its valley floor during floods. It meanders across this plain, its course shifting, either gradually as it erodes its banks or abruptly after severe floods. Most river basins—but not all (see p.197)—empty into an ocean or a lake. A river's mouth usually features an estuary where freshwater and saltwater mix, and often a delta, formed by the accumulation of sediments carried by the river.

◁ **SOURCE**
Locating the source of a great river is often contentious. Since 2014, scientists have believed the Amazon's source to be in the Mantaro river basin, high in the Peruvian Andes.

Watershed
The boundary of a river basin, called the watershed, commonly runs along a mountain ridge. The Andes form most of the western watershed of the Amazon.

Pacific Ocean

Upper course
In the upper course of a river, it tends to descend steeply and is joined by many small tributaries. In the Amazon's upper course, rivers and streams cascade down steep slopes of the Andes.

region of seasonal flooding

▶ **CATCHMENT AREA**
A river basin, also called its catchment area, encompasses the steep upper courses of a river system as well as the gentle gradients of the river's lower courses. In the Amazon's lower course, the river flows over such flat terrain that it falls only 230 ft (70 m) from the city of Manaus to the sea–a distance of some 820 miles (1,400 km).

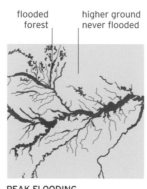

△ **AQUATIC ACROBAT**
Unlike oceanic dolphins, the Amazon river dolphin has a flexible neck, which helps it negotiate vegetation in the flooded forest. The dolphins are seen lifting their heads clear of the water to check the surroundings, and occasionally they jump up to 3 ft (1 m) above the water's surface.

REGIONAL FLOODING

Every year, vast areas of the Amazon's floodplains become inundated with water, creating the most extensive system of river-flooded forests in the world. The weight of water is so great that GPS sensors show it depressing the land by 3 in (7.5 cm) during peak flood. Due to the Amazon Basin's vast size, and the uneven distribution of its seasonal rainfall, different regions in the basin are affected at different times.

water confined to river channels

flooded forest

higher ground never flooded

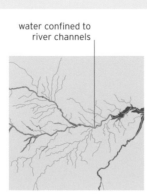

LOW-WATER PERIOD

PEAK FLOODING

It takes **Amazon river water** at least **one month** to flow **from the Andes to the Atlantic**

Main channel
The largest channel in a river system usually bears the name of the main river, but not always. Here, above the confluence with the Rio Negro, the Amazon's main channel is known locally as the Rio Solimões.

◁ **CONFLUENCE**
At a confluence known as the Meeting of Waters, just downstream of the city of Manaus, the dark waters of the Rio Negro ("black river") meet the sandy-colored waters of the Rio Solimões. The waters run side by side for several miles before mixing.

Plateau
A broad plateau, rather than a sharp mountain ridge, may divide river basins. The Guiana Highlands plateau forms the northeast section of the Amazon's watershed. Rain falling on the northwest of this plateau flows into the Orinoco.

Atlantic Ocean

Shelf area
Beyond a river's delta, river sediment continues to settle, building up a shelf of material on the seabed. On the Amazon's shelf area, a newly discovered reef system thrives, even though the river's vast sediment plume cuts out most sunlight.

△ **ESTUARY**
The form of a river's mouth depends on many factors, including the load of sediment and the volume of water it carries. The Amazon's broad mouth contains many islands, including Marajó Island, thought to be the world's largest island created by river sediment.

◁ **FLOODPLAIN**
Repeated flooding distributes a river's sediment over its valley floor, where it settles in perfectly flat layers, creating a floodplain. In the rainy season, the Amazon's floodplain can be covered by up to 28 ft (9 m) of water. To cope with this, many floodplain inhabitants live in stilt houses.

Deforestation
Changes in vegetation affect the speed of water runoff into a river system and the amount of soil erosion. Deforestation in the Amazon Basin makes flood waters peak more quickly and at a higher level. It also increases soil erosion and sediment load in the river.

Upon seeing Iguaçu Falls for the first time, **U.S. First Lady Eleanor Roosevelt** is reputed to have exclaimed **"Poor Niagara!"**

◁ **WATER VOLUME**
The amount of water in a waterfall determines its erosive power. On average, more than 26.4 million gallons (100 million liters) of water flows over Iguaçu Falls every minute—one of the largest volumes of any waterfall on Earth.

Devil's Throat

▷ **FEEDING OPPORTUNITY**
Waterfalls may be home to a variety of plants and animals. Black-fronted piping guans live in the Atlantic Forest and sometimes venture out onto Iguaçu Falls to search for mollusks and insects.

viewing gallery on Brazil side of falls

△ **BASALT OUTCROPS**
The hard caprock forming the crest of the falls was created during flood basalt eruptions that covered the region in layers of lava more than 120 million years ago.

underlying layer of soft rock

plunge pool

Main flow of falls
This cross-section downstream of the Devil's Throat shows the channel through which half of the Iguaçu's waters flow. A further series of falls lie below and to the right, on the Argentina side.

Retreating falls
Waterfalls retreat upstream as the caprock is eroded. The edge of the basalt cap of Iguaçu Falls is receding by 1/8 in (3 mm) a year.

E. South America

◁ **IGUAÇU RIVER**
The Iguaçu River is one of the major tributaries of the Paraná River, South America's second longest river after the Amazon, which it joins 14 miles (23 km) downstream of the falls.

viewing gallery on Argentine side of falls

◀ **GREAT WATER**
Iguaçu Falls is about three times the width of North America's Niagara Falls (see p.52) and around 50 percent taller. The name of the falls and river is from a word in the indigenous Guaraní language that translates as "great water."

Atlantic Forest
The tropical forest surrounding the falls is a remnant of the once vast Atlantic Forest. The Iguaçu National Park protects the thousands of species of plants and animals living within its borders.

Iguaçu Falls

One of the world's greatest waterfall systems, situated within a lush South American tropical forest

Spanning part of the border between northeastern Argentina and southern Brazil, Iguaçu Falls is one of the largest waterfall systems on Earth, rivaled in size only by Africa's Victoria Falls (see pp.194–95). It is located on a tight horseshoe bend of the Iguaçu River and drops a total of 269 ft (82 m).

Complex system

Iguaçu Falls is a vast and complex series of up to 300 separate waterfalls—depending on the water level of the river—stretching 1¾ miles (2.7 km) along a fragmented basalt escarpment. Above the falls, many islands split the river into the flows that feed the waterfalls, which vary in size. At the heart of the falls is the Devil's Throat, a narrow, semicircular chasm down which half of the river's waters plunge. Iguaçu Falls produces huge clouds of spray that constantly soak the surrounding area, creating a humid microclimate in which a diverse group of plants and animals flourish.

top layers of hard rock

△ **DROUGHT-AFFECTED FALLS**
Waterfalls can experience great fluctuations in water flow. During the dry season, when long-periods of droughts may occur, Iguaçu Falls can shrink to a fraction of its size, completely transforming its appearance.

HOW A WATERFALL FORMS

Waterfalls form where a river flows from hard to softer rock. The water erodes the soft rock faster than the hard rock, creating a step then a plunge pool. Undercut sections of the hard rock eventually collapse, and as this process continues, the waterfall retreats upstream.

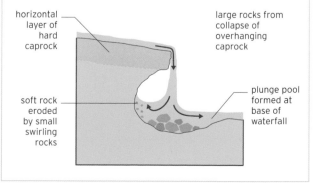

horizontal layer of hard caprock

large rocks from collapse of overhanging caprock

soft rock eroded by small swirling rocks

plunge pool formed at base of waterfall

C. South
America

The Pantanal

An immense freshwater wetland system with one of the greatest concentrations of plant and animal species on the planet

The world's largest freshwater wetland, the Pantanal covers more than 70,000 square miles (180,000 square km) of western Brazil and parts of Bolivia and Paraguay. Sitting in the upper Paraguay River basin, it is a gently sloping floodplain composed of a mixture of habitats, including grasslands, forests, marshes, and rivers, that are subject to seasonal flooding.

Thriving in floods

A wide variation in water levels across the Pantanal has led to a great diversity in plant life. In higher areas, drought-resistant woodland is common, while in lower areas plants adapted to seasonal flooding flourish. Areas that are permanently under water support a diverse community of aquatic plants, including lilies and hyacinths. Many thousands of waterbirds feed on the great numbers of fish that migrate and spawn each year when the floods arrive, as do the estimated 10 million caimans that make up the largest population of crocodilians in the world.

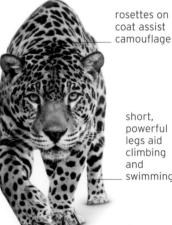

rosettes on coat assist camouflage

short, powerful legs aid climbing and swimming

△ **BIG CAT**
The Pantanal is a vital sanctuary for the jaguar, the world's third largest feline after the tiger and the lion. It is estimated that only 15,000 jaguars remain in the wild.

S. South America

General Carrera Lake

A large glacial lake in the shadow of the Andes with a remarkable complex of carved marble caves on its shores

General Carrera Lake, also known as Lake Buenos Aires, spans the border between Chile and Argentina in Patagonia. With a surface area of 715 square miles (1,850 square km), it is the second largest freshwater lake in South America, after Lake Titicaca (see p.105). General Carrera Lake is surrounded by the Andes, and its glacier-fed water drains into the Pacific Ocean through the Baker River, the largest river in Chile. Its most famous feature, known as the Marble Cathedral, or Marble Caves, is a series of pure marble cliffs that have been carved into elaborate caves, tunnels, and columns by 10,000 years of wave action. The smoothly formed light-gray rock takes on a blue tint from the surrounding water, which is colored turquoise by glacial silt.

▷▽ **SCULPTURED MARBLE**
The Marble Cathedral system includes an island monolith known as the Marble Chapel. The clarity of the water of General Carrera Lake reveals how the rock formations continue below the surface.

CLIMATE

The Pantanal has a tropical climate with a hot rainy season and a drier season during the cooler months. The extent and depth of floodwaters vary greatly throughout the year.

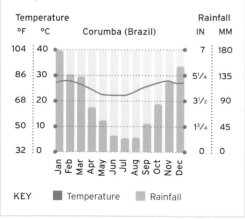

Temperature		Corumba (Brazil)	Rainfall	
°F	°C		IN	MM
104	40		7	180
86	30		5¼	135
68	20		3½	90
50	10		1¾	45
32	0	Jan Feb Mar Apr May Jun Jul Aug Sep Oct Nov Dec	0	0

KEY ■ Temperature ■ Rainfall

△ **FLOODED PLAINS**
Around 80 percent of the Pantanal's vast plains are inundated during the wet season, as water levels rise by up to 16 ft (5 m).

C. America

The Great Blue Hole

A stunningly beautiful submerged sinkhole in a reef off the coast of Belize

The Great Blue Hole is situated on Lighthouse Reef, an atoll lying 35 miles (55 km) east of the huge Belize Barrier Reef. It is a large, almost perfectly circular sinkhole within the limestone mass of the reef.

Formation and exploration

Analysis of stalactites found in the Great Blue Hole shows that it formed in stages going back at least 153,000 years, during periods when sea level was lower than it is today. Erosion by streams during these times produced a complex of air-filled caves and tunnels in a block of limestone on dry land. Eventually, the ceiling of one cave collapsed, creating what is now the entrance to the Great Blue Hole. Since its formation, sea level has gone up and down, most recently rising to flood the blue hole and a cave system linked to it. Today, both the sinkhole and caves are accessible only to adventurous scuba divers but are regarded as one of the world's most exciting dive sites. Diving into the hole is not recommended for novice divers because perfect buoyancy control is required. Caribbean reef sharks are often seen swimming around the entrance.

PREVIOUS SEA LEVELS

The Great Blue Hole has not always been underwater. It formed over several tens of thousands of years through erosion of a mass of limestone on dry land. The sea level has risen over the past 18,000 years, since the last ice age, causing the Great Blue Hole to become submerged.

Depth
FT	M	
10	3	
50	15	6,000 years ago
		8,000 years ago
150	45	10,000 years ago
260	80	14,000 years ago
380	120	18,000 years ago

◁ STALACTITE EVIDENCE
The presence of stalactites on several locations around the walls of the hole are proof that it formed above sea level, since stalactites and stalagmites cannot form underwater.

South grotto cave
Large stalactites hang from the cave roof and stalagmites project from its floor, giving the entrance the appearance of an open shark's mouth.

Sediment dune
This is created by coral fragments raining down from the lip of the shelf some 360 ft (110 m)

△ WALL CAVITY
The sinkhole's walls are pockmarked with small cavities. These provide ideal hiding places for some marine animals, except at great depth, where there is a lack of oxygen in the water.

▲ THE GREAT BLUE HOLE
With a diameter of 1,043 ft (318 m), the Great Blue Hole is 407 ft (124 m) deep and has rough-textured limestone walls indented by ledges, grottoes, and caves. Its entrance is almost completely surrounded by a ring of reef projecting slightly above the sea surface.

▷ AT THE BOTTOM
The floor of the blue hole is covered in chunks of limestone and fallen, broken stalactites, as first reported by the underwater adventurer Jacques Cousteau, who used a small submersible to explore the sinkhole in 1971.

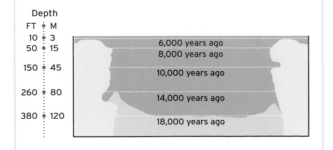

Surrounding reefs
Smaller coral reefs surround the Great Blue Hole and are home to a great variety of marine life.

West wall cave
At a depth of 165 ft (50 m), a cave system extends horizontally for more than 150 ft (46 m) and contains sea turtle skeletons.

Steep shelf
This slopes down to a depth of 50–60 ft (15–18 m), to the rim of the sinkhole, where there is a vertical dropoff.

△ CORAL RESERVE
The rim of the hole, close to the surface, is covered in a profuse array of both stony, reef-building corals (hexacorals), such as brain coral, and many colorful soft corals (octocorals).

Notch
This indentation is the result of dissolved hydrogen sulfide in the water at the bottom of the hole dissolving the limestone walls.

Thermocline level
At a depth of around 90 ft (27 m), warm surface water abruptly gives way to colder, clearer water lower down.

Sediment layers
At the bottom of the hole are layers of carbonate deposits—the remains of plankton and corals—and sediment washed from outside the reef and into the sinkhole.

Hydrogen sulfide layer
A turbid, brownish layer of water at about 310 ft (95 m) depth contains hydrogen sulfide gas from decomposing organic matter.

△ GIANT SINKHOLE
Like the pupil of an eye, the blue hole appears from above as a dark inky-blue circle. This contrasts with the mottled turquoise shades of the surrounding shallow water covering Lighthouse Reef.

Stalactites hanging over the entrance to a cave on the sinkhole's wall are up to **25 ft (8 m) long**

Lençóis Maranhenses

A region of sweeping white dunes in northeastern Brazil, seasonally adorned with thousands of crystal-clear turquoise surface pools

NE. South America

Covering some 600 square miles (1,500 square km), Lençóis Maranhenses is one of the most unusual areas of coastal sand dunes in the world. For part of the year, it looks like a desert, but from March to June it receives large amounts of rain, which collects in depressions between the dunes and creates many aquamarine pools. From the air, the vast expanse of dunes and blue pools looks like white bedsheets hung out to dry on a windy day. In fact, the word *lençóis* is Portuguese for "linen."

▷ **SLEEPING FISH**
In the dry season, tiger fish, or trahira, burrow down into the sand and lie dormant, reemerging in the pools during the next rainy season.

Blue lagoons

The pools, or lagoons, reach their maximum size around the end of the rainy season, in July, and remain so until September. Some are over 295 ft (90 m) long, with temperatures of around 86°F (30°C). Although they exist for only a few months each year, the pools are not devoid of life. Fish can usually access some of the pools via temporary channels that connect them to nearby rivers, such as the Rio Negro. The watery world of Lençóis Maranhenses is a brief one, however. Once the dry season returns, the hot equatorial sun heats the region, causing the pools to quickly dry up.

In places, the dunes stretch 30 miles (50 km) back from the coast

Chilean Fiordland

A mazelike network of fiords, islands, and magnificent, glacier-sculpted rocky landscapes in southern Chile

S. South America

▽ **GLACIAL SKYLINE**
In the Beagle Channel, the southern Andes soar behind a lonely lighthouse. Clear skies are rare here as moisture-laden Pacific air cools and forms clouds as it rises.

The Chilean Fiordland is a large region of southern Chile, including part of Tierra del Fuego, which was covered by a massive ice-sheet 10,000 years ago. Today, following the retreat of that ice-sheet, it is an extensive network of fiords—long, narrow sea inlets formed by the flooding of formerly glaciated valleys—and intervening islands. In total, it covers an area of about 21,500 square miles (55,000 square km) to the west of the southern Andes. At the terrestrial ends of some of the fiords, descending from ice-fields that still remain in the southern Andes, are the tongues of huge glaciers, which continuously calve enormous icebergs into the water. On the edges of the fiords, waterfalls cascade down granite walls, while hundreds of bird species feed around the often mist-shrouded coast and islands. Mammals that live along this coast include marine otters, sea lions, and elephant seals.

⊲ **CLEAR WATERS**
The pools are typically crescent-shaped with crystal-clear water. They can be up to 10 ft (3 m) deep.

SEASONAL CHANGES

In the dry season (September to November), strong winds blowing over Lençóis Maranhenses dry its surface and form sand dunes, with small valleys in between. In the rainy season, the valleys fill with water, which is prevented from percolating downward by a layer of impermeable rock lying under the sand.

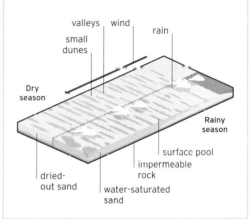

Cape Horn

A headland at the southern tip of South America, notorious for the atrocious weather conditions that often develop nearby

S. South America

⊲ **BREEDING AT THE CAPE**
Some breeding colonies of Magellanic Penguin are found on coasts around the cape.

Cape Horn is a rocky headland at the southern end of a small island called Hornos, in the Chilean part of Tierra del Fuego. It was named after the city of Hoorn in the Netherlands, the birthplace of the Dutch navigator Willem Corneliszoon Schouten, who rounded it in 1616.

Its importance and mystique derives from the fact that it is a navigational landmark, marking the northern edge of the Drake Passage, the strait between the southern tip of South America and the northern tip of the Antarctic Peninsula. "Rounding the Horn" is one of the main ways of moving between the Atlantic and Pacific Oceans, and because the sea around the headland is noted for its bad weather—including powerful winds, large waves, and a risk of icebergs—sailors rightly consider it a major challenge. The number of ships rounding Cape Horn from the Atlantic Ocean to the Pacific Ocean dropped sharply when the Panama Canal opened in 1914.

⊲ **HIGH POINT**
The rocky, roughly pyramid-shaped landmass known as Cape Horn stands 1,390 ft (425 m) tall at its most elevated point.

The Monteverde Cloud Forest

One of the world's most biodiverse and protected habitats, where sea, moisture, and mountains create a rain forest constantly shrouded in clouds

C. Central
America

Characterized by foglike clouds and humidity levels of close to 100 percent throughout the year, cloud forests make up just 1 percent of the world's woodlands. Costa Rica's Monteverde Cloud Forest, encompassing 40 square miles (105 square km), is nonetheless one of the most biodiverse habitats on Earth.

A safe haven
The highly protected Monteverde Cloud Forest Reserve is located around 4,720 ft (1,440 m) above sea level in the Cordillera de Tilarán, part of a larger mountain chain running along the center of Costa Rica.

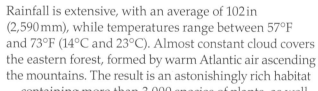

Rainfall is extensive, with an average of 102 in (2,590 mm), while temperatures range between 57°F and 73°F (14°C and 23°C). Almost constant cloud covers the eastern forest, formed by warm Atlantic air ascending the mountains. The result is an astonishingly rich habitat containing more than 3,000 species of plants, as well as thousands of insect, 400 bird, and around 100 mammal species.

◁ **SURVIVAL TACTIC**
Algae-covered fur helps the brown-throated three-toed sloth blend in with its lush forest surroundings—vital for such a slow-moving mammal.

△ **ONE FOREST, MANY TREES**
At least 755 tree species grow in the Monteverde reserve. They are taller on the cloud-covered eastern slope and shorter on the drier western side.

Costa Rica is home to **4 percent** of Earth's **plant** and **animal species**

The Andean Yungas

A unique band of dense forest, home to many species-rich ecosystems on the side of a mountain range

W. South America

The Yungas covers the eastern slopes of the Andes from northern Argentina through Bolivia and into Peru. At an altitude of 3,300–11,500ft (1,000–3,500m), it forms a transitional area between the Andean highlands and the eastern lowlands. In this area are many temperature-sensitive climatic zones, ranging from high, subtropical cloud forests to moist temperate woodlands. Because many animal species are confined to "shoestring distributions"—areas that extend for several miles horizontally but only a few hundred feet vertically—the biodiversity of the Yungas is very high.

▷ **LIFE ON THE EDGE**
Steep-sided ravines like those near the ruins of Peru's Machu Picchu create many complex microclimates. The resulting ecosystems are vital for native species.

TROPICAL COVER
Many species of mosses, epiphytes, and orchids thrive in the upper reaches of the cloud forest.

TRADE WINDS AND CLOUD FORESTS

Warm, moisture-laden trade winds from the Atlantic form Costa Rica's cloud forests. As warm air rises, its moisture condenses into fog clouds, creating a rainy tropical climate. Mountains prevent the clouds from entering the country's Pacific side, so the climate is drier and less humid there.

northeasterly trade winds bring warm air from the Atlantic Ocean

clouds form on eastern side of mountains

rain-shadow area to west receives little rainfall

The Valdivian Temperate Forest

Earth's second-largest temperate rain forest, where almost 90 percent of plant species are found nowhere else

SW. South America

South America's only temperate rain forest is also one of the largest in the world, second only to the coastal temperate rain forest of the Pacific Northwest. Bordered to the west by the Pacific Ocean and to the east by the Andes, the Valdivian runs mainly along the Chilean coast, with parts stretching into Argentina. Having evolved separately from plants in the Northern Hemisphere, many species are unique to the region, including the rare Patagonian cypress (a giant conifer) and the prehistoric monkey puzzle tree.

▷ **PREHISTORIC SURVIVOR**
The monkey puzzle is a primitive evergreen conifer species thought to have been in existence for 200 million years.

N. South
America

The Amazon Rainforest

*Earth's largest tropical rain forest, home to
390 billion trees and 2.5 million insect species*

Covering an area of almost 2 million square miles
(6 million square km), the Amazon is the world's
largest tropical rain forest, situated within the planet's
biggest river drainage basin. The bulk of the rain forest
is located in northern Brazil, but parts extend into eight
other South American countries. Rainfall is high—as
much as 120 in (3,000 mm) per year—with humidity
averaging over 80 percent, and temperatures rarely
falling below 79°F (26°C).

This constantly moist, warm environment hosts an
incredibly rich ecosystem, where an estimated 16,000
tree species coexist alongside 1,300 bird species and
a staggering 2.5 million insect species. Scientists
believe even more species have yet to be discovered.
However, deforestation, due mainly to cattle ranching
and agriculture, remains a huge threat to the Amazon
Rain forest. From 1970 to 2016, 296,887 square miles
(768,935 square km), or roughly 20 percent, of the
Brazilian rain forest was destroyed.

More than **half** of the world's **species live** in **tropical rain forests**

◁ CANOPY FEEDING
Many rain forest plants
depend on animals to
disperse their seeds.
The scarlet macaw
spreads seeds of the açaí
palm by eating its fruit.

Kapok
This giant of the rain
forest towers above the
canopy, reaching heights
of up to 60 m (200 ft).
Kapok flowers are
pollinated by bats.

▶ RAIN FOREST DYNAMICS
A tropical rain forest is both
highly structured and densely
populated–300 species of
woody plant species can be
found in just 2½ acres
(1 hectare). The many
species of plants and
animals interact with and
depend on each other
to survive.

Açaí palm
A slender palm of the
canopy layer, the açaí
grows up to 50-100 ft
(15-30 m) tall.

Ground level habitat
Larger animals search the
forest floor for food. Vines
growing from the floor create
"highways" to upper layers.

Fallen tree
If an old tree falls over, it
creates a gap in the canopy,
allowing in sunlight. Nearby
trees undergo a growth surge.

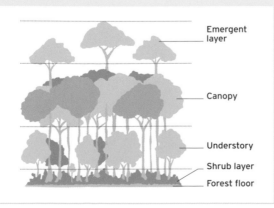

TROPICAL RAIN FOREST LAYERS

Tropical rain forests grow in five
main layers. The forest floor is
the hottest, darkest, and most
humid. Small shrub-layer plants
come next, competing for light
and food. In the understory, ferns
and vines grow over taller trees,
where most birds nest. Eighty
percent of rain forest life lives in
the hot, dry canopy at heights of
around 100 ft (30 m). Only the
tops of the tallest trees, the
emergents, project above it.

Emergent
layer

Canopy

Understory

Shrub layer

Forest floor

▷ **LIFE ON AIR**
Epiphytes, or air plants, like this *Cattleya* orchid, require no soil. Instead, they anchor themselves to trees, drawing water and nutrients from moist air.

▷ **CLOSED CANOPY**
Most rain forest plants and animals live in the canopy, where trees compete for sunlight via maximum leaf growth and spread. The overlapping branches form an almost impenetrable "roof."

Brazil nut tree
This emergent tree reaches heights of 130–165 ft (40–50 m). Most Brazil nuts come from wild rain forest trees.

Rubber tree
A rubber tree grows very fast so will soon establish itself if a gap appears in the canopy.

Cacao tree
The cacao tree grows in the understory in the shade of other trees. Each of its pods contains up to 60 cocoa beans.

Leaf litter
Microorganisms feed on and recycle decaying plant material lying on the forest floor.

Shallow roots
An extensive shallow root system allows a tree to absorb vital topsoil nutrients before they are washed away by rain.

▷ **FOOD ON THE FLOOR**
Some rain forest trees, like the Brazil nut, rely on ground-dwelling animals to spread their seeds. The agouti gnaws holes in tough, fallen seed pods to reach the nuts inside. It buries some but often forgets them, so new Brazil nut trees grow instead.

Topsoil
Rain forests have just a thin layer of topsoil. The soil contains all available organic matter and nutrients for plant growth.

Buttress roots
These large roots grow at angles from the trunk to stabilize a tall tree in poor-quality soil.

CLIMATE

The Pampas has a temperate climate, with almost constant breezes that help offset high humidity levels of 75 percent or more. Although winter temperatures occasionally drop below freezing, the yearly average is 54–64°F (12–18°C). Precipitation is higher in coastal areas, decreasing inland, and annual rainfall ranges from 20 in to 71 in (500 mm to 1,800 mm), decreasing toward the southwest.

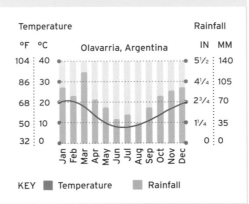

Temperature

Olavarria, Argentina

Rainfall

°F	°C		IN	MM
104	40		5½	140
86	30		4¼	105
68	20		2¾	70
50	10		1¼	35
32	0		0	0

Jan Feb Mar Apr May Jun Jul Aug Sep Oct Nov Dec

KEY ■ Temperature ■ Rainfall

△ UMBRELLA TREE

The evergreen ombu is one of the few tree species that grow in more arid Pampas regions. Its umbrella-like canopy spreads up to 50 ft (15 m) across.

SE. South America

The Pampas

Wide, flat grasslands stretching across the southeastern corner of South America, forming some of the richest grazing land in the world

South America's huge expanse of grassland known as pampas extends from the foothills of the Andes east to the Atlantic coast. Because *pampa* means "flat surface," the term is used for many smaller plains all over the continent, but the largest and most well known is the Argentine Pampas.

Covering around 295,000 square miles (760,000 square km), the Pampas consists mainly of a flat, unbroken plain that slopes gently from an elevation of 1,640 ft (500 m) in the northwest to just 65 ft (20 m) in the southeast. This expanse of land can be divided climatically into two main sections. The larger western dry zone of central Argentina is dominated by coarse grass species such as stipa and is dotted with marshes. The smaller, more fertile, but heavily populated humid zone in the east features a combination of grassland and dry woodland. While much of the Pampas' original coarse grass has been replaced with grass species more suitable to domestic livestock that have grazed here for two centuries, the region remains a vital habitat for endangered species such as the Pampas deer and the Argentine tortoise.

▷ **STEALTHY MOVEMENT**
Geoffroy's cat is a small, efficient, and opportunistic predator, hunting throughout the Pampas for rodents, reptiles, birds, and fish.

△ **FEATHERY TUFTS**
Pampas grass reaches heights of up to 12 ft (3.6 m) or more. The featherlike flower heads hold up to 100,000 seeds each.

▷ **HIGH VIEWPOINT**
Standing 5 ft (1.5 m) tall, the greater rhea cannot fly, but it can run as fast as 37 mph (60 kph). It is often seen grazing with herds of Pampas deer.

Few trees can **survive** the frequent **wildfires** of the Pampas, but **grasses regenerate easily**

HIGH DUNES
Vast sand dunes collect at the base of the Andean foothills. One, called Cerro Medanoso, rises as high as 1,800 ft (550 m).

△ **PENITENTES**
Snow rarely melts in the Atacama. Instead, it forms clusters of jagged, ice-encrusted penitentes, named after the spiky hats of Spanish penitents.

▷ **CAPTIVE CLOUDS**
The Pacific's Humboldt Current pushes cold water toward the ocean's surface, creating fog banks. When trapped near the coast, these bring moisture ashore.

W. South America

The Atacama Desert

The world's highest desert and one of the driest places on Earth, a sterile, stony plateau full of salt flats, sand, and active volcanoes

Lying to the west of the Andes and parallel to the Pacific coastline, the Atacama forms a narrow strip of arid desert in northern Chile, averaging less than 100 miles (160 km) in width but extending about 620 miles (1,000 km) south from the border with Peru. With an average elevation of about 13,000 ft (4,000 m), this high plateau is, according to NASA studies, the driest place in the world. Some weather stations have never recorded a single drop of rain; even at lower levels, a dry spell of four years is common. The average rainfall in the desert's estimated 40,600 square miles (105,200 square km) range is just ¹⁄₁₆ in (1 mm) a year.

In mountain shadows

The Atacama is a double rain-shadow desert, because two high mountain ranges, the Chilean Coast Range and the Andes, block incoming moisture from both Pacific and Atlantic Oceans. In this hyper-arid climate, nothing decays; Earth's oldest mummies, dating from around 7020 BCE, were found here, still beautifully preserved. However, on the eastern side, where the foothills of the Andes rise up, there is a small amount of rainfall in winter.

Although parts of the Atacama are considered "absolute desert," other areas are more hospitable. Fog zones called *lomas*, created at lower levels when clouds are trapped by mountains or coastal slopes, bring in enough moisture to sustain lichens, algae, and hardy cacti and shrubs. Birds, which are able to fly in and out as conditions dictate, are the most abundant wildlife: three species of flamingoes are found in the Atacama, feeding on algae growing in salt lakes, while hummingbirds and Peruvian song sparrows are seasonal visitors.

However, due to the high content of sulfur, copper, and mineral salts such as sodium nitrate in its soils, the desert is hostile to most animals apart from lizards, insects, and rodents—and the South American gray fox, which feeds on them. In fact, the Atacama's soil composition is so similar to samples taken from Mars that NASA uses it for testing space rovers.

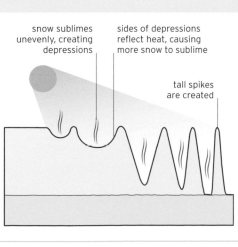

◁ **HARDY MAMMAL**
The highly adaptive South American gray fox, or *chilla*, copes with the Atacama's challenging conditions by hunting small animals in lower-lying fog zones.

HOW PENITENTES FORM

Penitentes are ice structures that usually form at altitudes of 13,000 ft (4,000 m) or more. Their formation is believed to be due to a combination of factors. When the Sun shines on snow, freezing temperatures prevent it from melting. Instead, it sublimes (changes directly from solid to gas without melting). Troughs form and reflect sunlight, raising the temperature enough to melt the surrounding snow. Strong winds freeze the melting snow into spikes.

snow sublimes unevenly, creating depressions

sides of depressions reflect heat, causing more snow to sublime

tall spikes are created

Studies of sediments indicate that parts of the Atacama have had **no rain for** more than **20 million years**

The Salar de Uyuni

The largest salt flat on Earth, the legacy of a huge prehistoric South American lake

W. South America

Southwestern Bolivia's Salar de Uyuni is considered the planet's largest and least-polluted salt flat. Situated on a high plateau, around 4,100 square miles (10,600 square km) of salt, so fine it is almost flourlike, floats in a crust on top of a lake of brine thought to contain 50–70 percent of the world's lithium reserves. While around 25,000 tons (22,700 tonnes) of salt per year are harvested, a staggering 10 billion tons (9 billion tonnes) remain.

Islands in a white sea

The Salar de Uyuni's salt crust varies from a few inches to 33 ft (10 m) thick, reflecting the 6–65 ft (2–20 m) depth of the saltwater lake below it. Despite these variations, the crust's surface is remarkably flat, with less than 3 feet of difference in level from one side to the other. The vast expanse of white is not entirely unbroken, however. A few elevated areas, such as the cactus-covered Isla Incahuasi (Inca House Island), pierce the crust in places. For the most part, though, there is little vegetation. Some wildlife is present, mainly in the form of flamingos that migrate here to breed.

▽ **NATURAL GEOMETRY**
When saline water evaporates undisturbed, hexagonal patterns are formed on the salt crusts. Widespread similar hexagons indicate calm conditions throughout the salar.

▽ **ROCKY OASIS**
One of the few oases in an almost pure-white landscape, the rocky Isla Incahuasi supports the growth of tree-sized pasacana cacti, which can reach 33 ft (10 m) in height.

Ponds formed by rain **dissolve bumps** on the **salt flat's** surface

HOW SALT CRUSTS FORM

Salt pans form when the rate of evaporation of mineral-rich water is higher than the rate of precipitation. As the water dries up, the previously dissolved salt ions solidify, forming a hard white crust, which thickens as the process continues.

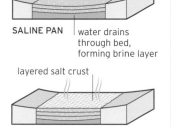

saline lake — evaporation

SALINE LAKE — lake bed

dry saline pan

SALINE PAN — water drains through bed, forming brine layer

layered salt crust

SALT CRUST

W. South America

The Valle de la Luna

A maze of canyons and badlands sculpted into bizarre and otherworldly rock formations by millions of years of wind and water erosion

The region in northwest Argentina popularly known as the Valley of the Moon is actually a series of sandstone and mudstone rock formations. Otherwise known as Ischigualasto—"the place where the Moon rests"—this rugged desert basin was, in prehistoric times, a volcanically active plain. Today, stone monoliths and contorted rock structures are all that remain of the mountains that once occupied this landscape. Pink, yellow, and purple bands reveal the mineral content of sediments in the rocks. Huge mudstone spheres and petrified tree trunks share space with sparse scrub and cacti. Ischigualasto also has some of the oldest dinosaur fossils, dating from 230 million years ago.

HOW ROCK PEDESTALS FORM

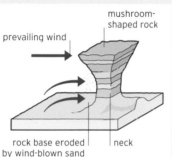

A pedestal-like rock formation is created by abrasion, when wind blows sand particles against a rocky outcrop. The force of the particles erodes softer sedimentary layers swiftly. The bottom section wears away faster than the top layers.

prevailing wind

mushroom-shaped rock

rock base eroded by wind-blown sand

neck

▷ **THE MUSHROOM**
One of Ischigualasto's most well-known rock formations is *El Hongo*, Spanish for "the Mushroom."

The Patagonian Desert

Argentina's largest desert, a cold, windswept region that undergoes seven months of winter and five of summer

SE. South America

coarse hairs extend beyond bands of skin-covered bony armor

△ **HAIRY ARMOR**
The center of the desert is too harsh for most animals, but some, such as the big hairy armadillo, survive in vegetated border areas.

Bordered by the Atlantic to the east and the Andes to the west, the Patagonian Desert covers 260,000 square miles (673,000 square km) of southern Argentina and just crosses the border into Chile. The region has two distinct climatic zones: in the semiarid north, top temperatures average 54–68°F (12–20°C), while the south is cooler, with highs of 39–55°F (4–13°C). Average precipitation is 8 in (200 mm). The center is mainly barren rocky plateau, but near the desert borders, mixed northern shrubland changes to sparser grassy vegetation in the south.

▷ **PETRIFIED FOREST**
Once a vast prehistoric forest, the Patagonian Desert today contains only the fossilized remains of huge trees at its center–some up to 89 ft (27 m) long.

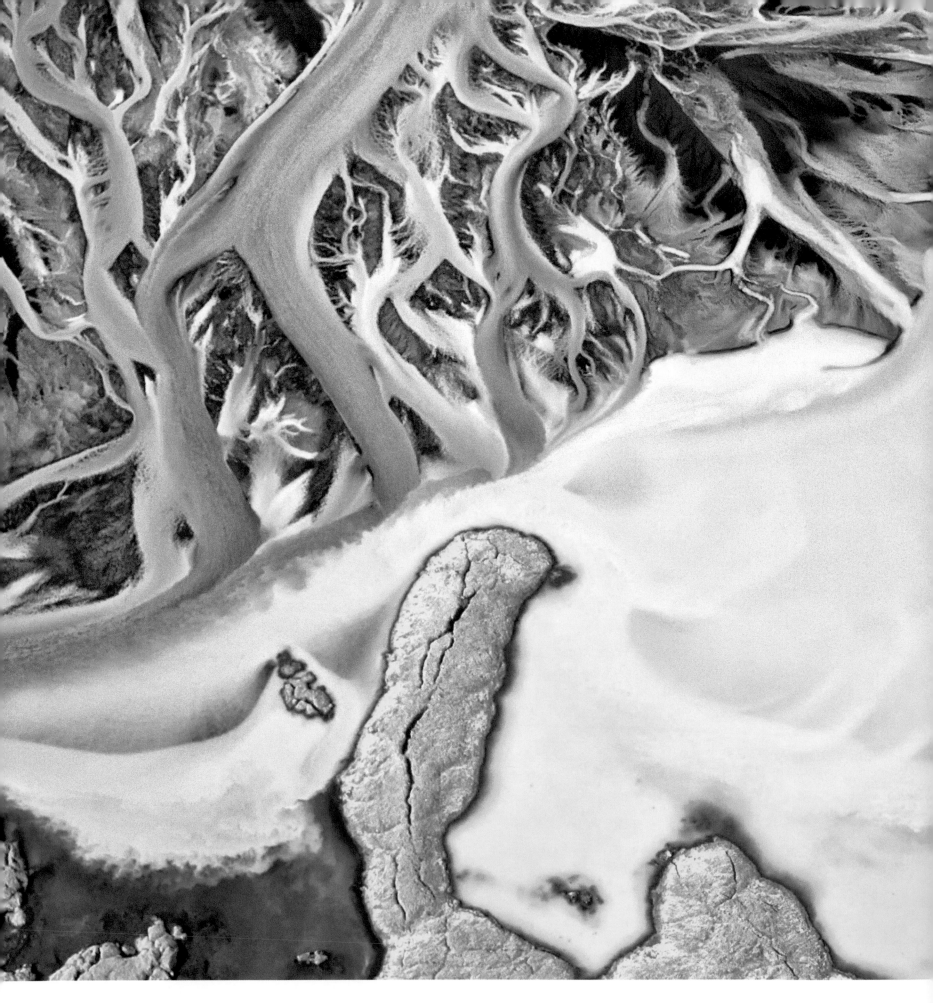

Transport network
Rivers deliver huge volumes of summer meltwater from
Iceland's glaciers to the Atlantic Ocean. On reaching flat
ground close to the coast, the rivers branch out to form
complex networks of channels called distributaries.

Europe

LAND OF RIVERS AND PLAINS
Europe

Europe is the world's second smallest continent; only Australia is smaller. Yet it has an astonishing variety of landscapes, sliced through by the continent's plentiful rivers, from the east-flowing Danube to the north-flowing Rhine.

Mainland Europe is attached to Asia all the way down its eastern boundary along the line of the ancient Ural Mountains in Russia. To the north, it is fringed by the chilly Arctic Ocean, to the west by the Atlantic Ocean, and to the south by the warm, almost landlocked Mediterranean Sea. The Mediterranean south of the continent is warm and the landscape dry, but it is separated from the damper, cooler north by chain after chain of jagged, high mountains, including the Alps and the Caucasus, which have been thrust up by the northward pressure of the subducting African Plate. To the north of the mountains are rolling, sediment-covered plains, and farther north still are ancient mountains worn smooth and sculpted with deep valleys by glaciers during the last ice ages.

KEY DATA

▲ **Highest point** Mount El'brus, Russia: 18,510 ft (5,642 m)

▼ **Lowest point** Caspian Depression, Russia: -92 ft (-28 m)

● **Hottest record** Murcia, Spain: 117°F (47°C)

● **Coldest record** Ust'Shchugor, Russia: -73°F (-58°C)

CLIMATE

Europe lies mostly within the mild temperate zone, with plentiful rains brought by winds off the Atlantic. Temperatures range from the warmth of the Mediterranean to the chill of the Arctic north.

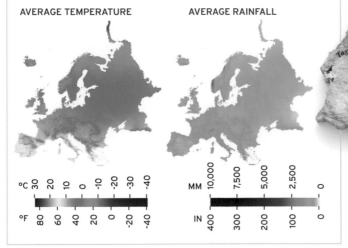

AVERAGE TEMPERATURE

AVERAGE RAINFALL

°C 30 20 10 0 -10 -20 -30 -40

°F 80 60 40 20 0 -20 -40

MM 10,000 7,500 5,000 2,500 0

IN 400 300 200 100 0

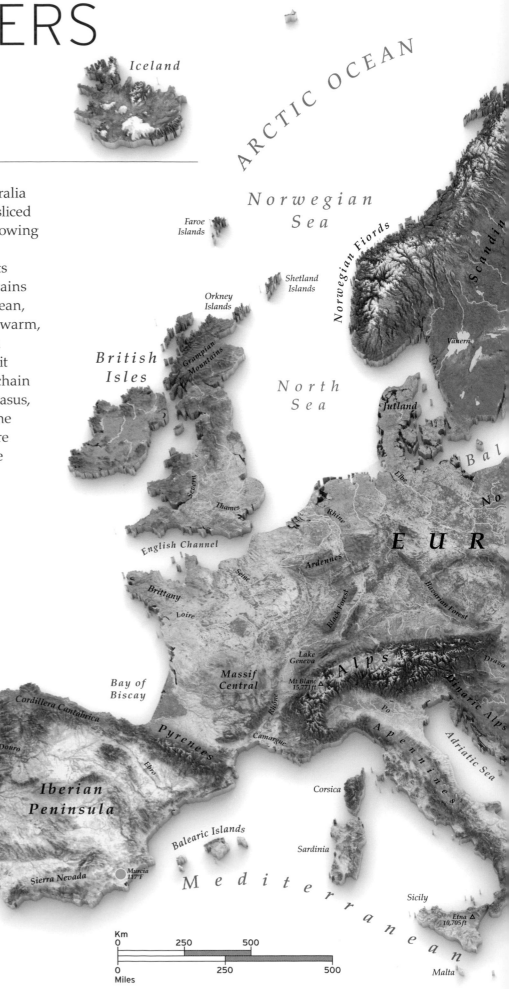

Iceland

ARCTIC OCEAN

Norwegian Sea

Faroe Islands

Norwegian Fiords

Shetland Islands

Orkney Islands

Scandin

Vänern

British Isles

Grampian Mountains

North Sea

Jutland

Bal

Elbe

No

Rhine

Severn

Thames

English Channel

Seine

Ardennes

Bavarian Forest

Brittany

Black Forest

EUR

Loire

Lake Geneva

Alps

Drava

Bay of Biscay

Massif Central

Mt Blanc 15,771 ft ▲

Dinaric Alps

Cordillera Cantábrica

Rhône

Po

Apennines

Adriatic Sea

Pyrenees

Camargue

Douro

Corsica

Ebro

Tagus

Iberian Peninsula

Balearic Islands

Sardinia

Sierra Nevada

Murcia 117°F

Mediterranean

Sicily

Etna ▲ 10,705 ft

Km
0 250 500

0 250 500
Miles

Malta

Barents Sea

Ural Mountains

Kola Peninsula

White Sea

Gulf of Bothnia

Lake Onega

Volga

Lake Ladoga

Ust' Shchugor
-73 ft

ia

tic Sea

Gotland

rth European Plain

N O R T H   E U R O P E

Central Russian Upland

Volga Uplands

Don

Vistula

Dnieper

Volga

Caspian Depression
-92 ft

Caspian Sea

Carpathian Mountains

Danube

Tisza

Great Hungarian Plain

Danube

Balkan Mountains

Caucasus

△ Mount El'brus
18,510 ft

Black Sea

Aegean Sea

Ionian Sea

Peloponnese

Crete

S e a

ECOSYSTEMS

Temperate forests dominate the continent, with steppe grasslands to the east, boreal forest and tundra to the north, and scrubby Mediterranean vegetation in the south.

KEY

- ■ Temperate broadleaf forest
- ■ Temperate coniferous forest
- ■ Desert and scrub
- ■ Temperate grassland
- ■ Mediterranean woodland and scrub
- ■ Montane grassland
- ■ Boreal forest and taiga
- ■ Tundra
- ■ Ice

GEOLOGY

The geological core of Europe is formed by the ancient rocks beneath Russia, Scandinavia, and the Caledonian Mountains of Scotland. Piled on top are the younger rocks of the southern mountains.

B

A

KEY

- ■ Precambrian (pre-541 million years ago)
- ■ Paleozoic (541–252 million years ago)
- ■ Mesozoic (252–66 million years ago)
- ■ Cenozoic (66 million years ago to present day)

ELEVATION

FT	M
6,600	2,000
0	0
-6,600	-2,000

Sicilian mountains

Mediterranean Sea

Apennines

Adriatic Sea

Alps

North European Plain

North Sea

Norwegian mountains

A ◄ ---------------- CROSS-SECTION ---------------- ► B

▷ **RISING PEAKS**
As the African Plate drives north, it has crushed southern Europe and thrown up the Alps and other high mountain ranges such as the Carpathians and Caucasus.

▷ **DEEP VALLEY**
A collision between tectonic plates 400 million years ago opened a gash across the Scottish Highlands called the Great Glen—which is now filled by Loch Ness.

THE SHAPING OF EUROPE

Europe is a relatively small continent, yet has its own unique geology. Whereas most continents are formed from several stable sections of continental crust called cratons, Europe was forged as fragments formed and reformed around just one—the East European Craton.

Europe rises
The East European Craton lies buried under sediments across the North European Plain that stretch from the Netherlands to the Ural Mountains. But in Sweden and Finland, the Craton is exposed.

At first, the East European Craton formed the continent of Baltica. Then, 430 million years ago, Baltica collided with ancient landmasses geologists call Laurentia (now North America) and Avalonia (now the British Isles and part of North America) to create the supercontinent Euramerica. Along the collision zone, mountains were thrown up across Norway and Britain—the Scottish Highlands are remnants of this range. Much of Euramerica was a desert, and it is sometimes known as the Old Red Continent for the red sandstones that formed there.

In time, Euramerica crashed into the southern supercontinent of Gondwana to help form the giant supercontinent of Pangaea, piling up new mountains, such as the Vosges and the Harz, which have now worn down. Pangaea eventually broke up to form the continents we know today, with North America splitting away from Europe as the Atlantic Ocean opened up. But the separated continent of Africa has for the last 40 million years been driving relentlessly north, forcing the southern edge of Europe up into a series of new mountain ranges from the Pyrenees to the Alps and the Caucasus.

Deep freeze
For much of the last two million years, in extremely cold periods known as ice ages, vast ice sheets covered northern Europe. Their grinding movement scoured large areas of lowland Europe flat, and left behind extensive deposits of debris, known as moraines, creating a bumpy landscape. In Scotland and the Lake District, glaciers carved deep, troughlike valleys. In Norway, they gouged out

5.6 million years ago, the Mediterranean Basin was a desert

KEY EVENTS

70 million years ago The Mid-Atlantic Ridge is spreading. In one area, hot magma from Earth's mantle cools into new crust, creating Iceland.

5.3 million years ago Tectonic activity lowers the Strait of Gibraltar to below sea level, causing the Atlantic Ocean to pour into the Mediterranean Basin, forming the Mediterranean Sea.

18,000-11,000 years ago During the coldest periods of the last ice age, glaciers carve out many of Europe's most dramatic landscapes.

10,000 years ago Vast forests start to cover Europe, replacing the grassy steppes of the ice age; remnants such as the Białowieza Forest survive today.

90 million years ago Tectonic activity in what will be southern Europe begins and will continue to eventually form the Alps.

30-20 million years ago Europe starts to form almost recognizable landmasses; most of western Europe is now above land, and a shallow ocean that separates Europe from Asia disappears.

9,000-8,000 years ago A tsunami floods the land link between what will become the British Isles and mainland Europe, forming the English Channel.

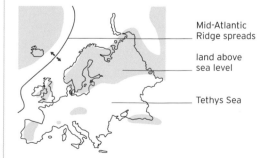

◁ CARVING ICE
The Jostedal icefield in Norway, the largest in Europe, is a remnant of the ice sheet that once carved out the deep valleys of the Norwegian fjords.

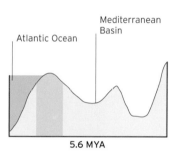

◁ NARROW CHANNEL
The Mediterranean Sea is linked to the Atlantic Ocean through the Straits of Gibraltar. The channel is so narrow that Morocco is visible from Spain.

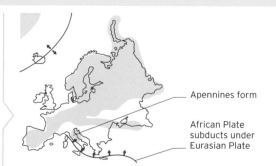

▷ NEW ISLAND
Rising sea levels just 9,000 years ago cut the British Isles off from mainland Europe.

valleys so deep that once the ice melted they were flooded by the sea to form the country's famous fjords.

Sea levels fluctuate

During the ice age, Britain was joined to Europe by a long block of limestone—this acted as a dam against rising glacial meltwater. About 450,000 years ago, this natural dam burst, leaving behind the White Cliffs of Dover. Sea levels continued to rise and fall during each glacial and interglacial period, until 9,000–8,000 years ago, when the last ice sheet melted and Britain was cut off from Europe for good.

The Mediterranean Sea has also come and gone. It was cut off from the Atlantic Ocean and most of it dried out to a salty desert. Then, the jostling of the African and the Eurasian plates caused the land underneath where the Straits of Gibraltar are now to suddenly fall. This let water flood through as a torrent from the Atlantic Ocean. At first the water pooled into the basin slowly over a period of a few thousand years. However, it is thought that 90 percent of the Mediterranean Sea's water eventually came rushing into the basin in a matter of months to years.

THE FORMATION OF THE MEDITERRANEAN SEA

About 5.6 million years ago, the Mediterranean Basin was cut off from the Atlantic Ocean and almost entirely dry. But 300,000 years later, the Straits of Gibraltar fell, and Atlantic waters flooded in and refilled the basin.

Atlantic Ocean

Mediterranean Basin

5.6 MYA

part of crust falls due to tectonic activity

water from Atlantic rushes in

5.3 MYA

erosion deepens Strait of Gibraltar

water from Mediterranean Sea mixes with water from Atlantic Ocean

PRESENT DAY

BUILDING THE EUROPEAN CONTINENT

KEY ▬ Convergent boundary ▬ Divergent boundary

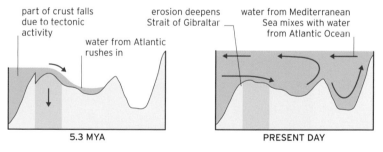

Mid-Atlantic Ridge spreads

land above sea level

Tethys Sea

Apennines form

African Plate subducts under Eurasian Plate

Europe linked to Asia by land

North Sea above sea level

African Plate continues to subduct

94 MILLION YEARS AGO The North Atlantic is spreading to push Europe and North America apart, while southern Europe is entirely inundated by the Tethys Sea between Europe and Africa.

50-40 MILLION YEARS AGO Africa drives north toward Europe. The Tethys Sea is squeezed out, and sediments pile up in southern Europe to form high mountain ranges such as the Apennines.

18,000 YEARS AGO As oceans recede, the Black, Baltic, and North Seas dry out, and Europe has become linked to Asia by land. Europe is at its greatest extent.

△ **SKYE LANDSCAPE**
Skye, Scotland's second largest island, boasts some dramatic landscapes. The large rock pinnacle at the center of this image is known as the Old Man of Storr.

The Scottish Highlands

A region of ancient mountains and extinct volcanoes, chiseled by glaciers and rainfall for many centuries, with a wild, rugged beauty

NW. Europe

HOW THE CALEDONIAN MOUNTAINS FORMED

During the Caledonian Orogeny, tectonic movements caused ancient continents to collide and push up mountains. Later, the combined landmass broke up, leaving the mountains scattered across the North Atlantic region.

490–390 MILLION YEARS AGO

TODAY

The Scottish Highlands are part of the Caledonian Mountains—an ancient, much larger mountain chain that was formed in a series of events called the Caledonian Orogeny (see panel, left). They include the Northwest Highlands and the Grampians—which are separated by a 60-mile- (100-km-) long fault called the Great Glen Fault—as well as the mountainous islands of western Scotland. The Grampian Mountains are the highest in the British Isles, their tallest peak being Ben Nevis, at 4,413 ft (1,345 m).

Volcanic remnants

Scattered among the highlands are remnants of past volcanic activity. Ben Nevis was once a large volcano that collapsed in on itself following an eruption 350 million years ago. Some highland subranges, such as the Cairngorms, are made of granites formed from solidification of magma underground, later exposed by erosion. The highlands have a characteristic rugged appearance. Forest still covers some areas, providing a home to wildlife such as the capercaillie, pine marten, and wildcat.

Glen Coe, a valley in the Highlands, **was once a supervolcano**

◁△ **HIGHLAND HEATHER**
Heather is an emblem of Scotland. In late summer, purple heather clothes the lower slopes of the Scottish Highlands, where highland cattle can often be seen grazing.

N. Europe

Strokkur Geyser

A famous Icelandic geyser, which erupts roughly every eight minutes

Strokkur geyser is situated in an area of hot springs about 50 miles (80 km) east of Iceland's capital, Reykjavík. Nearby is another famous but less frequently erupting geyser, named Geysir, which was first described in 1294 and gave the world the word "geyser."

Strokkur first became active following an earthquake in 1789. The geyser erupts from a pool of water, and its typical eruption height is about 50–65 ft (15–20 m), but occasionally the fountain of hot water reaches 130 ft (40 m). Each blast lasts for just a few seconds and is accompanied by a dramatic explosion of sound.

△ **REGULAR CYCLE**
The geyser's eight-minute cycle owes its regularity to human intervention. In 1963, its conduit was cleaned to ensure regular eruptions.

◁ **READY TO BLOW**
When Strokkur erupts, first a dome of water bulges up as a steam bubble rises from below. A plume of boiling water then blasts violently skyward.

W. Europe

The Massif Central

A highland region of France with a superlative scenery of extinct volcanoes

Covering about one-seventh of France, the Massif Central is separated from the Alps by the valley of the Rhône. Volcanism began here about 65 million years ago—possibly due to crustal thinning and upwelling of hot mantle rock—and is apparent today as volcanic remnants, known as the Chaîne des Puys, across a northern part of the region. The southern part of the massif features raised limestone plateaus cut through by deep canyons. The most famous of these is the magnificent Tarn Gorge, which is 33 miles (53 km) long and in places up to 2,000 ft (600 m) deep.

INSIDE A CINDER CONE

Pyroclastic, or cinder, cones are relatively small, steep-sided volcanoes built mainly from loose volcanic cinders (glassy fragments of solidified lava). Some also contain volcanic ash. Most have just one main eruptive phase when they first appear. After erupting and growing for a few months to years, they go permanently quiet.

cinders with layers of ash and lava

bowl-shaped crater

simple conical shape

single conduit

◁ **THE CHAÎNE DES PUYS**
This line of small, extinct volcanoes, mostly in the form of cinder cones, is about 25 miles (40 km) long. Puy de Côme is shown in the foreground with a winter sprinkling of snow.

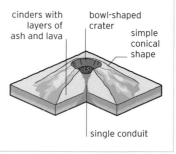

One face of **Marmolada**, the **highest peak** in the Dolomites, is a near-vertical cliff **2,000 ft (600 m) high**

▽ **BOLD CLIFFS**
The topography of the Dolomites features massive rock walls, towers, and jagged pinnacles.

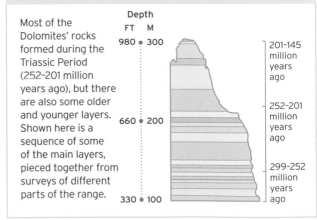

S.Europe

The Dolomites

A dramatic range in northeastern Italy, with 18 peaks rising above 10,000 ft (3,000 m)

A subrange of the Alps, the Dolomites consist of multiple layers of sedimentary rock, thousands of feet thick, that originally formed as coral reefs in shallow seas hundreds of millions of years ago. Later, these rock layers were lifted up as a result of tectonic plate movements, then eroded by glaciers to form the dramatic landscape seen today. The mountains are renowned for the fossils of coral reef animals they contain.

Pale mountains

The Dolomites are named after an 18th-century French mineralogist, Deodat de Dolomie, who was the first to describe the rock—dolomite—from which they are largely made. Dolomite is similar to limestone and typically white, cream, or light gray. Because of this, the Dolomites are also known as the Pale Mountains. Their highest peak is Marmolada at 10,968 ft (3,343 m). The spectacular scenery draws numerous climbers, skiers, paragliders, and hikers.

dolomite crystals

▷ **BUILDING BLOCK OF THE DOLOMITES**
Dolomite is the name given to a mineral, as well as to a rock. In this specimen, cream-colored dolomite crystals are seen mixed with crystals of quartz (pink) and a third, darker mineral.

ROCK LAYERS OF THE DOLOMITES

Most of the Dolomites' rocks formed during the Triassic Period (252–201 million years ago), but there are also some older and younger layers. Shown here is a sequence of some of the main layers, pieced together from surveys of different parts of the range.

Depth

FT	M
980	300
660	200
330	100

201–145 million years ago

252–201 million years ago

299–252 million years ago

The Pyrenees

A high natural barrier between France and Spain that has played a significant role in the history of both countries

W. Europe

△ **BLUE RIDGES**
This early morning view shows the western part of the French Pyrenees. The distinctive steep-sided peak in the foreground is Pic du Midi d'Óssau.

The Pyrenees mainly formed in the last 55 million years (see panel, below). Over time, the uplifted land was eroded, resulting in a series of parallel mountain ranges, together with some flat-topped massifs, that in places reach over 10,000 ft (3,000 m) in height.

Steep valleys and waterfalls

There are only a few places where the Pyrenees can be crossed through passes lower than 6,600 ft (2,000 m). During the last ice age, the Pyrenees experienced extensive glaciation. The mountains are characterized by jagged peaks that are steepest on the French side. They contain some of Europe's most spectacular waterfalls, the highest of which, at the Cirque de Gavarnie, plunges 1,385 ft (422 m). Major rivers, such as the Garonne of France and tributaries of the Ebro of Spain, flow out of the mountains. Wildlife in the region includes the rare lammergeier vulture, eagles, chamois, wild boars, and a small population of brown bears.

△ **SOURCE OF LEAD**
The Pyrenees contains deposits of some valuable minerals, including galena (an ore of lead). The crystals seen here have unusual shapes called truncated octahedrons.

HOW THE PYRENEES FORMED

In the distant past, Iberia was an island being pushed away from Eurasia by a spreading midocean ridge. But about 70 million years ago, a new convergent plate boundary formed between Iberia and Eurasia, which began to approach each other. This led to the collision that pushed up the Pyrenees.

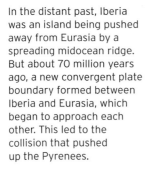

Mid-Atlantic Ridge
spreading ridge

Eurasia

Iberia

130 MILLION YEARS AGO

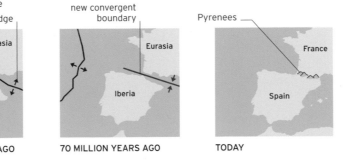

new convergent boundary

Eurasia

Iberia

70 MILLION YEARS AGO

Pyrenees

France

Spain

TODAY

The Alps

The highest and most extensive of Europe's mountain ranges, containing many famous peaks

C. Europe

Some 750 miles (1,200 km) long and more than 120 miles (200 km) wide at their broadest point, the Alps cover more than 80,000 square miles (207,000 square km) of western Europe. Pushed up over the last 100 million years by the coming together of the African and Eurasian plates, they form a curved belt, with more than 50 peaks rising to over 12,500 ft (3,800 m).

Shaped by glaciers

The current landscape of the Alps is a result of glaciation over the past 2 million years. The glaciers that once covered these peaks sculpted out enormous amphitheatre-like cirques (mountain hollows), sharp arêtes (knifelike ridges), and majestic pyramidal peaks such as the Matterhorn on the Swiss–Italian border and Austria's highest mountain, Grossglockner. As the glaciers melted, they left behind some U-shaped valleys, immense waterfalls pouring from hanging valleys, and elongated lakes of great depth. The longest remaining glacier is the Aletsch (see p.151).

European playground

The Alps include famous mountains such as Mont Blanc on the French–Italian border, the Matterhorn, and the Eiger in Switzerland, whose north face has long been considered one of the world's climbing challenges. During summer, the region attracts mountaineers and people who enjoy hiking, mountain biking, and paragliding. Winter sports such as skiing and snowboarding are also popular.

◁ **LIVING AT ALTITUDE**
A type of large squirrel, the marmot is a common sight in high-altitude pastures during summer. Marmots hibernate for the rest of the year in burrows underground.

In 1991, a **5,000-year-old mummy**, since named **Ötzi**, was **found** in the **Ötztal Alps**

SUBRANGES OF THE ALPS

The Alps contain many subranges, some of which are mapped below. The loftiest are the Pennine Alps, which contain 13 of the highest 20 alpine peaks; the Bernese Alps contain four; and the Mont Blanc Massif in the Graian Alps has three.

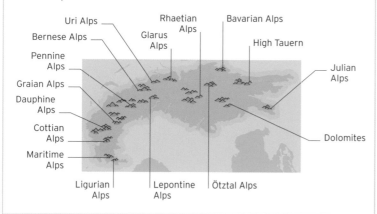

Uri Alps
Bernese Alps
Pennine Alps
Graian Alps
Dauphine Alps
Cottian Alps
Maritime Alps
Ligurian Alps
Glarus Alps
Rhaetian Alps
Bavarian Alps
High Tauern
Julian Alps
Dolomites
Lepontine Alps
Ötztal Alps

▽ **FLORAL MIGRATION**
As global temperatures have risen and glaciers retreated, flowering plants of all kinds have spread to much higher latitudes and altitudes in the Alps.

ICONIC PEAK
With its precipitous, glacier-sculpted walls, the Matterhorn is the quintessential alpine peak. Its summit, at 14,692 ft (4,478 m), was first conquered in 1865.

Mount Etna

The tallest and by far the largest active volcano in Europe, with a long history of major, often spectacular eruptions

S. Europe

Occupying 459 square miles (1,190 square km) of eastern Sicily, Mount Etna is one of the world's largest, most famous, and most active volcanoes. Towering above Catania, it has one of the world's longest documented records of volcanic activity, dating back to 1500 BCE. Etna's origins are probably connected to it being close to the boundary between the African and Eurasian Plates; it may also lie over a mantle hotspot.

Complex structure

Etna is a stratovolcano, with a complex structure that includes four separate summit craters—and more than 300 smaller parasitic volcanic vents and cones on its flanks. Its summit is at 10,925 ft (3,330 m), and lava flows cover much of the surface of its flanks. Although Etna has an overall conical shape, there is a deep and prominent horseshoe-shaped depression on its eastern side, known as the Valle del Bove.

Eruption types

Over the past several thousand years, Etna has been almost continuously active, producing eruptions of two main types. Spectacular explosive eruptions flare up from one or more of its summit craters. These can produce tall fountains of fiery lava, volcanic bombs, cinder showers, and large ash clouds. Quieter eruptions, with large-scale outpourings of lava, also emerge from flank vents and fissures. Etna's most destructive eruption occurred in March 1669. It produced massive lava flows that destroyed most of Catania's city walls. Despite the danger it poses, most of the inhabitants of Sicily regard Etna as a major asset.

◁ **LAVA LUMP**
The lava that erupts from Etna is mostly of a composition and temperature that makes it flow several miles before solidifying into lumps like this.

In **1999**, Etna erupted **lava fountains 1 ¼ miles (2 km) high**–the highest ever recorded

HISTORIC LAVA FLOWS

This map shows areas covered by Etna's lava flows over the centuries, going back to the 17th century. It also shows the extent of older flows. Many eruptions have come from Etna's flanks rather than its summit, and some flows have invaded urban areas. The pre-Etnian sediments are more than half a million years old—Etna's approximate age.

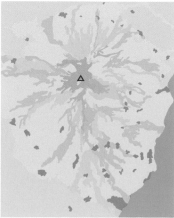

KEY

△ Summit of Mount Etna
■ Historic summit lavas

Flank lava flows
■ 21st century
■ 20th century
■ 19th century
■ 18th century
■ 17th century
■ Pre-16th century
■ Prehistoric lavas
■ Pre-Etnean sediments
■ Urban areas

△ **FIRE TUNNEL**
This cave under Etna's southeastern flank is a lava tube–a natural tunnel within a solidified former lava flow. Fiery magma once flowed through it.

ASH COLUMN
During an eruption in December 2015, Etna blasted ash to a height of 5 miles (8 km), where it flattened into a mushroom shape.

Inside a
STRATOVOLCANO

There are two main types of large volcano—shield volcanos and stratovolcanos. A shield volcano slopes gently, whereas a stratovolcano has steep sides and a conical shape built up from layers of eruptive products. The reason for these differences lies in the types of lava they emit. A shield volcano produces relatively runny lava that flows a long way before solidifying. A stratovolcano, such as Etna, also produces more viscous, water-rich lava that does not flow far. Instead, the lava tends to solidify around, and can sometimes block up, the volcano's main vent, or vents. As a result, much of the material erupted by a stratovolcano consists of fragmented solids—ash, pumice, and cinders—that are blasted into the air and later fall to the ground as the volcano clears its main vent. This helps explain not just the structure and shape of stratovolcanoes but also their long-term behavior, with phases of violent activity interspersed with quiet periods, lasting from a few years to several thousand years.

THE EVOLUTION OF A STRATOVOLCANO

Stratovolcanoes grow from the accumulation of their own eruptive products: solidified lava, ash, and cinders. The volcano initially grows rapidly, as eruptions add a lot of material relative to its existing size. A mature volcano gains height more slowly, partly due to erosion.

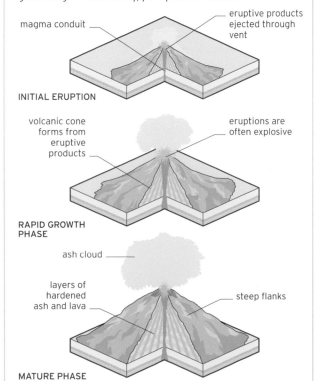

magma conduit — eruptive products ejected through vent

INITIAL ERUPTION

volcanic cone forms from eruptive products — eruptions are often explosive

RAPID GROWTH PHASE

ash cloud

layers of hardened ash and lava — steep flanks

MATURE PHASE

Ash cloud
An ash cloud is formed by gas and magma that is explosively fragmented into tiny particles as it is blasted into the air.

▷ **FLANK CRATERS**
Etna's flanks are covered in hundreds of secondary cones, vents, and craters, many of them now inactive. This is one of a pair called the Silvestri craters, dating from an eruption in 1892. It is 360 ft (110 m) across.

Solidified lava
Etna's surface is covered in solidified lava of various ages.

directly underneath Etna are layers of ancient volcanoes that form the foundation for the present-day cone

▷ **LAVA FLOWS**
Some eruptions occur from fissures that open up quite low down on the volcano's flanks. These can produce huge flows of lava that destroy agricultural land.

Lithospheric mantle
Made of the coarse-grained igneous rock peridotite, the lithospheric mantle underlies the crust.

lightning occurs in ash clouds due to friction between the ash particles and normal atmosphere

⊲ SUMMIT CRATERS
A stratovolcano may have one or several summit craters. Etna has four. The two shown here are known as Voragine (meaning "the chasm") and Bucca Nuova ("new mouth").

secondary pipe or conduit

Main pipe or conduit
Magma is carried up to the vents in the summit craters through the main pipe.

▽ EXPLOSIVE ERUPTIONS
These can occur from summit craters or sometimes from new vents that open near the summit. They typically produce lava fountains, ash plumes, and lava flows.

Crust
Beneath Etna, the crust is about 20 miles (30 km) deep and consists mainly of faulted and fractured layers of sedimentary rock with intrusions of molten and solidified magma.

magma rising through fractures in crust

Sill
Horizontal sheets of magma between rock layers are called sills.

Large magma chamber
Beneath Etna is a cavity full of magma (molten rock) and dissolved gases.

Dike
Vertical channels of magma are called dikes.

◀ MOUNT ETNA
Etna has many features of a typical stratovolcano, although it is more active than most in terms of eruption frequency. It produces spectacular explosive eruptions from craters at or near its summit and quieter eruptions, with outpourings of lava, from flank vents and fissures.

Stromboli

A small volcanic island, known for thousands of years as the Lighthouse of the Mediterranean because of the regularity, reliability, and brilliance of its eruptions

S. Europe

Stromboli is one of the few places on Earth that anyone can visit, on any day, with a high likelihood of being able to witness a volcanic eruption. Around every 20 minutes, one of its three summit vents blasts a fountain of lava fragments up to a height of 490 ft (150 m). These eruptions have been occurring for several thousand years and are so distinctive that geologists use the term Strombolian to describe similar eruptive activity at other volcanoes.

Visiting Stromboli

Roughly 3 miles (5 km) in diameter and 3,032 ft (924 m) high, Stromboli lies north of Sicily in the Mediterranean. A hike to the summit is permitted only in the company of a local guide. The eruptions can also be viewed from a boat off the island's northwest coast.

▷ **GIANT FIREWORK**
This photograph shows a typical eruption at Stromboli. Hundreds of glowing lava particles blast into the air, then fall to the ground in gentle arcs.

VOLCANIC ISLAND

Steep paths from two shoreline villages lead to vantage points for viewing eruptions. During forcible explosions, streams of lava cascade down a depression on one side of the volcano called La Sciara del Fuoco.

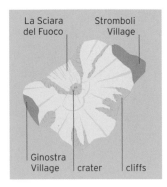

La Sciara del Fuoco — Stromboli Village — Ginostra Village — crater — cliffs

---

The Urals

An upland divide between Europe and Asia and one of the most ancient and mineral-rich mountain ranges on Earth

E. Europe

The Ural Mountains are about 1,600 miles (2,500 km) long and run north to south, from the Arctic Ocean to near the Aral Sea, encompassing a wide range of landscapes, from polar wasteland to semidesert. Located in Russia and Kazakhstan, they have traditionally been regarded as a boundary between Europe and Asia.

Ancient mountains

Partly due to the long period of erosion since they were formed, the Urals are not exceptionally high. The highest peak is Mount Narodnaya, with an altitude of 6,217 ft (1,895 m). Much of the area is forested, although the northern Urals are strewn with glaciers and also

feature alpine meadows and tundra. An extensive network of rivers and lakes covers the region, while karst topography, with many caves, is found on the western slopes. This mix makes the Urals one of the most beautiful areas in Russia. The richest and most varied fauna in the Urals are found in the forests and include brown bear, lynx, and wolverine. The mountains are particularly rich in natural resources, including lumber, coal, metal ores, and a range of gemstones.

◁ **CRYSTALLINE TREASURE**
Amethysts, emeralds, and topaz are among the precious gemstones mined in the Urals. This uncut hexagonal crystal is an emerald.

The Urals formed **300–250 million** years ago and are among the **world's oldest mountains**

Vesuvius

Europe's most dangerous volcano, famous for its large historic eruptions, including the massive and deadly eruption of 79 CE

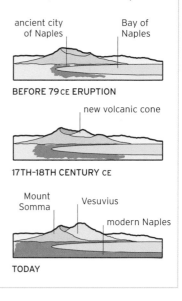
S. Europe

Vesuvius is a stratovolcano that lies just 5 miles (8 km) away from the Italian city of Naples, in the most densely populated volcanic area in the world. Its location and its capacity for especially violent eruptions are what make it so dangerous. Vesuvius has a steep cone, with a height of 4,203 ft (1,281 m), and sits within the caldera of an older volcano called Mount Somma. Its lower slopes are dotted with villages and vineyards.

Destructive past

Of Vesuvius's numerous large historic eruptions, the most infamous occurred in 79 CE. During the explosion, falls of volcanic ash and pyroclastic surges buried the towns of Pompeii and Herculaneum, killing about 2,100 people. Another enormous volcanic blast in 1631 killed more than 3,000 people, while an eruption in 1906 killed more than 200. There has been no eruption since 1944, suggesting that a large explosion is overdue. Plans have been drawn up for an evacuation in the event of warning signs, such as increased seismic activity.

△ **DECEPTIVE CALM**
Vesuvius's 1,300-ft- (400-m-) wide summit crater is outlined here. Part of the Bay of Naples is visible in the background, with the island of Capri in the far distance.

THE EVOLUTION OF VESUVIUS

Before the eruption in 79 CE, Vesuvius was a single volcanic cone with a huge crater. Later, a new cone grew within this crater. Today, this new, much enlarged, cone is known as Vesuvius. Much of the older cone, now called Mount Somma, has eroded away.

ancient city of Naples   Bay of Naples

BEFORE 79 CE ERUPTION

new volcanic cone

17TH–18TH CENTURY CE

Mount Somma   Vesuvius   modern Naples

TODAY

△ **WINTER FOREST**
A large part of the Urals is covered in taiga, the coniferous forest that stretches across northern Eurasia. In winter, much of the forest is snow-covered.

◁ **ROCK GIANTS**
Known as the Manpupuner rocks, these stone pillars, more than 100 ft (30 m) high, jut out of a plateau in the western Urals. Legend has it that they were giants turned to stone by a shaman.

△ BLUE CAVERN

Ice caves frequently form around the edges of the ice cap. They are carved by meltwater streams beneath the glacier that have since dried up. Their locations tend to shift from year to year.

◁ BEACHED ICE

Blocks of ice are often seen on the beaches of black sand near the edge of Vatnajökull. These are remnants of icebergs that reached the sea but were driven back by waves.

▷ PANOPLY OF COLORS

At sunrise, the undulating surface of Vatnajökull displays a glorious array of colors from soft pastel ochers and yellows to powder blue and bright turquoise.

N. Europe

Vatnajökull

A vast ice cap covering 8 percent of Iceland's surface area and the largest glacier by volume in Europe

One of several ice caps in Iceland, Vatnajökull derives its name from the Icelandic words *vatna*, meaning "water," and *jökull*, meaning "glacier." The ice cap completely covers the mountainous terrain of southern Iceland. Under the ice cap are valleys, mountains, and plateaus.

Shape and size
Roughly elliptical, the ice cap forms a frozen dome, with its highest point lying at an altitude of over 6,600 ft (2,000 m). It covers a total area of about 3,130 square miles (8,100 square km). Around 30 outlet glaciers drain the ice from Vatnajökull, carrying it toward the sea. Some of the ice breaks off as icebergs into a large glacial lagoon, Jökulsárlón, and from there makes its way via channels to the sea. Owing to the light-bending effects of atmospheric refraction, the ice cap is said to be visible sometimes from the top of the highest mountain in the Faroe Islands, which are 340 miles (550 km) away. This has been recorded as one of the world's longest sight lines.

Subglacial volcanoes
Three active volcanoes, as well as some volcanic fissures, lie under Vatnajökull. The active volcanoes are named Bardarbunga and Öræfajökull (both stratovolcanoes), and Grímsvötn, which is a caldera volcano. When one of these volcanoes erupts, along with the usual effects and dangers of a volcanic eruption, there is the added risk of large amounts of ice melting and causing a catastrophic flood (see panel, below). Events of this sort, which are widely referred to by the Icelandic term *jökulhlaups* (meaning "glacier floods"), are most often associated with the Grímsvötn volcano, which has the highest eruption frequency of all Icelandic volcanoes. In November 1996, an eruption of Grímsvötn melted millions of tons of Vatnajökull's ice and triggered a flood that lasted two days. The floodplain between the glacier and the sea became littered with huge blocks of ice.

◁ **BORN SWIMMER**
One of six seal species found in Iceland, the common, or harbor, seal is often seen in a meltwater lagoon, Jökulsárlón, situated between Vatnajökull and the sea.

FLASH FLOODING

When the Grímsvötn volcano under Vatnajökull erupts, large quantities of the overlying ice melt. This can create such pressure that the ice cap lifts up slightly and vast amounts of water flood out from beneath it. This can devastate the coastal plain next to the glacier.

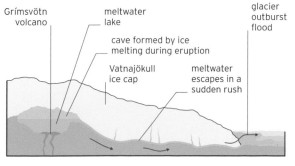

Grímsvötn volcano

meltwater lake

glacier outburst flood

cave formed by ice melting during eruption

Vatnajökull ice cap

meltwater escapes in a sudden rush

The **ice cap** has a maximum thickness of **3,300 ft (1,000 m)** and is **1,300 ft (400 m) thick** on average

N. Europe

The Monaco Glacier

An immense glacier on Spitsbergen, the largest island of the Svalbard archipelago

Spitsbergen, in northern Norway, is around 1,150 miles (1,850 km) from the North Pole, and about 80 percent of it is covered in glaciers. The Monaco Glacier, situated on an area of the island called Haakon VII Land, is one of the largest. The glacier discharges copious amounts of ice into a long fjord, Liefdefiord, at its front end.

A princely river of ice

Some 25 miles (40 km) long, the glacier is named after Prince Albert I of Monaco, a pioneering polar explorer of the early 20th century, who organized the first expeditions to explore and map the glacier. According to mapping studies, the glacier has retreated by more than 2 miles (3 km) in the last 50 years. Ringed seals, bearded seals, various seabird species, and polar bears are the main inhabitants of the neighboring ice floes, coasts, and islands.

△ SEA OF ICE
Ice calving off the Monaco and other glaciers nearby is the source of imposing, ship-sized icebergs that float in the middle of Liefdefiord.

▷ FROZEN WALL
A flock of kittiwakes feed in front of the Monaco Glacier's enormous terminal wall, which is more than 165 ft (50 m) high and 2 miles (4 km) wide.

The **thundering sound** of ice breaking off the glacier can be heard for **tens of miles** around

The Jostedalsbreen Icefield

The largest glacier in mainland Europe, a remnant of a vast ice sheet that covered the whole of Norway until about 10,000 years ago

N. Europe

The Jostedal Icefield (*breen* is Norwegian for "glacier") lies in southwestern Norway and has an area of about 185 square miles (480 square km), with its highest point at 6,421 ft (1,957 m) above sea level. In places, its ice is more than 1,300 ft (400 m) thick. The icefield survives primarily because of high regional snowfall rather than particularly cold temperatures.

Locked-up water

The top, or central, part of the icefield is a slightly undulating and largely featureless expanse of white ice and the source of numerous outlet glaciers. The total mass of ice corresponds to approximately 300 billion bathtubs of water, or all of Norway's water use for 100 years. However, this mass is diminishing, as reflected by a high rate of melting at the snouts of Jostedalsbreen's outlet glaciers. Some of these, such as Nigardsbreen, contain spectacular blue ice caves.

△ AZURE GROTTO
Ice of such a deep blue color as seen in this cave is very dense and strong, containing fewer air bubbles than white ice.

▽ GLACIER TOUR
A line of climbers, equipped with poles and crampons, make their way up Briksdalsbreen, one of the most accessible of Jostedalsbreen's outlet glaciers.

JOSTEDALSBREEN'S OUTLET GLACIERS

The Jostedalsbreen Icefield has a long, thin shape, with a maximum length of some 40 miles (65 km). About 50 outlet glaciers flow out of the icefield and down to the valleys below. Four of the largest are marked below.

Briksdalsbreen
Lodalsbreen
Austerdalsbreen
Nigardsbreen

W. Europe

The Mer de Glace

A glacier on the northern, or French, side of Mont Blanc,
notable for the spectacular banding pattern on its surface

The Mer de Glace is a valley glacier—a long glacier
with mountains on both sides—that extends down
from the Mont Blanc Massif toward the Chamonix
Valley in eastern France. It is some 7 miles (11 km) long.

The Mer de Glace, "Sea of Ice," gets its name
from the undulating, banded pattern seen on part
of the glacier below an icefall (a steeply descending
section) situated at an altitude of around 10,500 ft
(3,200 m). This surface pattern has been described as
waves frozen in the midst of a storm. The "waves"
are actually alternating bands of thicker, lighter ice
and thinner, darker ice, created due to seasonal
factors (see panel, right). The bands, known as
ogives, are curved—the result of the ice in the
middle of the glacier moving downhill slightly faster
than the ice at its sides.

▽ SNAKING DOWNWARD
The Mer de Glace descends some 8,200 ft (2,500 m)
vertically from its ice accumulation area, which lies
below Mont Blanc, visible here at top left, to its snout.

THE MER DE GLACE'S BANDING PATTERN

In summer, ice passing down
the icefall loses volume by
melting, and forms a trough
at the bottom, darkened by
the concentration of dust
inside it. In winter, thicker and
lighter-looking ice emerges
at the bottom.

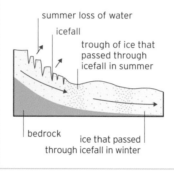

summer loss of water

icefall

trough of ice that
passed through
icefall in summer

bedrock

ice that passed
through icefall in winter

△ ICY HIGHWAY
Dark medial moraines (longitudinal stripes)
run down the glacier, reinforcing the appearance
of a winding highway of ice.

TYPES OF MORAINE

Debris carried along by a glacier and later
deposited is called a moraine. Lateral
moraines occur at the sides of a glacier,
whereas medial moraines form from smaller
glaciers merging. Subglacial moraines are
found at the base, and englacial moraines
are trapped within the ice. A terminal
moraine (not shown here) is rock material
dropped at the melting snout of a glacier.

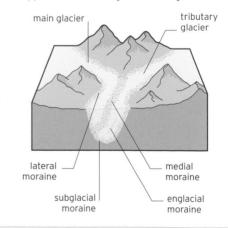

main glacier

tributary
glacier

lateral
moraine

medial
moraine

subglacial
moraine

englacial
moraine

The Aletsch Glacier

The longest and largest valley glacier in Europe, some 14 miles (23 km) long, which descends from a particularly scenic area of the Bernese Alps, in Switzerland

C. Europe

The Aletsch Glacier forms from the merger of four smaller glaciers that originate in the southern flanks of the impressive Jungfrau and Monch mountains. The four glaciers converge at a flat ice plateau called Kondordiaplatz at an elevation of 9,022 ft (2,750 m), where the ice is as much as 2,950 ft (900 m) deep.

Movement and retreat

The glacier moves at speeds varying from about 660 ft (200 m) a year in the Konkordiaplatz to 33 ft (10 m) a year near its snout. The pattern of terminal moraines below the snout indicates that it has retreated about 3 miles (5 km) since 1860. Spectacular views down the glacier can be seen from Jungfraujoch, a saddle area in the mountains that is accessible via a steep-grade railroad.

▷ **INSIDE THE ALETSCH**
Several ice caves have been carved out by meltwater streams. These are accessible from the glacier's edges and it is possible to enter them and view the brilliant blue interior.

The Aletsch glacier **covers an area** of about **31 square miles (80 square km)**

Litlanesfoss

A stunning Icelandic waterfall cascading over a remarkable outcrop of volcanic columns

N. Europe

Situated on the Hengifossá River in eastern Iceland, Litlanesfoss is a spectacular waterfall framed by a series of tall and exceptionally regular hexagonal basalt columns. Its total height of over 100 ft (30 m) includes two separate falls. The upper, smaller tier sits at an angle to the lower, near-vertical section, which falls into a small, bright-blue plunge pool. Litlanesfoss is one of several waterfalls dotted across the volcanic landscape of this part of Iceland, including Hengifoss, one of Iceland's tallest falls, just ½ mile (1 km) upstream.

Ancient lava flow

The Litlanesfoss Falls effectively cascade down the cross section of a thick ancient lava flow. The wide columns in the basalt rock are the result of the lava cooling relatively slowly over time. As it cooled and contracted, it formed joints, or fractures, that grew perpendicular to the base and surface of the lava flow. An array of straight and regular columnar jointing, as seen at Litlanesfoss, is known as a colonnade and is thought to have formed due to cooling from the base of the flow upward. Other well-known examples of columnar jointing include the Giant's Causeway in Ireland (see pp.164–65) and Fingal's Cave on the Scottish island of Staffa (see p.166).

△ **HORSETAIL FALLS**
The lower, taller section of the waterfall cascades down in what is known as a horsetail formation, with the water fanning out as it falls.

▽ **UPPER TIER**
The upper tier of Litlanesfoss is much shorter and less steep. Water pools at the bottom before it plunges down the lower section.

RADIAL DRAINAGE

The broadly circular Lake District comprises a core of central mountains from which a series of valleys, lakes, and waterways radiates outward. This pattern has been produced by the uplifting of the underlying dome of rock millions of years ago.

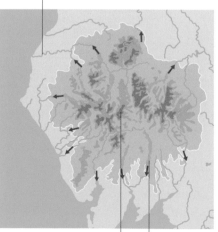

river flows downhill toward sea

rivers radiate outward from high central point

Lake Windermere

The Lake District

England's highest and arguably most picturesque region, a landscape shaped by glaciers and loved by tourists

NW. Europe

△ QUEEN OF THE LAKES
Seen here with a cloud of mist hanging over it, Derwentwater is thought by many to be the most beautiful of the Lake District's lakes and is referred to as the Queen of the Lakes.

The Lake District is a scenic area of mountains, moorlands, deep valleys, and lakes located within the county of Cumbria in northwest England. It contains most of the country's highest land, including Scafell Pike, its tallest peak, as well as England's longest and deepest lakes—Windermere and Wastwater, respectively. The region is named after the 16 lakes and many smaller water bodies, called tarns, which are found in higher areas. However, only one of the lakes, Bassenthwaite Lake, is actually called a lake; the others are known as waters or meres. Commonly defined by the boundary of the UK's largest national park, the Lake District covers an area of 912 square miles (2,362 square km).

Glacial landscape

The underlying geological structure of the Lake District is a huge granite dome, on top of which sits a series of three broad bands of rock of different ages and origins. Variations in bedrock can be seen in the different features across the region, from volcanic outcrops to slate scree slopes.

However, much of the landscape of the Lake District has been shaped by episodes of glaciation over the last 2 million years. Indeed, the region is known for its classic examples of glacial landforms. The most obvious of these are the broad valleys with their U-shaped cross sections, at the bottom of which often sit ribbon lakes.

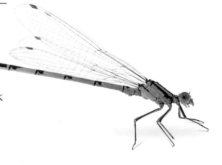

△ COMMON PREDATOR
The blue damselfly is a common predator on many UK waterways and large lakes. Unlike dragonflies, damselflies fold their wings along their body when resting.

The **Lake District National Park** is the **most visited** national park **in the UK**

The Camargue

One of Europe's major coastal wetlands, famous for its semi-wild horses and cattle, designated a wetland of international importance

W. Europe

The Camargue is a vast wetland region located within the Rhône Delta—the largest river delta in Western Europe—on the south coast of France. It comprises more than 360 square miles (930 square km) of diverse habitats including salt marshes, sandbars, brackish lagoons, freshwater ponds, reedbeds, and low-lying islands. Many of these landforms are in a state of constant change, and the gradual accumulation of silt means the delta is steadily growing.

Historically, the Camargue has been prone to flooding, and a combination of dykes and canals has been built to help control the flow of water through the area. Large parts of the original wetlands at the periphery of the region have been drained for agricultural use, including the creation of rice paddies, which have been a feature of the

region since the Middle Ages. The salt marshes in the southeast section of the Camargue are a major center of commercial salt production.

Famous wildlife

The Camargue's best-known animals include its semi-wild white horses and black cattle. However, its remarkable birdlife is one of the region's greatest attractions. More than 300 species are found in the Camargue, and it is an important stopover for birds migrating between Europe and Africa.

◁ **REEDBED SPECIALIST**
Attracted to reedbeds, the bearded reedling is also called the bearded tit. In summer it eats insects and larvae, and in winter it eats reed seeds.

COASTAL WETLAND

The Camargue is situated in a roughly triangular region between two arms of the Rhône Delta—the Petit-Rhône and Grand-Rhône—and the Mediterranean Sea. An area to the west of the western arm is known as Petite-Camargue.

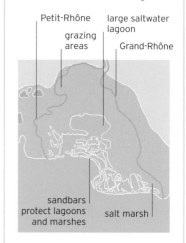

Petit-Rhône

large saltwater lagoon

grazing areas

Grand-Rhône

sandbars protect lagoons and marshes

salt marsh

▽ **FLOCKING FLAMINGOS**
The Camargue is home to a large population of greater flamingos, which feed on brine shrimps that thrive in brackish waters.

The **Camargue horse** is an **indigenous breed** that has lived here for **thousands of years**

The Verdon Gorge

A spectacular river-cut gorge in the South of France, referred to by some as the European Grand Canyon

W. Europe

The Verdon Gorge is a deep river canyon in the limestone massifs of the alpine foothills straddling the Alpes-de-Haute-Provence and Var regions of southeastern France. The gorge lies on a 15-mile (25-km) stretch of the Verdon river, which has carved the gorge over the last several thousand years. The river's striking turquoise-green color, produced by glacial rock flour suspended in the water, gives the river and gorge their name: *vert* means "green" in French. At the end of the canyon, the river flows into the artificial Lake Sainte-Croix.

Limestone walls

The width of the gorge at river level ranges from 20 ft (6 m) to 330 ft (100 m), and on each side steep cliffs reach up to 2,300 ft (700 m) tall. The limestone rock in which the Verdon Gorge formed was laid down during the Triassic Period (252–201 million years ago) when the area was under an ocean. Subsequent episodes of tectonic uplift and rock fracturing, followed by a period of glaciation, prepared the landscape for the river to carve out the gorge as it is today.

◁ **LUSH VALLEY**
The bottom of the gorge, which has a mild and humid microclimate, is covered in lush vegetation that diminishes with elevation up the cliff face.

Lake Geneva

W. Europe

A large alpine lake on the French–Swiss border, once hit by a huge tsunami

Known as Lac Léman in France, Lake Geneva is a crescent-shaped lake on the northern side of the Alps, stretching along the border between southeastern France and southwestern Switzerland. The largest of the alpine lakes, it is 45 miles (73 km) long, up to 9 miles (14 km) wide, and has a shoreline of just over 120 miles (200 km).

Lake of two halves

Lake Geneva lies on the course of the Rhône River, which enters through a delta at its eastern end and leaves at the western end through the city of Geneva. The lake is geographically divided into two basins separated by the Strait of Promenthoux. The Large Lake, to the east, is the widest section, and reaches a maximum depth of 1,020 ft (310 m). The Small Lake, to the west, is much narrower and shallower. Lake Geneva is notable for its rhythmical fluctuations in water level, known as seiches, caused by strong winds and rapid changes in atmospheric pressure.

▷ **BUSY SHORES**
The shores of Lake Geneva are home to more than 1 million people, including the residents of Lausanne and Geneva.

THE LAKE GENEVA TSUNAMI

Historical accounts and evidence from the lakebed suggest that Lake Geneva was struck by a large tsunami in 563 CE. Thought to have been caused by a rock fall near the Rhône Delta, a wave up to 52 ft (16 m) high traveled the length of the lake, devastating the shorelines in its wake. Moving at speeds of up to 45 mph (70 kph), the wave would have reached Geneva in just over an hour.

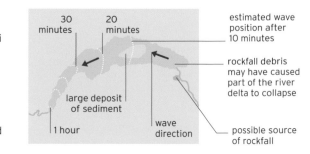

30 minutes
20 minutes
estimated wave position after 10 minutes
rockfall debris may have caused part of the river delta to collapse
large deposit of sediment
1 hour
wave direction
possible source of rockfall

C. Europe

The Danube

One of Europe's great waterways, linking the west of the continent to the east, crossing and forming many borders

Europe's second longest river, after the Volga, the Danube forms at the confluence of the Brege and Brigach Rivers in the Black Forest, in southwest Germany. Over a long journey eastward to its mouth at the Black Sea, the river flows a total of 1,770 miles (2,860 km). The Danube can lay claim to being the most international river in the world. It passes through, or partly borders, 10 countries—more than any other river—and its huge drainage basin includes parts of nine others.

Three courses

The course of the Danube is often divided into three major sections. The upper course runs from its source eastward across southern Germany and northern Austria to the Devin Gate, a natural gorge east of Vienna. At the start of its middle section, the river slows down and deposits large amounts of gravel and sand. Turning south, it flows through Budapest and across the Great Hungarian Plain to become a border between Romania

and Serbia, where its banks narrow into a series of towering limestone cliffs known as the Iron Gate. The Danube's lower course becomes wider and shallower as it starts its journey across a broad, flat plain. As the river approaches the Black Sea, it winds north around the Dobruja Hills before turning east, splitting into three main distributaries and forming Europe's second largest delta after the Volga.

Vital artery

The Danube has played a central role in the social and economic history of Central and Eastern Europe. At some times, it has formed the border between territories; at others, it has been a vital link allowing the flow of commerce between nations, enabling the countries it crosses to thrive.

◁ **BRIGHT FROG**
Pool frogs live in ponds, lakes, and slow-moving rivers over much of Europe. Those found in Central and Southern Europe are often a brighter green than their western cousins.

The Danube is the **only major European river** that flows **west to east** from Central to Eastern Europe

△ **CHANGING HABITATS**
Over its long course, the Danube passes through a variety of habitats, including deciduous woodland, coniferous forests, semiarid plains, and marshland.

HOW RIVERS TRANSPORT SEDIMENT

The amount of material that a river is able to transport depends on the energy of its waters and the size of the material. While dissolved materials are carried in solution, the smallest particles, such as clay and silt, are suspended in the water column when its currents are strong enough. Larger items of sediment that are not held in suspension are called its bedload.

When a river has enough energy, it can transport smaller bedload by pushing or bouncing it along the riverbed. Rivers with high energy levels and strong currents—during a flash flood, for example—can move boulders by rolling them downstream.

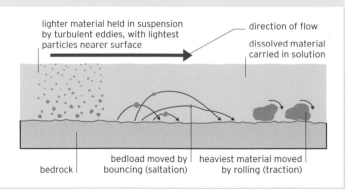

lighter material held in suspension by turbulent eddies, with lightest particles nearer surface

direction of flow

dissolved material carried in solution

bedrock

bedload moved by bouncing (saltation)

heaviest material moved by rolling (traction)

▷ **PELICAN HAVEN**
Home to 70 percent of the world's white pelican population, the Danube Delta is a key sanctuary for more than 300 bird species.

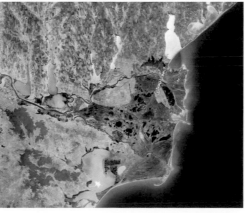

◁ **THE DANUBE DELTA**
The river's delta sprawls across an area of more than 1,500 square miles (4,000 square km). It is one of Europe's largest wetland areas, with one of the world's largest reedbeds.

C. Europe

The Biebrza Marshes

A large Central European wetland area with a wide variety of habitats supporting flourishing populations of birds and mammals

Centered on the Biebrza river valley in the far northeast of Poland, the Biebrza Marshes are a diverse wetland complex covering an area of more than 390 square miles (1,000 square km). The region features a mosaic of habitats including extensive marshes and reedbeds (see panel, right), grasslands, river channels, lakes, low islands, and wooded higher grounds. The area also includes one of the largest and best-preserved peat bogs in Central Europe. In many sections of the Biebrza Marshes, classic examples of ecological succession (see p.60) can be seen, as marshes grade first into raised peat bogs and then gradually into wet woodland.

The varied habitats of the Biebrza Marshes support a wide range of wildlife, and it has been recognized as a site of global significance. The area is one of the most important breeding sites for wetland birds in Central Europe, and more than 270 bird species have been recorded. Mammals include wetland specialists such as moose, beavers, and muskrats, as well as a few packs of wolves.

◁ **WATER LOVER**
Called moose in North America and elk in Europe, they usually live close to water and are excellent swimmers. Up to half of their diet consists of aquatic plants, although they also graze on trees and shrubs.

The Hortobágy Wetlands

A vast flooded grassland area in the Hungarian steppe region, regarded as one of the world's most important wetland reserves

C. Europe

Located on the upper reaches of the Tisza River in an area known as Puszta on the Great Hungarian Plain, the Hortobágy Wetlands is a steppe—a plain of grasslands and shrubs—that experiences regular flooding. This has turned an otherwise semiarid area into a wetland ecoregion of marshes, streams, and reedbed-lined lakes and lagoons. Part of the largest continuous natural grassland in Europe, the Hortobágy Wetlands is a key section of a national park that covers an area of 310 square miles (800 square km).

One of the best birdwatching sites in Europe, the Hortobágy Wetlands has a year-round wealth of birds. More than 340 bird species have been sighted here, and almost half have nested in the area.

▷ **STEALTHY PREDATOR**
Common in European wetland areas, the great egret wades through shallow water feeding mainly on fish, frogs, and small mammals, which it spears with its long, sharp bill.

In **fall**, up to **70,000 common cranes rest** and **feed** at Hortobágy

This cross section of a reedbed shows the succession from young reeds in open water to tightly packed old reeds on higher land. Old reeds give the dense cover needed by nesting birds, while young reeds provide shelter for small fish.

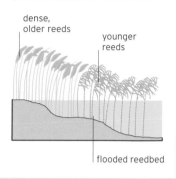

dense, older reeds

younger reeds

flooded reedbed

◁ **RICH PLANT LIFE**
The rich alluvial soils produced by regular flooding in the marshes support a wide variety of plants, including many rare species.

Slovak Karst Caves

An exceptionally large group of Eastern European caves with a great diversity of formations, a system of major geological and archaeological significance

C. Europe

The Slovak Karst cave system consists of more than 1,000 caves within an area of about 210 square miles (550 square km) along the border between southeastern Slovakia and northeastern Hungary. These diverse and complex caves formed as limestone and dolomite were dissolved by water.

Cave formations

Throughout the system, the caves are richly decorated by an array of structures and formations. The 15-mile- (25-km-) long Baradla-Domica cave complex, one of the most studied, has an elaborate display of stalagmites and stalactites of various shapes and colors, and a stream running through its main section. Like some of the other caves in the region, it shows evidence of ancient human occupation. The Slovak Karst system also includes many ice caves containing year-round ice formations. Perhaps the best known of these, the Dobšinská Ice Cave, was discovered in 1870 and is thought to be about 250,000 years old.

◁ **COMPLEX CAVE**
The ceiling of the Jasovská Cave is decorated with a variety of stalactites and curtainlike calcite flowstones.

△ **WOODED SHORELINE**
Much of the shoreline of Lake Ladoga is covered in coniferous and deciduous trees. The conifers include pine and spruce, and deciduous species include willows and birches.

Lake Ladoga

The largest lake in Europe, with a huge drainage basin made up of thousands of smaller lakes and rivers

NE. Europe

▷ **AQUATIC BLOOM**
The amphibious bistort, an aquatic plant in the knotweed family, is found in the shallows of Lake Ladoga and along the Neva River.

Located in northwestern Russia, near the border with southeastern Finland, Lake Ladoga is Europe's largest lake. Its total area is about 6,800 square miles (17,700 square km); from north to south it is 136 miles (219 km) long, and it is 86 miles (138 km) across at its widest point. The deepest part of the lake is near the tall, rocky cliffs of its northern shore, where it reaches 750 ft (230 m) deep just west of Valaam Island, one of the largest of more than 650 islands in the lake. The southern part of the lake is shallower and has a lower shoreline.

Lake and sea

Lake Ladoga formed in a graben, a depression between two fault lines, which has since been reshaped by glaciers. During the last ice age, it was part of the Baltic Ice Lake, which eventually formed the Baltic Sea when the ice retreated. Lake Ladoga is now separated from the Baltic Sea to the southwest by the Karelian Isthmus. The Neva, the only river that flows from the lake, leaves at its southwestern corner, passes through St. Petersburg, and drains into the Gulf of Finland (part of the Baltic Sea).

There are more than **50,000 lakes** and **3,500 rivers** in Lake Ladoga's drainage basin

SE. Europe

The Volga

The longest river in Europe and the core of its biggest river system, regarded as Russia's national waterway

The Volga is Europe's longest river, at the heart of the continent's largest river system. Rising in the Valdai Hills, northwest of Moscow, it flows 2,293 miles (3,690 km) southeastward through western Russia before emptying into the Caspian Sea. The lush Volga Delta, which spans 100 miles (160 km), is the largest river estuary in Europe. With a catchment area extending eastward to the Ural Mountains, the Volga drains most of European Russia. Along its course, it collects more than 200 tributaries, and its watershed contains 11 major cities. The river is traditionally referred to by Russians as Mother Volga.

△ **THICK BILL**
The largest species of tern, the Caspian tern is a common sight in the Volga Delta. This species has a distinctly chunkier bill than other terns.

▷ **ICE FLOES**
The Volga is affected by ice over much of its length for about 100 days each year.

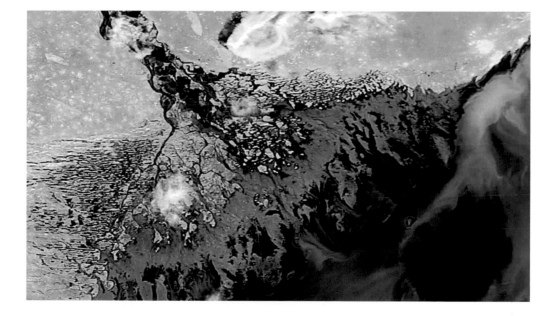

△ **FROZEN INLET**
The edges and inlets of the lake start to freeze in December, and by the end of February the entire surface of the lake is frozen over.

EUROPE'S LONGEST RIVERS

Four of the five longest rivers in Europe pass through Russia at some part of their course, and the Volga and the Don are contained within its borders. The Danube crosses 10 different countries on its course.

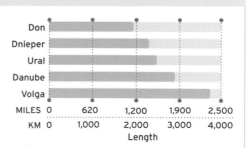

	MILES 0	620	1,200	1,900	2,500
Don					
Dnieper					
Ural					
Danube					
Volga					
	KM 0	1,000	2,000	3,000	4,000

Length

△ **GREEN DELTA**
The Volga brings life to an otherwise arid landscape as it flows into the saline Caspian Sea, the world's largest inland body of water.

NE. Europe

Norwegian Fjords

The most complex coastline in the world, a maze of long, deep marine channels, many overlooked by high cliffs

A fjord is an elongated ocean inlet, originally created by the action of glaciers over thousands or millions of years. While fjords are found in many countries, those of Norway are particularly magnificent, numerous (around 1,000), and in many cases exceptionally long.

A fjord-dominated landscape runs as a strip all around Norway's coast. This has a unique appearance, with a tangle of islands and peninsulas interspersed between the fjords themselves. Except in the area where they meet the sea, fjords are much deeper than typical coastal plain estuaries. Sognefjord, for example, reaches a maximum depth of 4,300 ft (1,300 m). Many Norwegian fjords have vertical cliffs on one or both sides, rising to a height of up to 3,300 ft (1,000 m).

Fjord life

Life in Norway's fjordland centers on picturesque fishing villages, which are connected by an extensive system of ferries. Wildlife includes marine mammals such as seals and porpoises, sea eagles, and millions of other seabirds, including colorful puffins. The harbor porpoise—one of the smallest marine mammals—is commonly seen here and feeds on small schooling fish.

Fjords **lengthen Norway's coast** by a factor of ten, to over **19,000 miles (30,000 km)**

HOW FJORDS FORM

During the last glacial period, huge glaciers extending down from a vast ice sheet scoured deep U-shaped valleys. As the glaciers melted and the sea level rose, seawater flooded into these valleys, forming fjords.

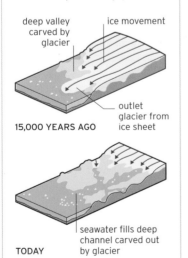

deep valley carved by glacier

ice movement

outlet glacier from ice sheet

15,000 YEARS AGO

seawater fills deep channel carved out by glacier

TODAY

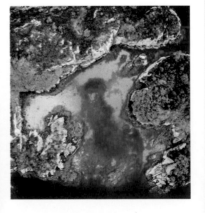

△ **TURQUOISE PASSAGE**
The waters of Hjeltefjord meander between islands in Hordaland County, southwestern Norway. Hjeltefjord is around 25 miles (40 km) long.

△ **SHEER DROP**
Soaring 2,000 ft (600 m) above Lysefjord, Pulpit Rock is one of Norway's most visited landmarks. It is popular for activities such as cliff jumping.

▷ **STUNNING ENTRANCE**
Traditional fishing huts can be seen near the mouth of Kierkfjord in the Lofoten Islands. A short distance from the mainland, the Lofotens are blessed with exquisite scenery.

NW.
Europe

The Giant's Causeway

An array of huge columnar rocks, famed worldwide for its spectacular geometric stepping-stone appearance

The Giant's Causeway is a group of thousands of interlocking columns of basalt rock at the foot of a cliff on the coast of Northern Ireland. The tops of the columns form a staircase that leads from the cliff foot up to a mound-shaped area, and then disappear under the sea.

Fiery origins

Legend has it that the Causeway was built by a giant, Fionn MacCoul, as a means of crossing the sea to fight a Scottish rival. It actually originated from lava poured out by volcanic fissures in the area roughly 55 million years ago. The lava cooled and solidified to form thick layers of rock, some of which fractured into columns beneath the surface. After millions of years of erosion by glaciers, and later the sea, these structures were revealed at the surface. Apart from the causeway itself, the surrounding area is excellent for studying sea birds and also for botanical studies, since some rare plants thrive here.

◁ **FINE-GRAINED ROCK**
The Giant's Causeway is composed of basalt, a dark rock that forms when a particular type of runny lava cools and solidifies rapidly.

Some of the **40,000 basalt columns** that form the causeway are more than **36 ft (11 m)** high

◁ △ **STEPPING-STONE COAST**
These columns were once a mass of cooling lava, which solidified, shrank, and fractured to form geometric patterns similar to those sometimes seen in drying mud.

▷ **ERODED CLIFFS**
Behind the Giant's Causeway are cliffs eroded out of the plateau of northern Antrim. The basalt columns jut out of these cliff faces, forming jagged steps that creep into the sea.

NW. Europe

Fingal's Cave

An atmospheric sea cave on the uninhabited Scottish island of Staffa, known for its wonderful acoustics

Staffa is one of the Inner Hebridean islands off the west coast of Scotland, close to the much larger island of Mull. Its cave, named after the hero of an epic poem, is about 65 ft (20 m) high and extends about 200 ft (60 m) into the island.

In summer, local boating companies organize regular boat trips to Staffa. In calm conditions, visitors can explore Fingal's Cave, including its interior, via a walkway. The cave's size and arched roof, and the echoes of lapping waves, give it a special acoustic quality. The German composer Felix Mendelssohn, after visiting the cave in 1829, was so moved that he wrote an overture, "The Hebrides," in remembrance. This piece of music is now known as the "Fingal's Cave overture."

◁ **HEXAGONAL FRAMING**
The cave's arched entrance is framed by an array of hexagonal basalt columns. The sea covers the cave floor, with waves regularly washing and splashing in and out.

The Cliffs of Moher

An imposing procession of high, banded gray cliffs on the west coast of Ireland, home to tens of thousands of seabirds during the annual nesting season

NW. Europe

The Cliffs of Moher extend along part of the coast of County Clare, Ireland, at the edge of a region called the Burren, which is dominated by karst topography (partially dissolved limestone). They take their name from an ancient fort called Moher, or Mothar, that once stood on Hag's Head, their southernmost point.

Ancient heights

The cliffs consist of layers of the sedimentary rocks shale and sandstone, which originally formed more than 300 million years ago. In parts, they rise more than 660 ft (200 m) above the Atlantic Ocean, and their tops offer fantastic views of various offshore islands and mountain ranges of western Ireland. These include the three Aran Islands in Galway Bay and the Maumturks mountain range in Connemara. Between April and July, more than 30,000 pairs of seabirds—including murres, razorbills, puffins, and cormorants—nest here, while numerous other species, including ravens, choughs, and peregrine falcons, can be seen all year round.

△ **SUMMER VISITOR**
From May to July, thousands of pairs of puffins, with their brightly colored beaks, nest on the cliffs.

◁ **BANDED GIANTS**
The cliffs wind in and out over a stretch of more than 5 miles (8 km). The banding pattern is due to color variation in the sedimentary rock layers.

The Cliffs of Dover

One of England's most famous natural landmarks, a parade of tall, brilliant white chalk cliffs, facing in the direction of France

NW. Europe

▽ **WHITER THAN WHITE**
The cliffs owe their gleaming whiteness to the almost 100 percent purity of the chalk.

The Cliffs of Dover, which form part of the coast of the English county of Kent, run along the northwestern side of the narrowest part of the English Channel. They are made from chalk formed during the Cretaceous period, between 100 and 70 million years ago. At that time, a large part of what is now northwestern Europe was underwater. The shells of tiny marine organisms accumulated on the sea floor and became compressed into a layer of chalk hundreds of feet thick. Later, after sea levels fell, this chalk formed a land bridge between Britain and mainland Europe. Around 8,500 years ago, this bridge was swept away in a catastrophic flood, leaving behind the cliffs of Dover and similar white cliffs at Cap Blanc Nez in France.

▷ **BRILLIANT BLUE**
The Adonis Blue butterfly favors chalk downland, such as the area of Kent behind the cliffs, and is found there in modest numbers.

Fossils found in the **cliffs** include **sharks' teeth, sponges, and corals**

The Dune du Pilat

The tallest sand dune in Europe, located close to Arcachon Bay in southwestern France, with magnificent views from the top

W. Europe

EVOLUTION OF THE DUNE

Analysis of sand samples from the dune and the underlying soil has shown how the dune has grown over several centuries. Wind power is currently moving it a few feet per year toward the forest to its east.

Containing more than 2 billion cubic feet (60 million cubic meters) of sand and running parallel to the shoreline, the Dune du Pilat rises gently to a height of 360 ft (110 m) above the west coast of France. It is nearly 2 miles (3 km) long and 1,600 ft (500 m) wide. The word

Pilat comes from the Gascon word *Pilhar*, which means "heap" or "mound." It takes about 30 minutes for a fit person to climb to the dune's top, where the wind can be very strong. It is a great vantage point in fall for viewing flocks of migratory birds.

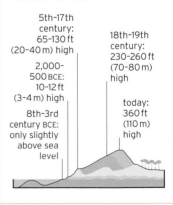

5th–17th century: 65–130 ft (20–40 m) high

18th–19th century: 230–260 ft (70–80 m) high

2,000–500 BCE: 10–12 ft (3–4 m) high

today: 360 ft (110 m) high

8th–3rd century BCE: only slightly above sea level

◁ **SHIFTING SAND**
The dune is slowly migrating away from the sea. Unless this movement can be stopped, the sand will eventually engulf a small forest on its land side.

▷ **COASTAL TREASURE**
The species of ammonoid fossilized in this rock specimen lived 195 million years ago. Called *Promicroceras*, its fossils are very common in parts of the Jurassic Coast.

limestone groundmass

WHAT ARE AMMONITES?

The rock strata of the Jurassic Coast are especially rich in ammonites–spiral-shaped fossils of the shells of extinct sea creatures called ammonoids. These predatory mollusks, alive 200-70 million years ago, were swimming organisms whose soft parts somewhat resembled those of present-day squids.

living ammonoid has long tentacles for catching prey

bouyant shell helped with swimming and defense

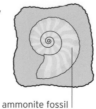
ammonite fossil embedded in rock

The **rock layers** of the Jurassic Coast cover **187 million years of** Earth's history

fossils are 1–2 in (2.5–5 cm) in diameter

NW. Europe

The Jurassic Coast

A stretch of English coastline famed for its fossils of ancient, extinct animals and plants, as well as some striking natural landforms

Approximately 96 miles (154 km) long, the Jurassic Coast occupies part of two English counties, Dorset and Devon. The coast has attracted worldwide attention mainly because of the cliffs, composed of layers of sedimentary rock, that run along most of its length.

A walk through Earth's past

Overall, the rock layers of the Jurassic Coast contain a near-complete record of Earth's Triassic, Jurassic, and Cretaceous periods. At different stages during this vast timespan, the area where the coastline now extends has been a desert, a tropical sea, an ancient forest, and a lush swamp. All these times and environments have been recorded in different rock layers. The area's significance in palaeontology only began to be realized in the early 19th century. In 1811, a local fossil collector, Mary Anning, and her brother discovered the complete fossil

of a large, extinct marine reptile called an ichthyosaur. Subsequently, she made more discoveries, including fossils of a pterosaur (an extinct flying reptile). This helped trigger an explosion of interest in the area's fossil legacy, which has continued to the present. Some of the more recent discoveries here include fossils of extinct crustaceans, insects, and amphibians, and a fossil forest of ancient conifers and tree-ferns.

In addition to its interest to paleontologists, the Jurassic Coast contains some superb examples of various types of natural landforms, including sea arches, rock pinnacles, and an enormously long gravel bank.

marcasite crystal

◁ **SHINY CLUSTER**
Nodules of the gleaming mineral marcasite, a form of iron sulfide, are commonly found in rocks on the Jurassic Coast.

chalk groundmass

densely packed fossils show that animals lived in great numbers

△ **SECLUDED BANK**
This remarkable gravel structure is Chesil Beach. It is 20 miles (30 km) long, and has the sea on one side and a lagoon on the other.

◁ **NATURAL ARCH**
Durdle Door is a limestone arch formed by wave erosion. The name Durdle comes from an Anglo-Saxon word *thirl* meaning "bore" or "drill."

The Algarve Coast

A coast famous for its breathtaking cliffs, golden beaches, limestone caves and rock formations, scalloped bays, and sandy islands

SE. Europe

△ **CAVE OF THE EYE**
This cave is close to Benagil Beach, near Lagoa, and can be visited by boat or by swimming. The enormous "eye" in its roof is 52 ft (16 m) across.

The Algarve is the southernmost region of mainland Portugal, and its coast has two sections. The first of these, some 100 miles (160 km) long, is south-facing and extends westward from Portugal's border with Spain out to the southwestern tip of the Iberian Peninsula, at Cape St. Vincent. The second section is west-facing and extends for 30 miles (50 km) northward from Cape St. Vincent.

Honey-colored cliffs

Bathed by the warm North Atlantic Current (an extension of the Gulf Stream), the Algarve Coast is notable for its picturesque, honey-colored limestone cliffs, small bays and coves, sheltered beaches of fine sand, and turquoise- or emerald-colored seas. Features typically formed by marine erosion can be seen along many sections. These include some large caves at the feet of cliffs, blowholes, and holes that have been excavated from headlands to create sea arches.

Parts of the shoreline are dotted with rocky sea stacks (isolated pillars of rock set off from headlands), some of which appear to be precariously balanced on the shoreline. Although limestone is a primary component of the landscape, other rocks, including sandstone and shale, are visible along scattered stretches of the coast. The region's warm climate and strikingly beautiful scenery have made this coast a popular vacation destination.

▷ **ERODED STACKS**
At Praia da Rocha, near Portimao, heavily eroded limestone sea stacks embellish the shoreline. Their coloration is typical of the entire Algarve Coast.

◁ **GOLD RUSH**
In Roman times, gold was mined across the Algarve region. Some new gold deposits have recently been discovered, so mining may restart.

The Dalmatian Coast

A fractured but highly scenic coast made up of hundreds of islands and straits, set in the azure waters of the Adriatic

S. Europe

The Dalmatian Coast extends down part of the eastern side of the Adriatic Sea (an arm of the Mediterranean Sea) and comprises a large section of the coastline of Croatia. From the seaport of Dubrovnik in the south, the coast meanders in a northwesterly direction for a distance of roughly 230 miles (375 km).

A drowned coast

The Dalmatian Coast is a classic of its type—a particular kind of drowned coast, formed from the submergence of what was once a landscape of many parallel mountain ridges close to a former coastline. As a result of the drowning, the higher parts of many of the old ridges are now long, narrow islands running parallel to the present-day coastline. The island of Dugi Otok (Croatian for "Long Island"), for example, is 27¾ miles (44.5 km) long, but only 3.0 miles (4.8 km) wide. Other parts of the former landscape are now long promontories, running parallel to the overall direction of the coastline, or have split up into innumerable small islands.

◁ **DOLPHIN-DOTTED WATERS**
Bottlenose dolphins are the only dolphin species commonly seen around the Dalmatian Coast. These sleek swimmers can reach speeds of more than 20 mph (30 kph).

The Dalmatian Coast is one of the **densest archipelagos** in the world

HOW THE DALMATIAN COAST FORMED

Around 10,000 years ago, the region contained mountains and valleys parallel to the coast. Over time, the rising sea level drowned the area, producing the coastal features seen today.

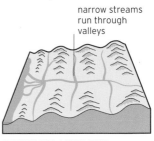
narrow streams run through valleys

BEFORE SEA-LEVEL RISE OF PAST 10,000 YEARS

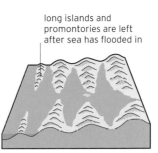
long islands and promontories are left after sea has flooded in

AFTER SEA-LEVEL RISE OF PAST 10,000 YEARS

△ **PARCHED ISLANDS**
This string of islands forms part of the Kornati Archipelago in northern Dalmatia. Their appearance is typical of many small islands of the region.

▷ **PRISTINE WATERS**
The clear waters and scenic beauty of Stara Baska in the southern part of Krk make it a popular tourist destination.

NW. Europe

The Hallerbos

A centuries-old Belgian woodland, where one species of wildflower produces a seasonal transformation

Every spring, the Belgian woodland called the Hallerbos undergoes a spectacular color change. From mid-April to early May, thousands of bluebells burst into bloom, covering its 2 square miles (5 square km) of forest floor in an intense blue-violet. The change is so striking that it has given rise to the woodland's other name: the Blue Forest.

Old and new

Usually, native bluebells grow in old woodland, but most trees in the Hallerbos date from a replanting effort conducted between 1930 and 1950. Its few ancient beeches and oaks are the only ones to survive World War I, when occupying forces devastated what had been a centuries-old forest. The first mention of the Hallerbos dates from 686 CE, and there are references throughout the medieval period; for this reason, official literature considers it "an ancient forest with young trees."

△ **DISTINCTIVE DROOP**
Bell-shaped flowers of native bluebells produce flowering stalks up to 20 in (50 cm) high, which sag at the top.

▽ **TRANSIENT CARPET**
Hallerbos bluebells display their most intense color just as beech leaves unfurl. Once the leaves open fully, less light hits the flowers beneath.

▷ **TOWERING FOLIAGE**
Spruce, including the towering Norway spruce, shares the Bavarian Forest with beech and fir on hillsides, and alder, beech, and willow in valleys.

▽ **TRAPPED AIR**
Fog is common in the Bavarian Forest and contributes to the area's precipitation. In fall and winter, damp valley floors also produce cold air lakes—areas of trapped frigid, foggy air.

The Bavarian Forest

A mixed mountain forest, dominated by spruce, fir, and beech, where plants and animals must contend with acidic soil, high rainfall, and months of cold winter weather

C. Europe

Southeastern Germany's Bavarian Forest covers the high region between the Danube River valley and the Bohemian Forest along Bavaria's border with the Czech Republic. Together, these two forests form Europe's largest area of continuous woodland. Bavarian Forest terrain consists mainly of tree-covered granite and gneiss hills and mountain domes, interspersed with valleys and elevated plains. Native spruce is the dominant tree species, mixed with fir and beech; all must cope with the area's highly acidic, nutrient-poor soils. According to a local saying, the climate is "three-quarters winter and one-quarter cold," which aptly describes the annual average temperature range of 37–46°F (3–7.5°C). Depending on altitude, yearly precipitation is 43–98 in (1,100–2,500 mm), 30–40 percent of which falls as snow.

Around **10,000 animal species** are believed to live in the Bavarian Forest

△ **SEED SEEKER**
The Eurasian red squirrel makes the most of the Bavarian Forest's plentiful supply of spruce seeds. It emerges just after sunrise to begin daily foraging.

The Białowieża Forest

The only old-growth forest in lowland Europe and a refuge for the largest population of free-ranging European bison

E. Europe

Covering more than 580 square miles (1,500 square km), the Białowieża Forest is shared between Poland and Belarus. It is the only large remnant of the primeval forest that once covered northeastern Europe after the last ice age and is still home to some of the continent's tallest trees; old-growth spruce reaches more than 165 ft (50 m) here. Part of the Central European mixed-forest ecoregion that extends from

eastern Germany to northeastern Romania, Białowieża contains tree species from both northern (boreal) and lowland mixed deciduous forests. The latter include English oaks—some of which are thought to be 150–500 years old—as well as the small-leaved linden and hornbeam. Mixed broadleaf and conifer stands are also interspersed with pocket ecosystems, such as the increasingly rare boreal spruce bog forest.

Dead wood, new life

A high percentage of standing and fallen deadwood supports thousands of fungi, insects, and bird species, including the protected three-toed woodpecker. Wild boars, elks, wolves, and lynxes are common forest residents, and Białowieża is an important refuge for the continent's largest mammal, the European bison or wisent, with the largest free-ranging herd of this species.

◁ **STRIPED COATS**
Wild boar young, known as boarlets or piglets, have striped coats to blend in with their dappled woodland surroundings.

TREES AND FUNGI

Many trees and fungi share a mutually beneficial relationship called a mycorrhiza. A fungus sends out hundreds of thin, rootlike threads called hyphae that grow around a tree's rootlets, absorbing sugars from them. In return, the tree uses the hyphae network to absorb more nutrients from the soil.

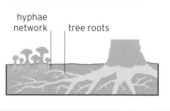
hyphae network   tree roots

SWAMPY SANCTUARY
The Białowieża Forest contains several different habitats, including sections of spruce bog forest that are home to many rare plant species.

Kola Peninsula Tundra

A treeless landscape of harsh beauty with permafrost, clear lakes, and rivers, and a vital home for Arctic plant and animal species

N. Europe

Northern Russia's Kola Peninsula is located almost completely within the Arctic Circle, between the Barents and White Seas. Tundra covers nearly all of its 40,000 square miles (100,000 square km), but approximately 22,700 square miles (58,800 square km) are classified as critically endangered coastal Arctic tundra. The landscape is covered with moss, lichen, grasses, and wildflowers, with shrubs such as dwarf Arctic birch and cloudberry; the frozen subsoil known as permafrost (see panel, right) prevents trees from growing. Kola winters are long and extreme; frost often occurs from August through to the following June, and strong winds blow up to 120 days a year. Despite its harsh environment, the Kola's tundra, lakes, and river systems are home to 200 species of birds and 32 mammal species, including migrating herds of reindeer. However, Soviet nuclear testing and mining have already harmed what was once a pristine environment.

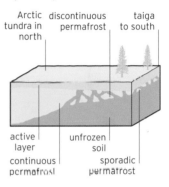

▷ **FRUITFUL SURVIVOR**
Cloudberries can survive temperatures as low as -40°F (-40°C). Their sweet fruit provides food for birds and mammals in fall.

PERMAFROST

Permafrost is ground that has been frozen for two years or more. It usually forms above a layer of unfrozen soil and beneath a layer, called the active layer, that freezes and thaws each year. The permafrost itself may be continuous or discontinuous, or form sporadic pockets.

Arctic tundra in north — discontinuous permafrost — taiga to south

active layer — continuous permafrost — unfrozen soil — sporadic permafrost

◁ **LIFE IN SMALL DOSES**
Tundra plants remain short to avoid freezing winds. Their shallow roots cannot penetrate permafrost.

The Western Steppe

A flat, windswept plain stretching thousands of miles and forming part of the largest temperate grassland on the planet

E. Europe

The Western Steppe lies chiefly in Ukraine, Russia, and Kazakhstan. With the Altai Mountains as its eastern boundary, it forms half of the vast Eurasian Steppe, the largest temperate grassland on Earth. Steppes get enough precipitation for grass growth but not enough for trees. The Western Steppe is 2,500 miles (4,000 km) long and is often called a "sea of grass," although where rivers and streams cut through the otherwise unbroken landscape, some trees do grow. Bordered to the north by taiga, and to the south by desert, the Western Steppe is relatively mild compared to eastern grasslands, but is still subjected to howling winds and temperatures ranging from 80°F (27°C) during the day to below freezing at night.

◁ **GRASSLAND FOOD CHAIN**
The grasses and wildflowers of the region abound with insects such as grasshoppers and beetles, which in turn attract many birds, including the western yellow wagtail.

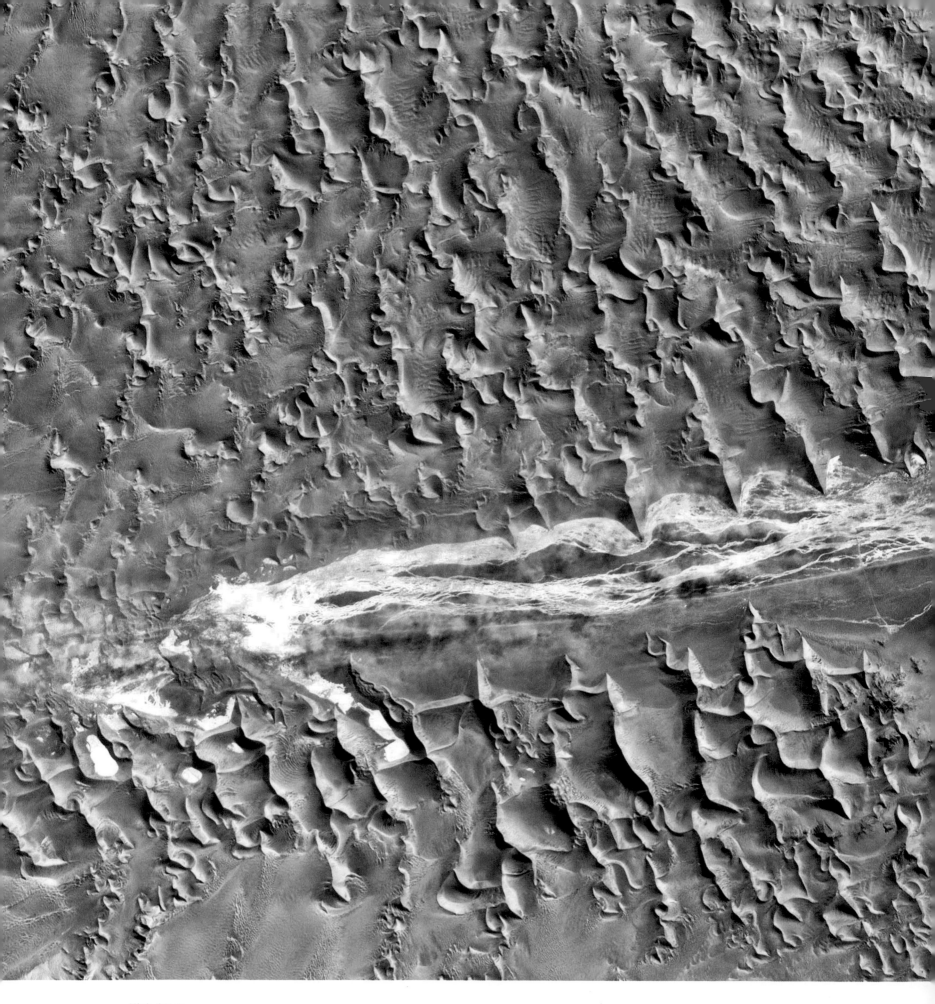

High dunes
Captured in visible light and infrared, these sand dunes are in Africa's ancient Namib Desert. Winds blowing off the Atlantic Ocean produce some of the world's highest dunes here, some reaching heights of 1,000 ft (300 m).

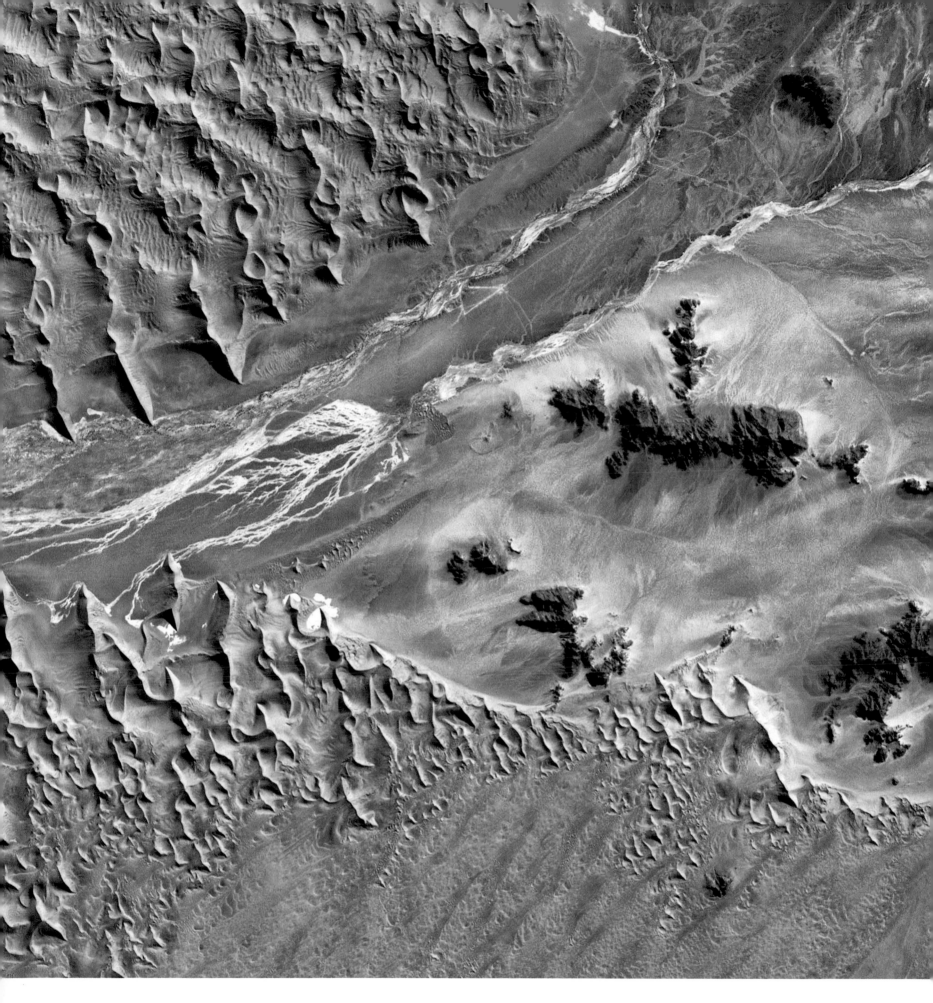

Africa

LAND OF THE RIFT

Africa

Africa is a gigantic continent; only Asia is larger. It stretches nearly 5,000 miles (8,000 km) from its most northerly point, Ras ben Sakka in Tunisia, to its most southerly point, Cape Agulhas in South Africa. It straddles the equator, which divides the continent in half, and consists of two large lobes—with the northern lobe extending far out west into the Atlantic Ocean, and the southern lobe running south between the Atlantic and Indian Oceans. The northern half is dominated by the world's largest desert—the Sahara. It is such an overwhelming geographical feature that Africa is often divided into the

Sahara and sub-Saharan Africa, the latter dominated by the western river basins filled with lush, tropical forests, and the drier, bushy grasslands or savanna on the eastern plateaus.

Despite its enormous size, Africa has no high mountain ranges, except for the Atlas in the extreme northwest. Much of the east and south are dominated by huge, high plateaus. In the northeast, the plateaus are split apart by the deep trench of the East African Rift valley, dotted by ribbons of huge lakes. The rift was created by tectonic forces beneath the crust cleaving the African Plate apart.

KEY DATA

▲ **Highest point** Mount Kilimanjaro, Tanzania: 19,341 ft (5,895 m)

▼ **Lowest point** Lake Assal, Djibouti: -512 ft (-156 m)

● **Hottest record** Kebili, Tunisia: 131°F (55°C)

● **Coldest record** Ilfrane, Morocco: -11°F (-24°C)

CLIMATE

Most of Africa lies within the warm tropics. Except in the forest areas of the west, rainfall is scarce with just one wet season in the savanna and almost none in the Sahara.

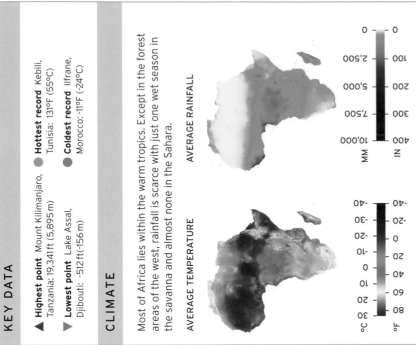

AVERAGE TEMPERATURE

AVERAGE RAINFALL

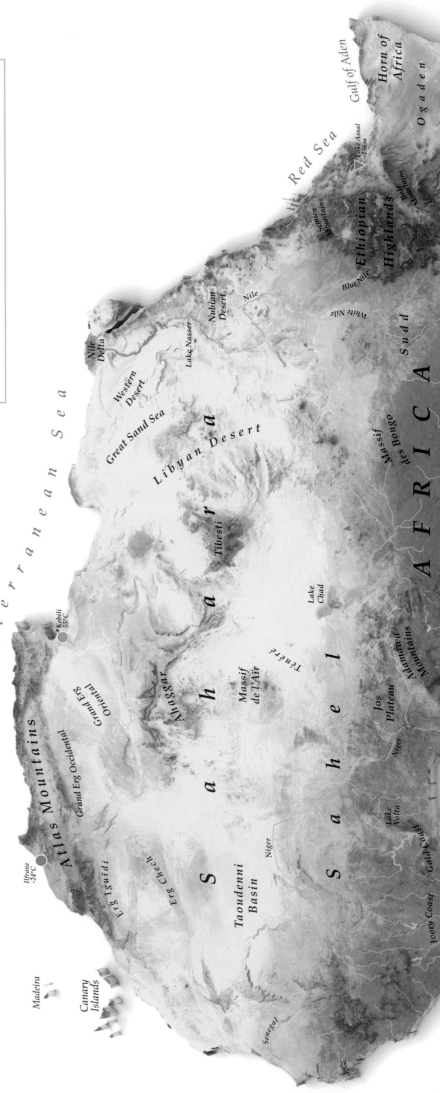

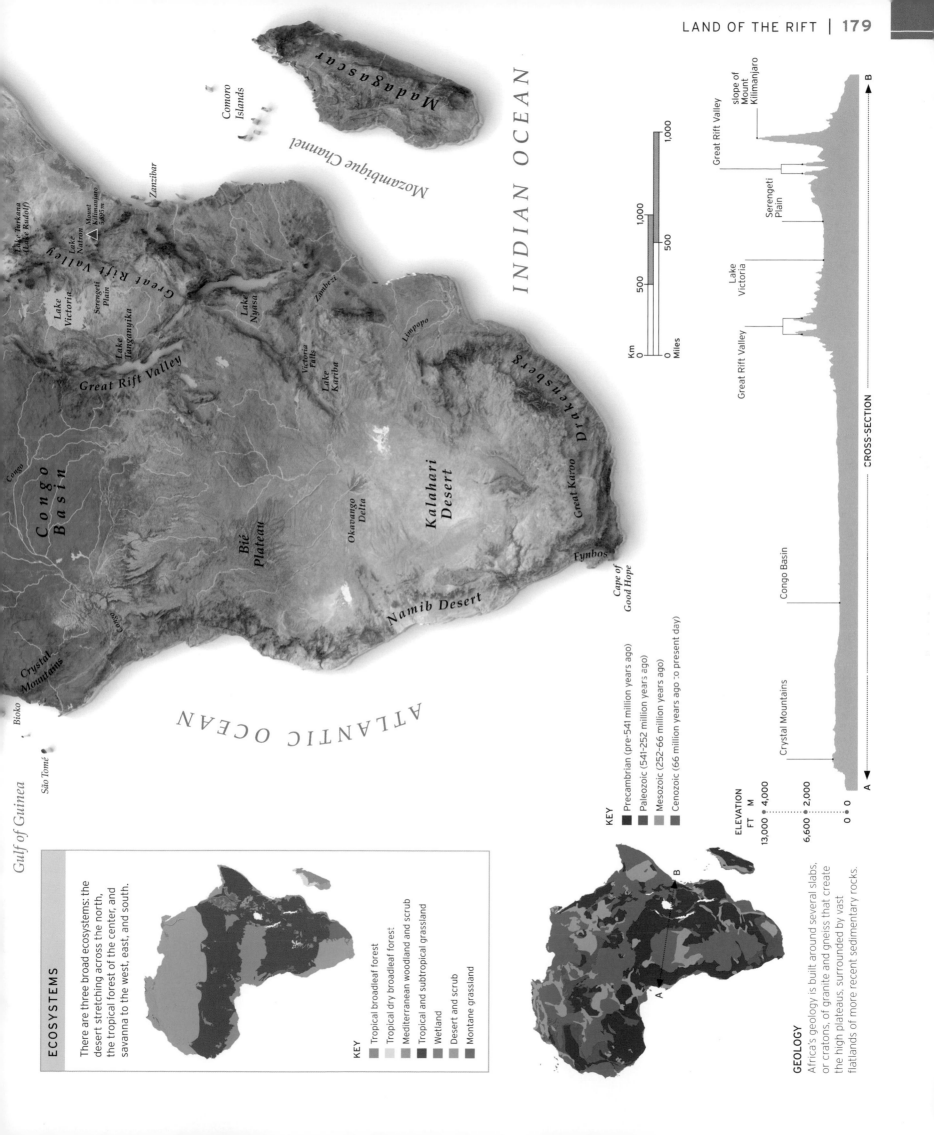

INDIAN OCEAN

ATLANTIC OCEAN

Mozambique Channel

Madagascar

Comoro
Islands

Gulf of Guinea

Bioko

São Tomé

Crystal
Mountains

Congo

Congo
Basin

Cuanza

Bié
Plateau

Okavango
Delta

Kalahari
Desert

Namib Desert

Cape of
Good Hope

Fynbos

Great Karoo

Drakensberg

Limpopo

Lake
Kariba

Victoria
Falls

Zambezi

Lake
Nyasa

Great Rift Valley

Lake
Tanganyika

Lake
Victoria

Serengeti
Plain

Great Rift Valley

Lake Natron

Lake Turkana
(Lake Rudolf)

Zanzibar

Mount
Kilimanjaro
5895 m

ECOSYSTEMS

There are three broad ecosystems: the desert stretching across the north, the tropical forest of the center, and savanna to the west, east, and south.

KEY
- Tropical broadleaf forest
- Tropical dry broadleaf forest
- Mediterranean woodland and scrub
- Tropical and subtropical grassland
- Wetland
- Desert and scrub
- Montane grassland

GEOLOGY

Africa's geology is built around several slabs, or cratons, of granite and gneiss that create the high plateaus, surrounded by vast flatlands of more recent sedimentary rocks.

KEY
- Precambrian (pre-541 million years ago)
- Paleozoic (541–252 million years ago)
- Mesozoic (252–66 million years ago)
- Cenozoic (66 million years ago to present day)

ELEVATION

FT	M
13,000	4,000
6,600	2,000
0	0

CROSS-SECTION

A — B

Crystal Mountains

Congo Basin

Great Rift Valley

Lake
Victoria

Serengeti
Plain

Great Rift Valley

slope of
Mount
Kilimanjaro

Km
0   500   1,000

Miles
0   500   1,000

▷ **CRATON REVEALED** In the Drakensberg, South Africa, the Blyde River has cut down to the ancient rocks of the Kaapvaal Craton.

THE SHAPING OF AFRICA

Africa is formed around five stable sections of Earth's crust called cratons. They were the heart of the giant southern continent Gondwana during the age of the dinosaurs. About 65 million years ago, the land around these cratons broke away to create Africa.

Ancient heart

The five African cratons were formed over a billion years ago when magma welled up from Earth's interior and solidified. In places, they have since been altered by heat and pressure, and in others, they have been covered by a veneer of sediments. The rocks they are made of are very tough—they have endured as other rocks came and went, and continents and oceans shifted around them. The cores of the cratons are ancient and date back to one of the oldest periods in Earth's history, the Archean Era, over 2.5 billion years ago. The oldest rocks are the granite, gneiss, and greenstone terrains of parts of South Africa, Zimbabwe, and Tanzania. The Kaapvaal Craton is noted for its distinctive komatiite rocks that formed when Earth was very hot in its earliest days. The area is also famous for its deposits of gold, nickel, and uranium.

BUILDING BLOCKS Africa is built around five major cratons of ancient rock, with the most ancient of all being the Kaapvaal Craton to the south. The Arabian-Nubian Craton is being rifted apart.

Arabian-Nubian Craton

craton

Kaapvaal Craton

Mountain fringes

For long periods in Earth's history, the African cratons were separate continents. A billion years ago, the craton to the west was attached to a landmass that would eventually hold the Amazon Basin within an ancient supercontinent, while the central and southernmost cratons were islands. Gradually, they were all glued together by mobile belts of rock and became part of the Pangaea supercontinent 250 million years ago. Meanwhile, the mobile belts were crunched between the cratons, folding up rocks to form mountain chains such as the Atlas and the Cape Fold Mountains, some of which survive in South Africa (others are in South America, Australia, and Antarctica).

KEY EVENTS

140 million years ago The landmass that will become Africa splits from what will become South America due to the opening of the Mid-Atlantic Rift, creating a new ocean—the Atlantic.

30 million years ago The Arabian Peninsula pushes away from Africa as magma forces through Earth's crust and cools, creating new land that is denser than the older, surrounding land. As a result of this higher density, it sinks—creating the Afar Depression.

1.8 million years ago Mount Kilimanjaro forms due to volcanic activity linked to the opening of the Great Rift Valley.

160 million years ago Madagascar (along with the Seychelles and India) splits from Africa, and then about 84 million years ago it splits again from the Seychelles and India—both events are due to rifting as the Indian Ocean opens up.

30 million years ago Many faults open between various plates in eastern Africa, and the land between three neighboring plates "drops" to form the Great Rift Valley.

2.5 million years ago North Africa becomes increasingly dry, eventually forming the Sahara Desert. Changes in the tilt of Earth's axis lead to brief periods of increased humidity and plant growth.

400,000 years ago Tilting of Earth's crust creates a basin. Water collects within it and Lake Victoria forms. During the ice age it drains away multiple times as the rate of rainfall rises and falls.

◁ **COLLIDING CONTINENTS**
The High Atlas mountains to the north of Ouarzazate were formed less than 50 million years ago when the African and Eurasian Plates started colliding.

▷ **PARCHED LAND**
Almost all of northern Africa is covered by the huge Sahara Desert. Half of the Sahara receives less than 1 in (2.5 cm) of rain per year— the rest gets less than 4 in (10 cm).

▷ **FIERY CRACKS**
The Erta Ale volcano lies between the splitting sides of the African and Arabian plates.

▷ **ACTIVE VOLCANO**
Oldoinyo Lengai in Tanzania is one of many erupting volcanoes along the widening crack in the African Plate that forms the Great Rift Valley. It is unique in that it erupts natrocarbonatite lava.

In time, volcanic activity forced cracks to open in Pangaea, and as a result Africa split from Europe and South America and started its own journey. Madagascar had split from Gondwana about 20 million years earlier.

A splitting continent

Around 25–22 million years ago, a big crack began to split east Africa in two as the tectonic plates began to pull either side apart and plumes of hot magma thrust their way up from underneath. The widening crack became one of the world's most dramatic geographical features—the Great Rift Valley—which runs over 3,700 miles (6,000 km) from the Dead Sea in Jordan to Mozambique. North of Ethiopia, the rift is splitting the Arabian-Nubian Craton apart and has already gaped wide enough to allow the Indian Ocean

to flood in and create both the Red Sea and the Gulf of Aden. In about 10 million years' time, the two sides of the Great Rift Valley will have pulled far enough apart to eventually open up a new ocean right through the middle of East Africa.

Africa dries out

Africa's location straddling the tropics has created three huge deserts—the Sahara in the north, along the Tropic of Cancer, and the Namib and Kalahari in the south along the Tropic of Capricorn. The Namib is the world's oldest desert and formed 80 million years ago. The Sahara is the world's largest hot desert, yet developed less than 2.5 million years ago—much of it was green until a dramatic climatic shift just 5,000 years ago scorched the north of Africa dry.

Africa was still attached to South America when Madagascar split away

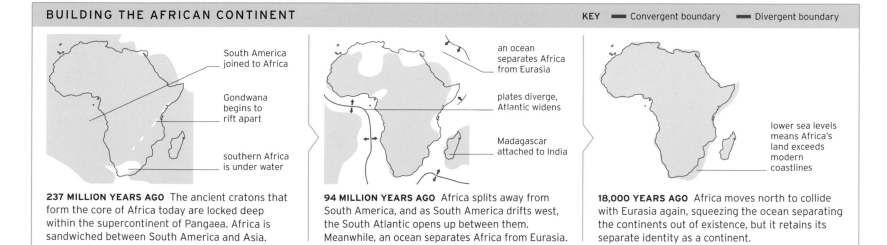

BUILDING THE AFRICAN CONTINENT

KEY ━ Convergent boundary ━ Divergent boundary

South America joined to Africa

Gondwana begins to rift apart

southern Africa is under water

237 MILLION YEARS AGO The ancient cratons that form the core of Africa today are locked deep within the supercontinent of Pangaea. Africa is sandwiched between South America and Asia.

an ocean separates Africa from Eurasia

plates diverge, Atlantic widens

Madagascar attached to India

94 MILLION YEARS AGO Africa splits away from South America, and as South America drifts west, the South Atlantic opens up between them. Meanwhile, an ocean separates Africa from Eurasia.

lower sea levels means Africa's land exceeds modern coastlines

18,000 YEARS AGO Africa moves north to collide with Eurasia again, squeezing the ocean separating the continents out of existence, but it retains its separate identity as a continent.

N. Africa

The Atlas Mountains

A barrier between the Mediterranean and Sahara, stretching from Morocco to Tunisia

Extending for around 1,600 miles (2,500 km), the Atlas Mountains consist of several distinct subranges, which are interspersed with plateaus and gorges. The subranges include the Anti-Atlas and High Atlas in Morocco and the Tell Atlas in Algeria.

Product of collision
The bulk of the Atlas Mountains were pushed up by a collision between the African and Eurasian plates, with the main uplift taking place 30–20 million years ago. However, certain parts were created during an earlier period of mountain building, from around 250 million years ago. Today, the appearance of the Atlas varies between forested slopes in the north and much more arid southerly subranges. The region contains some of the world's largest and most diverse mineral resources.

◁ MINERAL WEALTH
The Tell Atlas range of Algeria is mined in several locations for iron ore, as well as gold and phosphate. The chunk of iron ore shown here is of a type called hematite.

In **subtropical Africa,** the **tallest peaks** of the Atlas were **covered in glaciers** during **the last ice age**

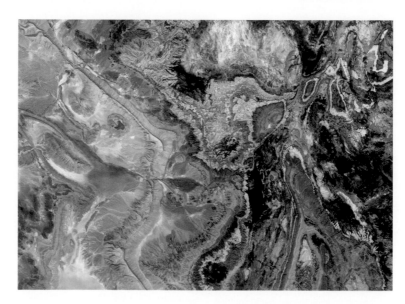

△ MOUNT TOUKBAL
This small village, seen here in winter, is at the base of Mount Toukbal, the highest peak in the Atlas. Although there is snow on the peaks, the lower slopes are hot and arid.

◁ INTRICATE FOLDING
The Anti-Atlas is geologically complex. This satellite image, taken with an infrared camera sensitive to differences in surface composition, reveals intricate folding among the rock outcrops.

E. Africa

The Afar Depression

One of the hottest and driest places on Earth, a region of volcanoes, hot springs, and salt lakes

Also called the Afar Triangle due to its shape, the Afar Depression is a low-lying region touching Ethiopia, Eritrea, and Djibouti. It is an area where Earth is thought to be rifting apart with thinning and stretching of the crust above an upwelling plume of hot mantle rocks. The presence of the magma underground is apparent in the form of many volcanoes at the surface, including a massive shield volcano (a broad volcano with shallow slopes), Erta Ale, which has a permanent lava lake. Partly because much of it lies below sea level, the Afar Depression is one of the hottest places on Earth.

THE AFAR TRIPLE JUNCTION

The Afar Triple Junction is a spot on Earth's surface where the Arabian Plate and two parts of the African Plate—the Nubian and Somalian Plates—are slowly separating. Projecting from this site are three arms, or rifts (down-faulted sections of crust). Two are occupied by the Red Sea and the Gulf of Aden. The northern part of the third arm is the Afar Depression.

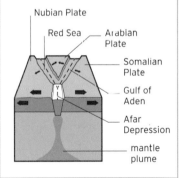

Nubian Plate
Red Sea
Arabian Plate
Somalian Plate
Gulf of Aden
Afar Depression
mantle plume

▽ **COLORFUL CRATERS**
These hot springs, with salt and sulfur deposits, lie 150 ft (45 m) below sea level at Dallol in the northern part of the Afar region.

◁ **NORTHERN SURVIVOR**
The only surviving primate in Africa north of the Sahara Desert, the Barbary macaque has adapted to living in the rocky cliffs and high-altitude oak and cedar forests of the Atlas range.

E. Africa

The Great Rift Valley

A system of faults in Earth's crust that extends for 3,700 miles (6,000 km)
through part of southwestern Asia and then down the length of East Africa

The Great Rift Valley marks a region where plumes of hot material from Earth's mantle are welling up and causing the crust to stretch and split apart along a series of fissures, or rifts. These rifts start in southern Lebanon and pass southward, via the Red Sea, to the Afar Depression (see p.183) and finally down through East Africa, ending on the coast of Mozambique. The average width of the Great Rift Valley is around 30 miles (50 km). In many parts, sections of crust have sunk down, leaving escarpments on either side, with steep

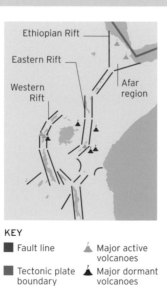

▷ **SOCIAL ANIMALS**
The Great Rift is home to many endemic species of African wildlife, including olive baboons. These live in groups of 15 to 150 individuals, with a complex social structure.

walls that are typically around 2,950 ft (900 m) high. The African section of the Great Rift, known as the East African Rift System, is gradually splitting the African tectonic plate in two.

Volcanoes and lakes

Dotted throughout the rift system are numerous volcanoes, formed where magma has welled up through fissures in the crust and erupted at the surface. Several, such as Ol Doinyo Lengai in Tanzania, have been active in the past 50 years. In the East African Rift System, many lakes—some very deep—fill the valley floors. Ancient fossils of various extinct relatives of modern humans have been discovered in different parts of the region.

The Great Rift Valley has **formed over** approximately the **last 35 million years**

EASTERN AND WESTERN RIFT VALLEYS

The oldest and best defined rift occurs in the Afar region of Ethiopia. This is usually referred to as the Ethiopian Rift. In East Africa, the Great Rift splits into two branches called the Western (or Albertine) and the Eastern (or Gregory) rifts. Each contains active volcanoes and a string of lakes. Some of these are highly alkaline and are termed soda lakes; others are very deep.

Ethiopian Rift

Eastern Rift

Western Rift

Afar region

KEY

■ Fault line
■ Tectonic plate boundary
▲ Major active volcanoes
▲ Major dormant volcanoes

△ **SALINE LAKE**
Kenya's Lake Bogoria is a saline and alkaline lake that is home to one of Earth's largest populations of lesser flamingos. There are also geysers and hot springs along the banks.

▷ **UNIQUE LANDSCAPE**
The crater of Ol Doinyo Lengai is filled with steep-sided cones and a unique dark lava that turns white as it comes into contact with moisture in the air and solidifies.

The formation of a
RIFT VALLEY

▷ DEEP AND HOT
The lava lake of Nyiragongo volcano is sometimes 2,000 ft (600 m) deep. It connects to a magma reservoir in the crust beneath it, which has intruded there due to the rifting process.

Continental rifting is a process in which part of one of Earth's tectonic plates, made up of the crust and upper mantle (together known as the lithosphere), becomes stretched and thinned to the point where fissures and faults appear. Blocks of crust collapse downward, leaving a series of depressions, called a rift valley, at the surface. The fissuring also allows magma (hot melted rock from the mantle) to intrude into the crust and create volcanic activity at the surface.

Rift valleys in East Africa

An excellent example of continental rifting can be seen in East Africa. The cause here is thought to be a mantle plume, or possibly plumes, under the region (see panel, below). The East African Rift system has two arms, east and west (see p.184), on either side of an area containing Lake Victoria. Rifting and volcanism started in a part of the eastern branch to the northeast of Lake Victoria around 35 million years ago, probably due to a mantle plume, then spread north and south. But as the rifting process encountered a stable core of lithosphere, around and to the south of Lake Victoria, which it could not go through, it is theorized to have diverged around it, leading to the two rift arms. Alternatively, it could be that separate mantle plumes exist, or once existed, under the eastern and western rift arms.

▶ EAST AFRICAN RIFT SYSTEM
In East Africa, two arms of a rift system are separated by a broad, stable plateau. Each arm is a curved depression in the landscape and the site of several remarkable natural features, many of them having a volcanic origin.

Lake Edward

Lake Kivu

Western Rift arm
This is marked by a string of four large lakes, which fill the lower parts of the rift's depressions.

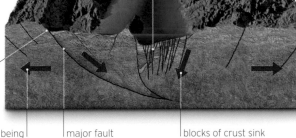

Lake Tanganyika
With a maximum depth of 4,820 ft (1,470 m), Tanganyika is the world's second deepest freshwater lake.

Lake sediments
Due to the lake's age, of up to 12 million years, its bottom sediments are several miles deep.

▷ RIFT FAULTING
In a rift, some blocks of crust drop down, along faults created by stretching forces. In the segment shown here, a long major fault has developed on one side. Along a whole rift, major faults may switch between sides, giving an overall sinuous appearance.

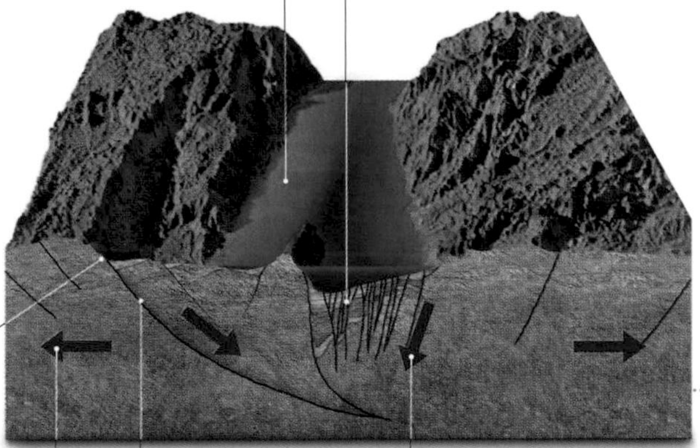

steep escarpment on shoulder of rift valley

crust being stretched

major fault

blocks of crust sink along fault lines

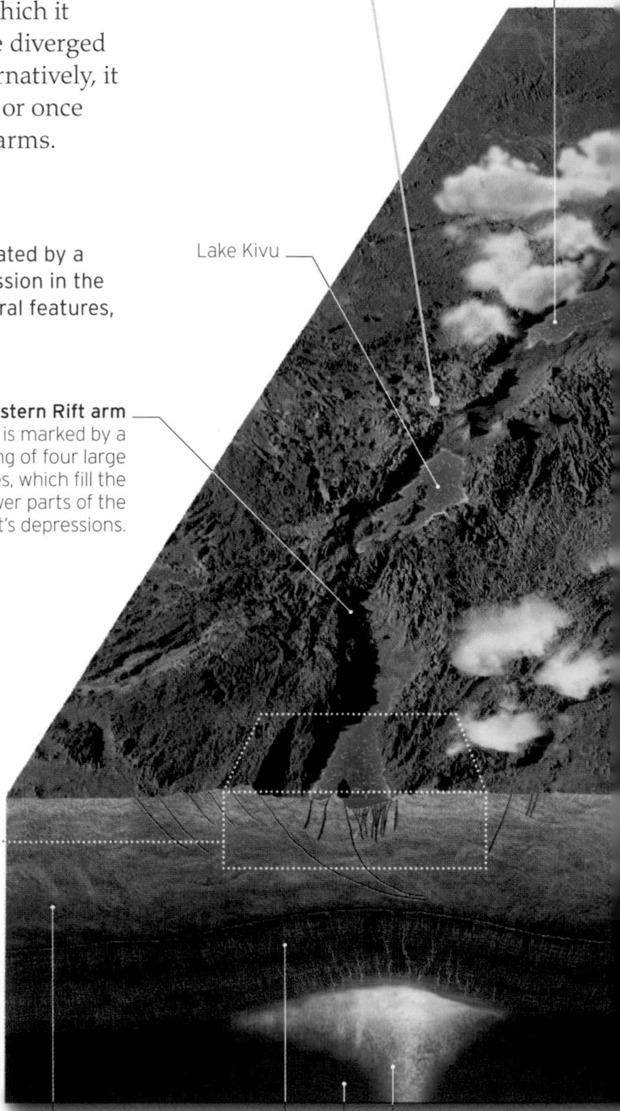

HOW THE EAST AFRICAN RIFT FORMED

The system of rift valleys in East Africa formed when a mantle plume, or possibly plumes, caused upward warping of the crust, then stretching and thinning as parts of the African Plate on either side began to move apart. Tensional faults appeared, and blocks of crust dropped down, producing depressions, which became partly filled by lakes, bordered by escarpments or mountains.

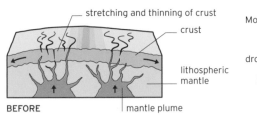

stretching and thinning of crust

crust

lithospheric mantle

BEFORE

mantle plume

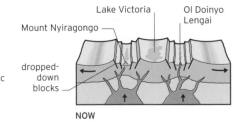

Lake Victoria

Mount Nyiragongo

Ol Doinyo Lengai

dropped-down blocks

NOW

Continental crust
This is laterally stretched and faulted in a rift region, with some crustal blocks collapsing downward. In places, magma intrudes into the crust.

Lithosperic mantle
The uppermost layer of mantle, with overlying crust, comprises the lithosphere, which is divided into plates.

Asthenosphere
This relatively deformable layer of Earth's upper mantle is warmer than the overlying lithosphere.

Speculative mantle plume
Volcanism in the Western Rift arm indicates that in some places magma has definitely intruded into the crust.

◁ **GREAT ELEVATION**
Mount Stanley in the Rwenzori Mountains is Africa's third highest mountain. The Rwenzoris originated on one side of the western rift arm as a gigantic horst—a crustal block that lifted up when land on either side subsided.

◁ **SCARLET GLOW**
The strange lava, called natrocarbonatite, exuded by the Ol Doinyo Lengai volcano glows an unusual scarlet color at night. The volcano developed as a result of rifting, but the reason it produces this unique lava is not yet known.

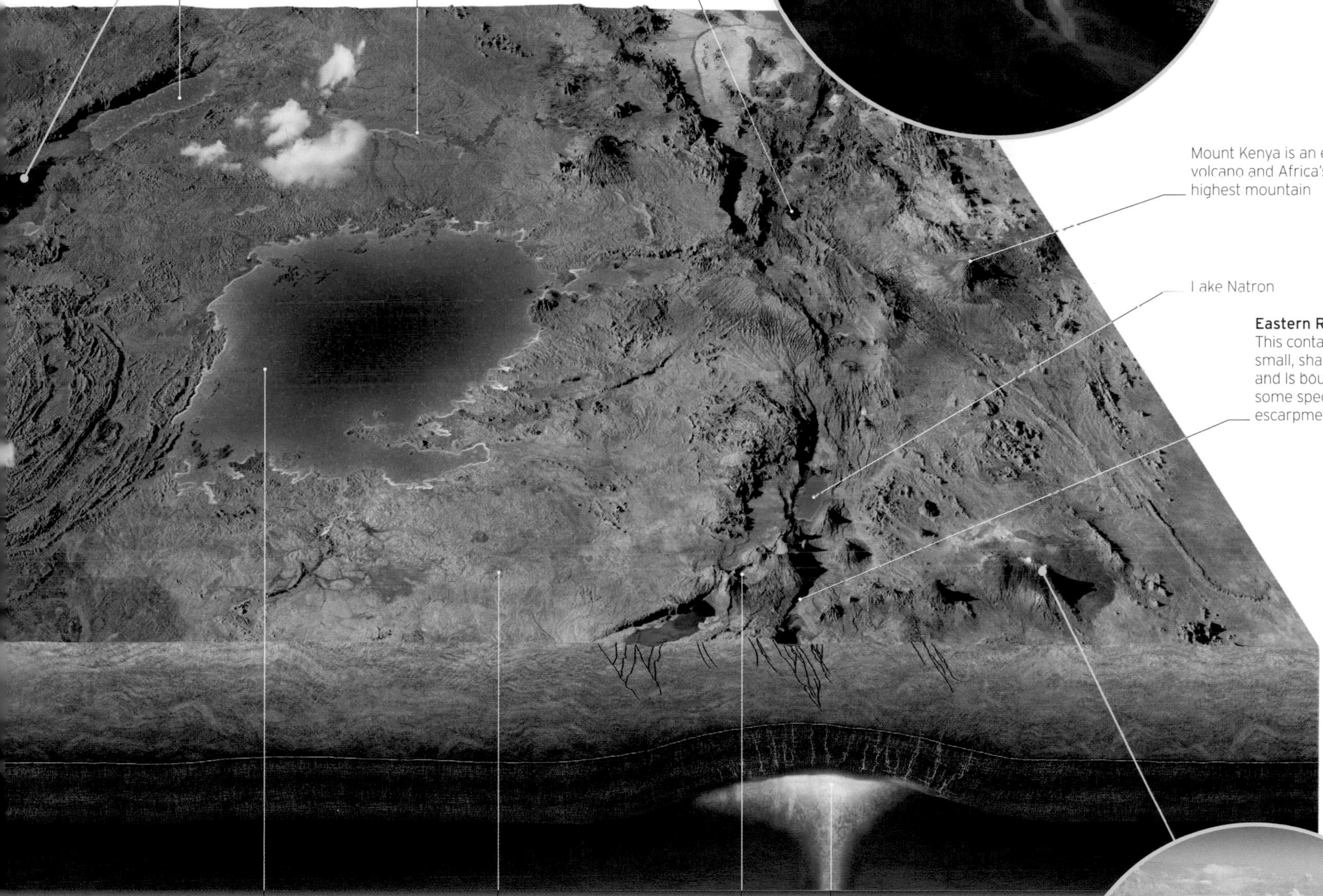

Lake Albert

Lake Kyoga

Lake Bogoria

Mount Kenya is an extinct volcano and Africa's second highest mountain

Lake Natron

Eastern Rift arm
This contains relatively small, shallow lakes and is bounded by some spectacular escarpments.

Lake Victoria
Africa's largest lake by area, Lake Victoria occupies a shallow depression in the central plateau.

Central plateau
The area between the two rift arms is 810 miles (1,300 km) wide, underlain by a stable ancient block of continental lithosphere.

Ngorongoro Crater
This large extinct volcanic caldera is famed for its wildlife, especially lions and other large mammals.

Mantle plume
A mantle plume is strongly suspected to be present under the Eastern Rift arm and has caused stretching and faulting.

▷ **SKY HIGH**
Kilimanjaro—a dormant volcano—is Africa's highest mountain at 19,341 ft (5,895 m). Its origins are connected to the development of the Eastern Rift arm.

There are over **50 volcanoes** in the **East African rift system**

The Brandberg Massif

An isolated, dome-shaped mountain, or inselberg, that towers over the Sun-scorched gravel plains of the northwestern Namib Desert

SW. Africa

The Brandberg formed around 130 million years ago when a large glob of magma from deep in Earth's crust pushed up through the surrounding rock layers before solidifying. Later, the granite mass was exposed by erosion of the surrounding rocks.

The burning mountain

The Brandberg is the highest mountain in Namibia, with its most elevated point at an altitude of 8,442 ft (2,573 m). The indigenous San people call it the "burning mountain" because it appears to glow red in the light of the setting Sun. With its considerable height and breadth, the Brandberg influences the local climate, drawing more rain to its flanks than the desert below. The rain slowly seeps out through springs. Unique animal and plant species thrive in its high-altitude environment, and prehistoric cave paintings decorate rock faces hidden within ravines around the base of the mountain.

◁ **RING OF ROCKS**
The Brandberg is around 14 miles (23 km) long and 12 miles (20 km) wide, and has a rugged, undulating surface. A ring of dark, steep-sided rocks encircle the granite mass.

S. Africa

Table Mountain

A long, flat-topped mountain that forms a dramatic backdrop to the city of Cape Town

Approximately 2 miles (3 km) long and edged by impressive cliffs, Table Mountain is one of the most iconic landmarks in Africa. With its high point at 3,563 ft (1,086 m), it lies at the northern end of a sandstone mountain range that terminates about 30 miles (50 km) to the south at the Cape of Good Hope. Table Mountain once had much thicker, softer layers of rock above it, but these eroded away to produce the flat top. This is often enveloped in a "tablecloth" of clouds, formed when onshore winds are directed up the mountain's slopes into colder air, where the moisture they carry condenses.

△ **SUNSET VIEW**
Table Mountain is seen here covered by its "tablecloth" of white clouds just before sunset. Below and to its left are the lights of Cape Town and to the right lies the suburb of Camps Bay.

◁ **ICONIC FLORA**
Also known as the Pride of Table Mountain, the spectacular Red Disa orchid typically grows near streams and waterfalls. It is the emblem of South Africa's Cape Province.

INSIDE TABLE MOUNTAIN

The mountain's upper part is a 2,000-ft- (600-m-) thick slab of hard sandstone, around 450 million years old. This lies above several layers of sedimentary rock, except at one end, where there is a large granite intrusion.

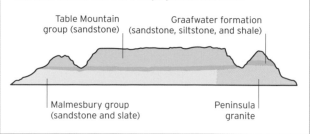

Table Mountain group (sandstone)

Graafwater formation (sandstone, siltstone, and shale)

Malmesbury group (sandstone and slate)

Peninsula granite

The Drakensberg

A gigantic, steplike, eroded landform that separates part of southern Africa's narrow coastal plain from its vast interior of savanna and desert

S. Africa

△ SAVAGE ARENA
In this part of the uKhahlamba-Drakensberg Park, the escarpment is shaped like an amphitheater. The pinnacles on the right are known as the Dragon's Teeth.

Meaning "Dragon's Mountain" when translated from Afrikaans, the Drakensberg is the eastern part of a long, winding escarpment, the Great Escarpment, which forms the edge of a high plateau of central southern Africa. The Drakensberg is South Africa's highest mountain range and runs for more than 620 miles (1,000 km) across eastern South Africa, with some peaks exceeding 10,000 ft (3,000 m) in height.

A giant, eroded step

From its base, the escarpment appears like a giant step. This is continuously worn down by gravity and the flow of water. The landscape is diverse, with steep cliffs, pinnacles, and caves. Rivers rising in the plateau behind the escarpment have carved some impressive gorges. The local fauna and flora includes vultures, rock hyrax, and 300 species of indigenous plants.

INSIDE THE DRAKENSBERG

The escarpment's steep sides consist of a thick layer of basalt (an igneous rock formed by cooling of lava), created as a result of ancient volcanic outpourings. Lower down are layers of older sedimentary rocks. Together, the basalt and sedimentary strata form part of a sequence of rocks found in a large part of Africa known as the Karoo Supergroup.

sedimentary rocks | basalt

Karoo Supergroup

150 million years ago (MYA)

190 MYA
200 MYA
210 MYA
220 MYA

240 MYA

▽ FREE FALL
The Drakensberg boasts some of the world's tallest waterfalls, and other beautiful cascades, such as the Berlin Falls, shown here.

◁ GROUND-DWELLER
The indigenous Drakensberg rockjumper is a ground-nesting bird, which is often seen perching on rocks.

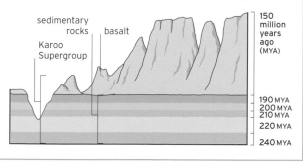

In places, the **escarpment** is **4,900 ft (1,500 m) high** from base to top

△ DESERT LIFELINE

Passing through arid landscapes over most of its course, the Nile provides a vital water supply for drinking and agricultural use. It is also an important transportation route used by millions of people each day.

THE WORLD'S LONGEST RIVERS

The accurate measurement of a river's length can be problematic. In some cases, it is difficult to identify a river's source or mouth, and over its course a river may split into many different channels. Although figures for the lengths of the longest rivers vary, the measurements below are the most widely recognized.

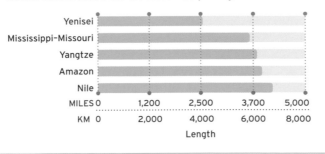

Yenisei		
Mississippi-Missouri		
Yangtze		
Amazon		
Nile		

MILES 0	1,200	2,500	3,700	5,000
KM 0	2,000	4,000	6,000	8,000

Length

△ VERSATILE PLANT

A species of aquatic sedge, papyrus forms tall, reedlike beds in the Nile's shallow waters. Since ancient Egyptian times, papyrus has been used to make various items, from scrolls to boats.

The Nile

The longest river on Earth, running from central Africa to the Mediterranean, and a lifeline for millions of people on its route

NE. Africa

The longest river in the world, the Nile flows more than 4,130 miles (6,650 km), from its source near the equator, through East and northeast Africa to its mouth on the Mediterranean Sea. The Nile basin drains more than 1 million square miles (3 million square km)—about 10 percent of Africa's land area.

White and Blue

The Nile has two major tributaries. Considered the river's primary stream, the White Nile flows from the north of Lake Victoria (see p.196), in Uganda, and descends into the Great Rift Valley (see pp.184–87) over the Murchison Falls. After crossing Lake Albert, it passes onto the huge wetland plain of the Sudd. At Khartoum, Sudan, it is met by the Blue Nile, which has its source at Lake Tana in the Ethiopian Highlands. From here, the Nile runs over a series of cataracts, or rapids, before passing through Egypt's historic valleys. North of Cairo, the river splits into two branches—the Rosetta to the west and the Damietta to the east— forming the Nile Delta as it enters the Mediterranean.

▽ **MISTY FALLS**
The Blue Nile Falls, in Ethiopia, is known locally as Tis Abay, meaning "great smoke." Spanning up to 1,300 ft (400 m) in the rainy season, its waters plunge 150 ft (45 m) to the bottom.

Welcome floods

The Nile Delta, one of the most populated and highly cultivated parts of Egypt, was originally formed of sediment brought down by the Blue Nile from the Ethiopian Plateau. Indeed, the flooding of the Nile has been an important annual cycle, delivering not only water but also fertile silt that replenishes the soil across the floodplain. However, since the completion of the Aswan High Dam in 1970, the flow of the Nile has been significantly moderated, and traditional farming practices have been greatly affected.

△ **SOLITARY FORAGER**
The shoebill is a large, storklike bird that typically forages alone for fish, which it snaps up with its huge, hook-tipped bill.

It takes **three months** for the **water leaving Lake Victoria** to reach the **Mediterranean**

Lake Retba

A West African lake with a bright pink hue during the dry season and very salty water

W. Africa

Also called Lac Rose, meaning "Pink Lake" in French, Lake Retba is a small, shallow lake in Senegal, West Africa, best known for its vivid color and the high salt content of its water. The color, which is deeper during the dry season (November to June), is created by an algae species called *Dunaliella salina* that produces a red pigment to help absorb sunlight for energy (photosynthesis). The salinity of Lake Retba's water is comparable to that of the Dead Sea (see p.236), and this has led to adaptations in the lake's fish, with many growing to only a quarter of the usual adult size.

▷ **SALTY SHORES**
Scores of salt collectors harvest the mineral from the lakebed by hand and export it across West Africa. They cover their skin in shea butter to protect it from the salty conditions.

The Congo

One of Earth's mightiest rivers, draining much of West Africa's vast tropical rain forest

W. Africa

Africa's second longest river after the Nile (see pp.190–91), the Congo carries more water to the sea than any other river on Earth except the Amazon (see pp.106–09). It is also the deepest river in the world, with sections reaching more than 660 ft (200 m) deep. From its source in the highlands of Zambia, it flows in a giant counterclockwise arc, draining a vast circular depression that covers most of the Democratic Republic of the Congo, before discharging into the Atlantic Ocean. Its route, which crosses the equator twice, covers more than 2,900 miles (4,700 km) and includes many waterfalls, rapids, broad waterways, and lakes.

▷ **MARSHY BANKS**
For most of its course, the Congo flows through tropical rain forest, where extensive marshes line its banks and run between tributaries.

ENDEMIC FISH
The tropical waters of Lake Malawi are thought to be home to as many as 1,000 species of cichlids, most of which are endemic to the lake.

LAKE MALAWI'S LAYERS

Lake Malawi has a definite stratification, with layers of water having different nutrient and oxygen contents that generally do not mix. However, from May to August, southern parts of the lake experience strong winds and cooling that create a large upwelling of water. This circulation sends nutrients from the mid-depths up the water column to mix with shallow, more oxygenated water, creating ideal conditions for fish to thrive.

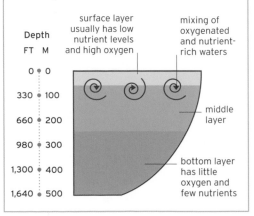

surface layer usually has low nutrient levels and high oxygen

mixing of oxygenated and nutrient-rich waters

Depth		
FT	M	
0	0	
330	100	
660	200	middle layer
980	300	
1,300	400	bottom layer has little oxygen and few nutrients
1,640	500	

Lake Malawi

A large, deep lake at the southern end of Africa's Great Rift Valley with a remarkably diverse population of fish

E. Africa

Also known as Nyasa and Niassa, Lake Malawi is the southernmost and third largest of the East African Great Lakes, lying between Malawi, Tanzania, and Mozambique. It is about 360 miles (580 km) long, up to 47 miles (75 km) wide, and reaches a depth of more than 2,300 ft (700 m). The lake is fed by the Ruhuhu River, from the north, and several smaller rivers. From the southern end flows its only outlet, the Shire River, a tributary of the Zambezi. Lake Malawi runs north-to-south, occupying a single basin in part of the southern end of the Great Rift Valley (see pp.184–87). Above its north and east shores rise steep mountain slopes, some covered by thick forest. The much flatter and shallower shores of the southern end of the lake feature many white, sandy beaches.

Multitudes of fish

Lake Malawi's waters and shores provide a variety of habitats and are home to many animals including crocodiles, hippopotamuses, and the African fish eagle. The lake is best known for containing as many as 3,000 species of fish—more than any other lake on Earth.

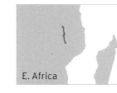

▷ **WINGED ARCHITECT**
The male village weaver builds a dense nest out of grass and leaves. Leaving an opening at the bottom, it suspends the nest from a tree branch.

△ **LAKE OF STARS**
The explorer David Livingstone called Lake Malawi the Lake of Stars, due to the dazzling reflections off its still waters. However, it is also known for the violent gales that blow through.

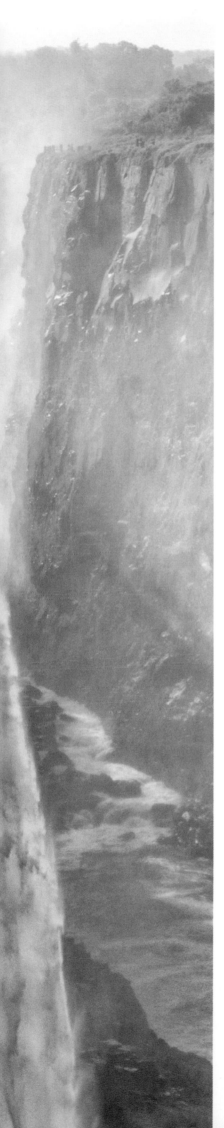

S. Africa

Victoria Falls

An immense African waterfall with a unique geological history, plunging into a deep, narrow chasm

Part of the mighty Zambezi River, Victoria Falls crosses the border between Zimbabwe and Zambia in southern Africa. Challenged only by South America's Iguaçu Falls (see pp.110–11) for the title of Earth's largest waterfall, it spans more than 5,600 ft (1,700 m) and reaches a height of 354 ft (108 m)—twice that of Niagara Falls (see p.52). It is often described as having the largest sheet of falling water in the world.

last 100,000 years, the river has eroded the sandstone to leave a series of narrow gorges, the latest of which the river plunges over (see panel, below). The zig-zagging of the gorges, which traces the falls' movement upstream, is due to a series of joints running across the ends of the gorges that have "captured" the river and changed its direction. The narrow chasm formation of the falls produces a spray that can be seen 30 miles (50 km) away.

Unique formation

The Zambezi runs across a basalt plateau that has large cracks, roughly perpendicular to the river's flow, filled with softer sandstone. Over the

◁ **FEARSOME PREDATOR**
The Nile crocodile, the largest crocodile species in Africa, is the most fearsome predator found along the Zambezi.

The falls' local name, **Mosi-oa-Tunya**, translates as "**the smoke that thunders**"

◁ **ON THE EDGE**
The falls form part of two national parks, both with large populations of elephants. Occasionally, one will venture out onto the falls to feed on trees on the islands.

△ **FALLING IN SECTIONS**
The islands in the river on the edge of the falls separate the flow into several recognized sections (from left): the Devil's Cataract, Main Falls, Rainbow Falls, and the Eastern Cataract.

THE VICTORIA FALLS GORGES

The steep, narrow gorges below Victoria Falls mark their changing positions over time. The Devil's Cataract, on the far west side, is the point at which a new waterfall is in the early stages of cutting back.

Zambezi River

position where next full-width falls will form

Devil's Cataract

Victoria Falls

2nd gorge

3rd gorge

4th gorge

5th gorge

E. Africa

Lake Victoria

Africa's largest lake and the second largest in the world, the major source of the Nile

Lake Victoria sits in a shallow depression on a plateau between two arms of the Great Rift Valley (see pp.184–87) in East Africa and is bordered by Uganda, Kenya, and Tanzania. It is the largest of the African Great Lakes and, with an area of more than 26,500 square miles (68,800 square km), it is the second largest freshwater lake in the world by area after Lake Superior (see pp.50–51). However, it is shallow for its size, with an average depth of only 130 ft (40 m).

Lake Victoria is the source of the White Nile (see pp.190–91), its only outlet, which is known as the Victoria Nile at its start. The Kagera is the largest river entering the lake, from the west, but the lake receives most of its water from rainfall.

▷ LONG TOES
The African jacana is a tropical wading bird with long toes and claws that enable it to walk across floating vegetation, such as lily pads, in shallow lakes.

△ INTRICATE SHORELINE
The lengthy shores of Lake Victoria, seen here from space, provide access to freshwater for millions of people, including those in several large cities such as Uganda's capital, Kampala.

▽ AQUATIC ANTELOPE
An antelope found in the wetlands of southern Africa, the red lechwe has long, widely splayed hooves that help it run on soft ground and a water-repellent coat that aids swift movement through water.

The Okavango Delta

A vast inland delta in southern Africa, an oasis within the Kalahari Desert that attracts millions of animals on annual migrations

S. Africa

The Okavango Delta is a gently sloping inland delta in northwestern Botswana. It is fed by the Okavango River, which flows southeastward from its source in Angola through the Kalahari Desert but never reaches the ocean. Instead, its water drains into a depression in the Kalahari Basin, where it spreads out to create an oasis comprising permanent and seasonally flooded marshlands, grassland plains, islands, and waterways. The annual flooding of the delta by water that fell as rain in the Angolan highlands, more than 620 miles (1,000 km) away, swells it to three times its usual size. Coinciding with Botswana's dry season, the flooding attracts a huge influx of animals, including herbivores such as buffaloes and elephants, creating one of the greatest concentrations of wildlife in Africa. Over 97 percent of the water that enters the delta evaporates or eventually seeps into the sand of the Kalahari Desert. What is left drains into Lake Ngami to the west.

◁ **BOLD MARKINGS**
Marbled reed frogs display a large range of colors and markings with a combination of stripes, spots, and dashes.

◁ **FLAT DELTA**
The Okavango Delta is mostly flat, with less than 6 ft (2 m) variation in height across the area. The pattern of its waterways changes every year as channels become blocked and the slow-moving water finds a new route.

ENDORHEIC BASINS

The Okavango Delta is part of an endorheic basin—a drainage system that has no outlet to the sea. These basins usually converge on a lake or swamp. Most endorheic basins are located in arid areas, and most water that enters evaporates.

Okavango River

Okavango Delta

NW. Africa

Los Gigantes Cliffs

A line of breathtakingly high cliffs on the western coast of Tenerife in the Canary Islands

Los Gigantes, or "the Giants," rise up to 2,620 ft (800 m) in places and are made of the igneous rock basalt. The cliffs continue under the sea to the seabed some 100 ft (30 m) deep, which harbors a rich variety of marine life.

Tenerife is built from three volcanoes that coalesced about 3 million years ago into a larger volcano. The Los Gigantes Cliffs are the result of the erosion of a huge mass of solidified lava.

▷ **GEMS IN THE CLIFFS**
Crystals of the olive-green mineral peridot are occasionally found in rocks and on beaches around Tenerife.

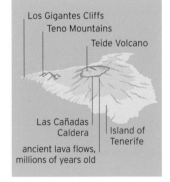

LOCATION OF LOS GIGANTES

The Los Gigantes Cliffs are located where multiple layers of basalt—solidified lava erupted by one of Tenerife's original volcanoes—have been eroded away by the sea. The mass of basalt makes up an area in the western part of Tenerife called the Teno Mountains.

Los Gigantes Cliffs
Teno Mountains
Teide Volcano
Las Cañadas Caldera
Island of Tenerife
ancient lava flows, millions of years old

△ **THE END OF THE WORLD**
In classical times, 3,000 years ago, mariners thought these cliffs marked the end of the dry world, beyond which there was only ocean.

ZONES OF A FRINGING CORAL REEF

A fringing reef has several zones. Facing the sea, the fore-reef is populated by different corals at different depths. At its top is the reef crest, which takes the brunt of wave action. On the shoreward side is the inner reef (or reef flat), which may become partly exposed at low tide.

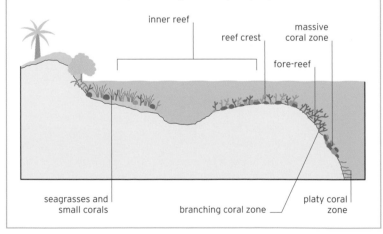

inner reef
reef crest
massive coral zone
fore-reef
seagrasses and small corals
branching coral zone
platy coral zone

The Red Sea Coast

The margins of the world's most northerly tropical sea with magnificent reefs and the site of a new ocean in the making

NE.
Africa

△ **RIOT OF COLOR**
Tiny green chromis fish surround scarlet and magenta soft corals—colonies of numerous minuscule animals called polyps—in a typical scene from a Red Sea reef.

The Red Sea is an inlet of the Indian Ocean and has coasts on seven countries—Egypt, Israel, Jordan, Saudi Arabia, Sudan, Eritrea, and Yemen. It stretches for 1,400 miles (2,250 km) over a tectonic plate boundary where the Arabian Plate is gradually moving away from the African Plate—a process that will eventually create a new ocean (see p.185).

Red Sea reefs
Reflecting the fact that it sits over an area where Earth's crust is rifting apart, the center of the Red Sea plunges to a maximum depth of 7,254 ft (2,211 m). However, most of its coast is rimmed by shallow submarine shelves and extensive fringing coral reefs (see panel, opposite). These are some of the most biologically well-endowed reefs, home to over 1,000 invertebrate species, 1,200 fish species, and several hundred types of hard and soft corals. With this rich marine life and high water clarity, the Red Sea Coast is a highly popular destination for scuba divers. Unfortunately, intense diving tourism, along with coastal development, has damaged the reefs in some areas.

▽ **VIVID CORAL**
This mushroom coral is one of over 300 different species of hard corals found on Red Sea reefs.

C. Africa

The Congo Rain Forest

The second-largest tropical rain forest on Earth, home to more than 11,000 plant species

One of the world's most important remaining wilderness areas, the Congo Rain Forest covers around 70,000 square miles (72 million square km) of central Africa's Congo River Basin, stretching across six countries.

Last refuge

Second in size to the Amazon, this warm, wet region is hugely biodiverse, with more than 600 tree species and 10,000 animal species. Vegetation is so thick in parts of the forest that it has become a last refuge for some of the most endangered species on Earth. Rare forest elephants, okapis, bonobos, and both eastern and western gorillas live in the Congo, as do hundreds of plant species found nowhere else.

▷ **RED FLAGS**
The tropical dogwood, or red flag bush, is often found beside streams in the rain forest.

THE WORLD'S TROPICAL RAIN FORESTS

Tropical rain forests occur around Earth's equatorial regions, between the latitude lines of the Tropic of Cancer (23° North) and the Tropic of Capricorn (23° South). The largest coincide with South America's Amazon and Africa's Congo river basins, while Southeast Asia and India also have significant rain forest coverage. Northeast Australia has a small area of rain forest on its eastern side, stretching from coastal hills down to the sea.

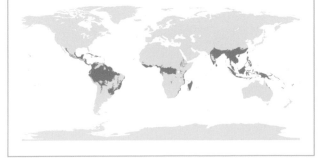

△ **VERSATILE MIMIC**
The African grey parrot feeds on nuts, seeds, and fruit. It mimics other bird and mammal calls, as well as human speech.

E. Africa

Madagascan Rain Forest

A narrow strip of rain forest, where isolation has led to an extraordinary number of unusual native species

Madagascar's rain forest runs along the island's eastern coastal region. Separated from the west by a chain of mountains—the Tsaratanana, Ankaratra, and Ivakoany massifs—and bordered to the east by the Indian Ocean, this area receives more than 100 in (2,540 mm) of rain fall a year. Constant warmth and humidity have created an ideal habitat for tropical plants to flourish, including around 1,000 orchid species, as well as unique animals such as lemurs. Madagascar's evolutionary isolation has also led to an extremely high number of native species: 80–90 percent of its rain forest plants and animals are found nowhere else on Earth.

△ **HIGH DINING**
Critically endangered mountain gorillas (a subspecies of eastern gorillas) live high in the Congo, at elevations of 8,000–13,000 ft (2,440–3,960 m). Their diet consists of plants such as wild celery and bamboo.

◁ **DENSE FOREST**
Rain forest vegetation in the Congo is so dense that many sections of the forest have never been explored by humans. In some areas, only 1 percent of sunlight reaches the ground.

△ **DAWN MIST**
Clouds of moisture often linger over Madagascar's rain forest. Around 12,000 plant species are native to the island, which has led to parts of the rain forest receiving UNESCO World Heritage status.

E. Africa

Madagascan Dry Forest

A distinctive, diverse, and rapidly disappearing ecosystem, with an extremely high percentage of native plant and animal species found nowhere else on Earth

Madagascar's tropical dry forest covers about 58,300 square miles (151,000 square km), mainly in small, fragmented blocks along the island's western and northwestern coasts. Characterized by a dry season that lasts up to eight months a year, it is a habitat full of unique, sometimes bizarre-looking plants that have evolved to cope with an extreme lack of water. Seven species of baobab tree grow here—six native to Madagascar—alongside shorter species that have also adapted to dry conditions, such as moringas and spiky succulents of the *Pachypodium* genus. The animals are also highly specialized, so much so that many—such as numerous lemurs, chameleons, and the island's largest predator, the catlike fossa—are found nowhere else. Slash-and-burn agriculture has already destroyed 97 percent of the forest and threatens the remainder.

◁ **CLINGING ON**
Numerous species of chameleons cling on to existence in what remains of Madagascar's dry tropical forest. The panther chameleon hunts by day in low trees or bushes.

ADAPTING TO DRY CONDITIONS

Drought-survival methods vary with tree species. Baobabs store water in their trunks. Their roots are shallow but widespread, absorbing moisture near the surface. Acacias have widespread surface roots, as well as a taproot that extracts deeper groundwater.

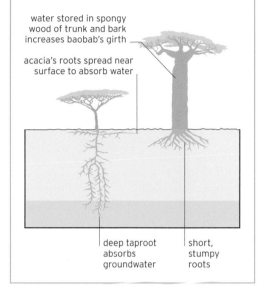

water stored in spongy wood of trunk and bark increases baobab's girth

acacia's roots spread near surface to absorb water

deep taproot absorbs groundwater

short, stumpy roots

△ **VERTICALLY MOBILE**
The golden-crowned sifaka's long, powerful back legs enable it to leap upright from tree to tree in search of seeds, fruit, and leaves.

◁ **DROUGHT SURVIVOR**
Madagascar's boadaka tree stores water in its trunk, like a cactus. Its trunk and branches can also photosynthesize, which means it can shed leaves during periods of drought.

LEAFLESS SILHOUETTES
Baobabs lose their leaves during the dry season, only growing them again when the rains come, so for much of the year they seem dead—or at least dormant.

Ethiopian Montane Grasslands

A region of rare high-altitude grasslands, situated amid some of Africa's tallest mountains and a refuge for many endangered species

NE. Africa

Northeastern Africa's montane grasslands are set mainly amid the eastern and western highlands of Ethiopia, with parts extending into the Eritrean highlands. Encompassing the Simien Mountains in the northwest, the Bale Mountains in the southeast, and part of the Great Rift Valley (see pp.184–87) between the two, this is a striking region, where jagged peaks surround high plateaus of grass and shrubland, as well as pockets of woodland, heather moorland, and alpine steppe. The most fertile montane grasslands occur at elevations of 5,900–10,000 ft (1,800–3,000 m). These attract not only a host of endemic reptiles, birds, and small mammals, but also a high human population, whose agricultural activities and removal of the area's natural resources have resulted in a loss of about 97 percent of its original vegetation. Even so, these grasslands are a vital habitat for some of the continent's most endangered creatures, including the world's only population of Walia ibex and the Ethiopian wolf—Africa's sole wolf species.

◁ **GRASS FORAGER**
Geladas live almost exclusively on grasses from high-altitude meadows. They are also called bleeding-heart baboons due to their red chest patches.

84 small mammal species, many endemic, are found throughout the Ethiopian highlands

The Sudanian Savannas

A vast expanse of dry woodland and grassland, and a region of great botanical biodiversity

W. and C. Africa

The band of mixed grasslands and trees known as the Sudanian Savannas stretches across western and central Africa from the Atlantic Ocean eastward to the Ethiopian highlands. To the north is the Sahel, the transition zone between the savannas and the Sahara Desert, while to the south lie the forests of Guinea and Congo.

West meets east

The ecoregion as a whole covers an estimated 354,000 square miles (917,600 square km) but divides geographically into two main sections, the West and East Sudanian savannas, which are separated by a much higher plateau in Cameroon. Large tracts of hot, dry woodland comprising acacia and other deciduous trees characterize this landscape. There is an understory of shrubs, such as bushwillow, and grasses including the bamboolike Napier grass, with an average height of up to 11 ft (3.5 m), as well as shorter species such as thatching grass, which grows up to 31 in (80 cm). More animals are found in the western savanna, while the eastern section contains a much higher degree of botanical biodiversity, including around 1,000 species of endemic plants. The entire savanna ecoregion forms an important habitat for elephants, wild dogs, cheetahs, and lions, as well as migrating birds.

△ **GREEN EXPANSE**
South Sudan's Boma National Park turns a lush green after the annual rains, heralding the arrival of more than a million savanna grazing animals such as the tiang and white-eared kob—two antelope species that have converged here for centuries.

CLIMATE

Ethiopian highland climate is classed as tropical monsoon, but due to the region's elevation, temperatures are cooler here than would be expected this close to the equator. Rainfall increases from April to October, with the southwestern section receiving the most, while the annual average for the rest of the region is around 63 in (1,600 mm).

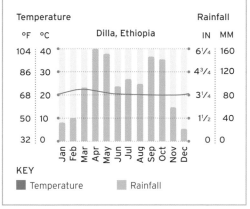

Temperature   Rainfall

| °F | °C | Dilla, Ethiopia | IN | MM |

Dilla, Ethiopia

104 : 40	6¼ : 160
86 : 30	4¾ : 120
68 : 20	3¼ : 80
50 : 10	1½ : 40
32 : 0	0 : 0

Jan Feb Mar Apr May Jun Jul Aug Sep Oct Nov Dec

KEY
■ Temperature   ■ Rainfall

◁ **JAGGED PEAKS**
The rugged basalt cliffs and gorges of Ethiopia's Simien Mountains were created 30–20 million years ago by volcanic eruptions.

The Cape Floral Region

A biological hotspot, one of the most botanically diverse regions in the world, containing thousands of species found nowhere else

S. Africa

Situated on the southwestern tip of South Africa, the Cape Floral Region is one of the country's eight World Heritage sites. This spectacular area covers just 4,227 square miles (10,947 square km) from the Cape Peninsula to the Eastern Cape—around 0.04 percent of Africa's landmass—yet it contains up to 20 percent of the entire continent's plant species. Of the Cape Floral Region's almost 9,000 plant species, a high percentage belong to the rare fynbos shrubland that makes up part of the area. Fynbos vegetation, which includes plants such as proteas and heathers, is found nowhere else on Earth outside South Africa; it thrives in the Cape's Mediterranean climate.

However, because many species are found only in localized, extremely small areas, this is a highly threatened, fragile environment, where just plowing a field or building a house can eradicate an entire species.

▽▷ **BLOOM TIME**
The term fynbos originates from a Dutch word meaning "fine-leaved plants." Fynbos plants include hundreds of protea species, such as the king protea (below), that bloom at varying times of year.

E. Africa

The Serengeti

The famous immense African grassland, the site of one of the most spectacular annual migrations on Earth

Covering 11,600 square miles (30,000 square km) across Kenya and Tanzania, the Serengeti lies near the equator and consists mainly of grassland in the southeast, open woodlands of acacia trees in the north, and a mix of grassland and acacias in the western regions. Ash from nearby volcanoes has added to the fertility of the region's rolling plains, nourishing its short, intermediate, and long grasses.

Mass migrations

"Serengeti" is derived from a Masai word meaning "endless plain," but even its vast grassland sections are occasionally broken up by kopjes—weathered granite outcrops—that not only provide much-needed shade for the savannas' wildlife species but also hold rainwater in rocky depressions, providing small water holes for animals. Hordes of wildlife, particularly herds of herbivores, are part of the fabric of the Serengeti. Each year, more than a million wildebeest and hundreds of thousands of zebras and gazelles come together to form a vast "super herd" in an amazing migration of 620 miles (1,000 km), following the seasonal rains that bring fresh grass and vital water supplies to this arid area of Africa.

◁ STANDING TALL
The tallest living land animals on Earth, giraffes grow to heights of 15-20 ft (4.5-6 m)—useful for reaching acacia leaves.

△ SYMBOL OF THE SERENGETI
Acacias dominate parts of the Serengeti. Among the few trees found in the region, acacias provide food and vital shade for many animal species.

△ NATURE'S RECYCLERS
Dung beetles provide waste removal in the Serengeti by rolling and burying dung to use as food and as a place to lay eggs.

◁ WATER BREAK
Water holes provide a lifeline to migrating herbivores, such as zebras—but also for the carnivores that prey on them.

TYPES OF SAVANNA

There are several types of African savanna vegetation, mainly influenced by the amount of rainfall, which varies with distance from the equator. Rain in the Serengeti falls in two main seasons: from March to May and November to December. In the more wooded northwestern Serengeti, the average annual rainfall is 41 in (1,050 mm), while in the grass-dominated open savannas of the southeast, it is just 22 in (550 mm).

rain forest (rain all the year) | parkland or closed savanna | savanna (summer rain; winter drought) | semidesert scrub or open savanna | desert (annual drought)

The **largest numbers** of **grazing and browsing animals** in the **African continent** are found in the Serengeti

N. Africa

The Sahara Desert

The world's largest hot desert, shaped by relentless winds and shifting sand dunes that are constantly on the move

Stretching from the Atlantic Ocean in the west to the Red Sea in the east, the Sahara covers almost all of North Africa. At 3.6 million square miles (9.4 million square km), it is slightly larger than the U.S., and surveys indicate that this vast, almost uninterrupted arid region is still expanding. Since 1962, the desert has grown by 250,000 square miles (647,500 square km).

More than sand

Although often associated with red sand dunes, most of the Sahara consists of barren, rocky plateaus called hamadas. These contain little sand, most having been swept away by the winds that continually transform the desert into a harsh landscape full of rocks, boulders, and gravel. In other areas, sand is piled up in dune fields. The largest of these are called ergs—vast areas spanning more than 50 square miles (125 square km). Many contain shifting dunes, some as high as 490 ft (150 m). Because ergs are highly active regions of loose sand, they are notoriously difficult to cross.

Several mountain ranges lie in the Sahara and many contain volcanoes. The desert also has gravel fields and salt flats, as well as dry river- and lakebeds. Despite such harsh terrain and little to no rainfall—half of the Sahara gets less than 1 in (25 mm) a year, the rest possibly up to 4 in (100 mm)—around 500 plant and 70 animal species are found here.

◁ **DESERT FOX**
The smallest of all foxes, the desert-dwelling fennec fox is the size of a house cat. Its massive ears keep it cool by releasing excess body heat.

Each year, **thousands of tons** of Saharan phosphorus dust reach the **Amazon rain forest**

DUST IN THE WIND

Saharan winds have a global impact, setting off weather events such as cyclones. The springtime scirocco blows damp, hot, dust-laden clouds and fog into southern Europe. The cool, dry, northeasterly harmattan carries tons of dust over the Atlantic during winter. The hot, dry khamsin blows into the Arabian Peninsula and southeast Europe from February to June.

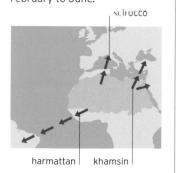

scirocco

harmattan | khamsin

◁ **ROCKY PEAKS**
At Tassili du Hoggar in southern Algeria, the central Saharan winds have sculpted a large sandstone plateau into jagged peaks, rock arches, and spires that resemble the ruins of a forgotten city.

△ **BULL'S-EYE**
The huge Richat Structure, known as the Eye of the Sahara, was thought to have been formed by a meteor strike. Scientists now believe it formed when igneous rocks were uplifted and later eroded.

◁ **SAND WAVES**
The red dunes of the Great Western Erg, near Taghit oasis, Algeria, resemble ocean waves. High winds here cause deadly sandstorms that can last for days.

The White Desert

A desert of chalk, shaped by wind-driven sand into an eerie landscape of stone monoliths

NE. Africa

Often compared to a lunar landscape, the White Desert is located northeast of the oasis town of Farafra, Egypt. This gleaming terrain was originally an ancient seabed.

From chalk to chickens

When the prehistoric sea that once covered this vast area dried up, the bodies of millions of marine creatures calcified, forming an enormous plateau of chalk and chalky limestone. Millions of years of wind erosion carved the upland into pinnacles, spires, boulders, and other shapes—some so bizarre that they have been likened to giant mushrooms, ice cream cones, and even an oversized chicken. Occasionally, chalky inselbergs (isolated, towerlike hills jutting out of the flat plain) are also found among the smaller formations. The sand surrounding the shapes glitters with quartz crystals, adding to the White Desert's overall frosty appearance, particularly when viewed under moonlight.

▽ **DESERT SCULPTURE**
Windblown sand has carved these striking stone formations that seem to balance dangerously on severely eroded bases.

NW. Africa

The Adrar Plateau

Mauritania's high, hot plateau, with a unique collection of desert landforms, and its own geological "eye"

The Adrar Plateau can be considered a miniature version of the Sahara Desert (see pp.208–209). With huge shifting sand dunes, rugged gorges, palm-filled oases, stone deserts, and rocky cliffs rising to about 790 ft (240 m) above their surroundings, this northwest African region has all the features normally associated with the Sahara as a whole. The center of the plateau, which lies in central Mauritania, is arid, with little to no vegetation, but enough water concentrates around the base of the surrounding uplands to allow the growth of some crops.

Desert eye

West of the Adrar, the plateau merges into the Richat Structure—a gigantic "bull's-eye" first noted from space. This geological dome, also known as the Eye of the Sahara (see p.209), is 25 miles (40 km) in diameter.

△ **CUSTOM BILL**
The greater hoopoe-lark uses its long, curved bill to probe for termites and other insects in the desert sands.

▽ **TABLETOP MOUNTAINS**
The dark tabletop mountains, or mesas, of the Adrar Plateau rise above almost barren sands.

The Kalahari Desert

Not a true desert but a massive sandy savanna with adequate rainfall and a covering of vegetation in parts, yet no water on the surface or in the soil

S. Africa

Despite summer temperatures that top 104°F (40°C) and seemingly endless sand sheets, the Kalahari is not, technically, a desert. Parts of this semiarid region, which covers most of Botswana and some of Namibia and South Africa, receive more rainfall annually than the 10 in (250mm) that is traditionally used to classify desert areas. While rainfall in the Kalahari's southwestern half falls within this range, the northeast gets almost double that figure—yet no surface water remains here. Any rain drains immediately through the deep sands, creating a soil completely devoid of moisture. Despite this and the huge dune-chains that characterize its western half, the Kalahari contains considerable plant life. Deep-rooted acacia trees, shrubs, and some grasses grow in the center, while the north boasts forested areas. More animals are found in the north, but in the arid south, grazing species such as kudu and gemsbok track down enough vegetation to survive.

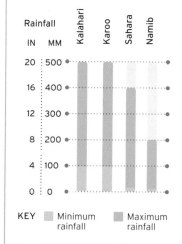

RAINFALL

Total rainfall is only one factor in classifying deserts. Although the Kalahari and Karoo receive over twice the precipitation of true deserts, their rainfall is seasonal, which means that there may be no rain in some areas for 6-8 months.

Rainfall: IN / MM — Kalahari, Karoo, Sahara, Namib

IN	MM
20	500
16	400
12	300
8	200
4	100
0	0

KEY ▪ Minimum rainfall ▪ Maximum rainfall

◁ RUST RED
A herd of gemsbok stands out against the red Kalahari sand, which gets its intense color from iron oxide.

The Karoo

A semiarid cape zone, containing many of the world's succulent plant species

S. Africa

Covering about a third of South Africa, the Karoo is one of the driest regions of the country. Noted for its vast open spaces interspersed with Karoo koppies (flat-topped hills), rugged mountains, and fossil deposits, this desert also contains a fertile zone along the coast that is home to one-third of the world's succulent plant species. In the west, spring rains transform the landscape, covering it with carpets of vivid orange and pink flowers.

The region is divided into sub-Karoos—the more desertlike Upper Karoo, the lower-lying plateau basin called the Great (Groot) Karoo, and the relatively fertile Little (Klein) Karoo.

▷ BRANCHING SUCCULENT
The quiver tree is a tree-sized succulent, with a stem that grows to a diameter of 3 ft (1 m). Desert bushmen hollow out its tubular branches to make quivers for arrows, giving it its name.

SW. Africa

The Namib Desert

A dramatic coastal desert, where terrain ranges from bare bedrock to high, mobile dunes that meet the sea

Stretching from the Atlantic to a high, inland plateau, the narrow Namib Desert extends for about 810 miles (1,300 km) along the coast of southwest Africa. Considered one of the world's oldest deserts, it is bordered to the north by the Kaokoveld Desert, which stretches into Angola, and to the south by South Africa's Karoo. In most places, the narrow Namib measures less than 100 miles (160 km) wide. Despite this, it is often viewed as having three sections: the Atlantic-influenced coastal strip; the Outer Namib, which takes up what remains of the western half; and the Inner Namib, comprising the desert's eastern half. Commonly called the Skeleton Coast, the coastal Namib receives almost no rainfall and instead relies on regular sea fogs for moisture. From sea level, the terrain rises, reaching about 2,950 ft (900 m) by the time the desert meets the base of the Great Escarpment to the east. In the Inner Namib, average annual rainfall is only around 2 in (50 mm).

Living the dry life

Dryness is such a feature of the Namib that it is believed to have been in an arid state for at least 55 million years. This makes it all the more surprising to find that the desert is home to animals ranging from adders and geckos to zebras and elephants. Parts of the Namib are also rich in plant life, including the ground-hugging, ragged-leaved tree tumbo (*Welwitschia mirabilis*), which captures moisture from sea fog and can live for more than 1,000 years.

▷ **SEA OF SAND**
At the heart of the desert lies the Namib Sand Sea, spanning an area of 12,000 square miles (31,000 square km). It is composed of two dune systems—an ancient, semistable one topped by a younger, active layer.

▽ **ADAPTED FEET**
The web-footed gecko's feet enable it to walk easily on top of desert sand, as well as to bury itself beneath the sand to avoid the daytime heat.

In the **southern Namib**, some **dunes** are up to **20 miles (32 km) long** and **790 ft (240 m) high**

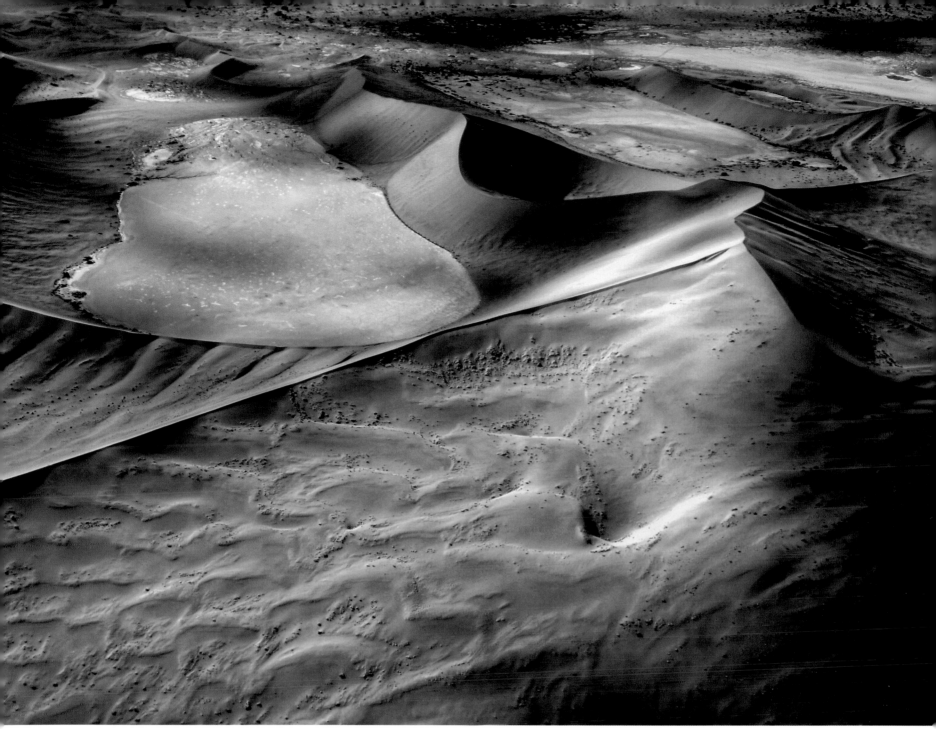

△ FOG HARVESTER

The darkling beetle survives in the Namib by using its own body to harvest water vapor. Microgrooves on its forewings collect water droplets, which are channeled to its mouth when it raises its back legs.

◁ A PLACE OF BONES

Dunes absorb myriad shipwrecks and bones that litter the Skeleton Coast. Here, the thick fogs that caused sailors to run aground bring precious moisture inland.

SKELETON COAST FOG

As the Atlantic's cold Benguela Current runs north along the coast of Namibia, it cools the moist air above the ocean. When this meets the hot desert air, dense fog banks form, rolling up to 60 miles (100 km) into the central Namib until they are burned off by the Sun.

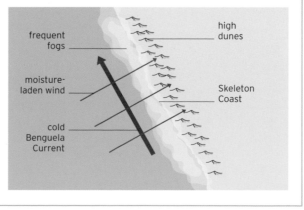

frequent fogs

high dunes

moisture-laden wind

Skeleton Coast

cold Benguela Current

Red pillars
Sandstone pillars rise up to 660 ft (200 m) at Zhangjiajie in southern China, close to the Tianzi Mountains. The pillars have been shaped by water erosion and weathering by ice and plant life.

Asia

TUNDRA TO TROPICS
Asia

Asia is the world's largest continent, taking up almost one-third of Earth's land surface. A large part of the north is the West Siberian Plain, the largest area of flat land in the world. It is dotted with huge swamps and blanketed by dark Siberian forests, ending in the far north with bleak, open tundra—which can often be colder than the Arctic in winter. Farther south are the extensive steppe grasslands of Central Asia, merging into the Gobi Desert. Asia's other great desert is the sand sea of the Rub' al Khali, scorching across the Arabian Peninsula.

In the heart of Asia is the Tibetan Plateau. The world's highest mountains soar around it: to the north, the Altai and Tian Shan; and to the south, the Himalayas, the highest of all, rising to the peak of Everest. Beyond the mountains, to the south and east, long stretches of land extend out into the Indian and Pacific oceans. They are cut by great rivers, filled by rain and melting snow from the mountains—the Indus and Ganges in Pakistan and India, the Mekong in Southeast Asia, and the Yellow and Yangtze in China.

GEOLOGY

Asia is divided between the huge, stable area of continental rock under Siberia and the more varied, unstable south. In between lie the mountains of Tibet, thrown up as India pushes north.

KEY

■ Precambrian (pre-541 million years ago)
■ Paleozoic (541–252 million years ago)
■ Mesozoic (252–66 million years ago)
■ Cenozoic (66 million years ago to present day)

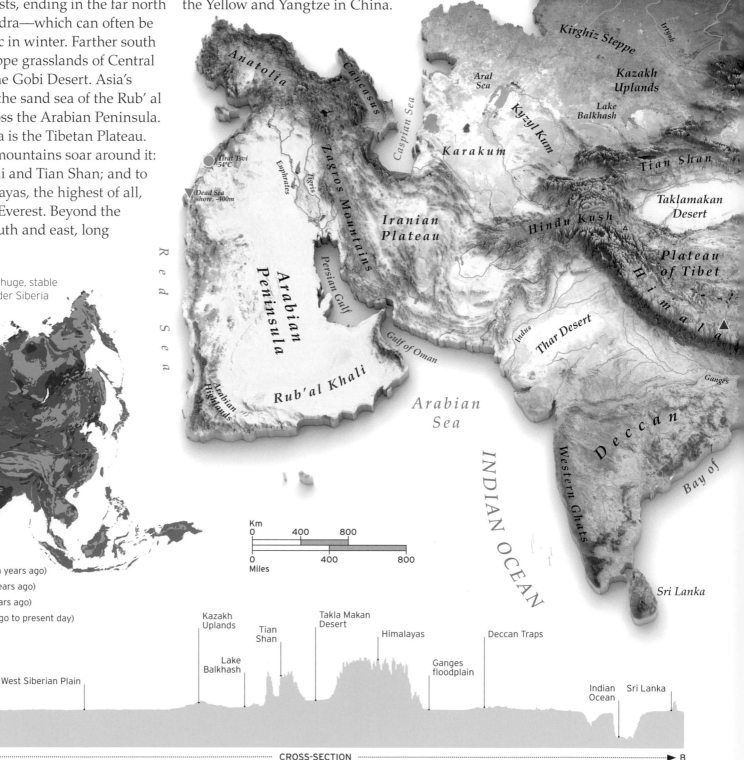

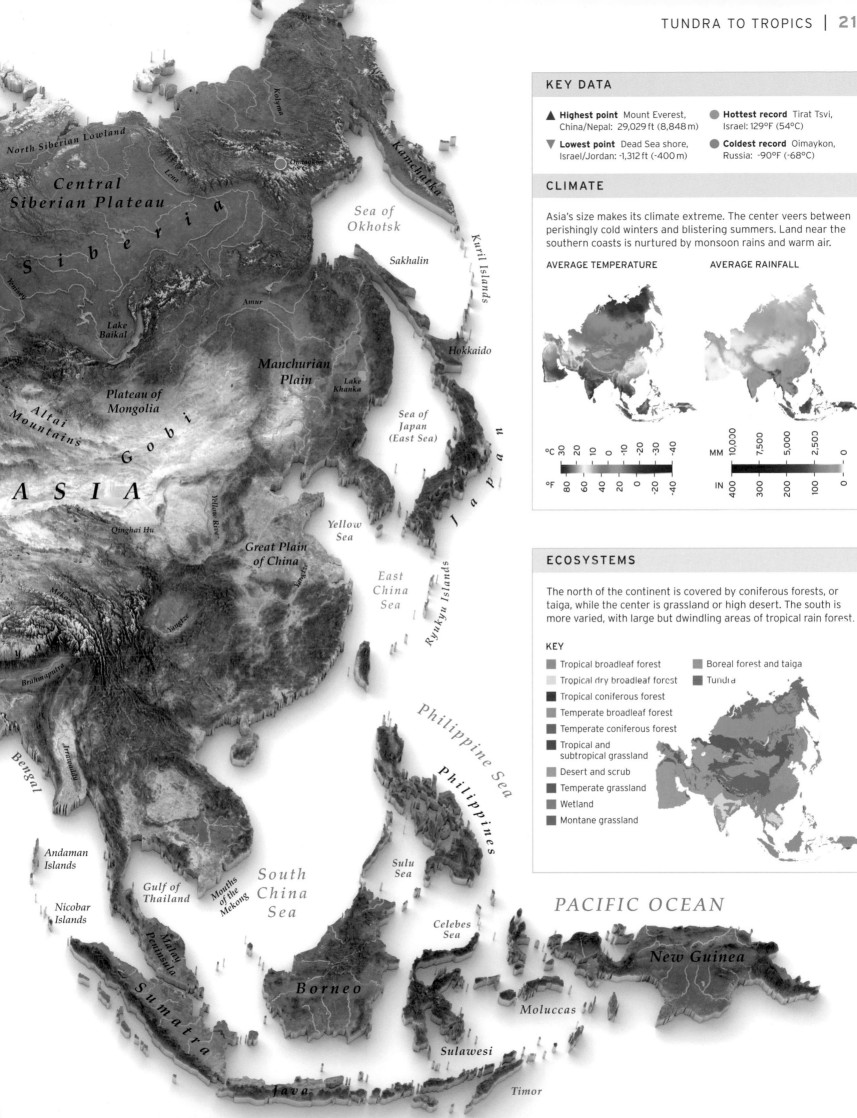

North Siberian Lowland

*Central
Siberian Plateau*

Kolyma

Siberia

Lena

Yenisey

Kamchatka

*Sea of
Okhotsk*

Amur

Sakhalin

Kuril Islands

*Altai
Mountains*

Lake
Baikal

*Plateau of
Mongolia*

Gobi

*Manchurian
Plain*

Lake
Khanka

Hokkaido

ASIA

Qinghai Hu

Yellow River

*Great Plain
of China*

Yellow
Sea

*Sea of
Japan
(East Sea)*

Japan

Mekong

Yangtze

Yangtze

*East
China
Sea*

Ryukyu Islands

Brahmaputra

Bengal

Irrawaddy

Philippine Sea

Philippines

*Andaman
Islands*

Sulu
Sea

*South
China
Sea*

Celebes
Sea

PACIFIC OCEAN

*Nicobar
Islands*

*Gulf of
Thailand*

Mouths
of the
Mekong

*Malay
Peninsula*

New Guinea

Borneo

Moluccas

Sumatra

Sulawesi

Java

Timor

KEY DATA

▲ **Highest point** Mount Everest, China/Nepal: 29,029 ft (8,848 m)

▼ **Lowest point** Dead Sea shore, Israel/Jordan: -1,312 ft (-400 m)

● **Hottest record** Tirat Tsvi, Israel: 129°F (54°C)

● **Coldest record** Oimaykon, Russia: -90°F (-68°C)

CLIMATE

Asia's size makes its climate extreme. The center veers between perishingly cold winters and blistering summers. Land near the southern coasts is nurtured by monsoon rains and warm air.

AVERAGE TEMPERATURE

AVERAGE RAINFALL

°C 30 20 10 0 -10 -20 -30 -40

°F 80 60 40 20 0 -20 -40

MM 10,000 7,500 5,000 2,500 0

IN 400 300 200 100 0

ECOSYSTEMS

The north of the continent is covered by coniferous forests, or taiga, while the center is grassland or high desert. The south is more varied, with large but dwindling areas of tropical rain forest.

KEY

■ Tropical broadleaf forest

■ Tropical dry broadleaf forest

■ Tropical coniferous forest

■ Temperate broadleaf forest

■ Temperate coniferous forest

■ Tropical and subtropical grassland

■ Desert and scrub

■ Temperate grassland

■ Wetland

■ Montane grassland

■ Boreal forest and taiga

■ Tundra

▷ **WORN HILLS**
The Urals are one of the world's oldest mountain ranges, although they have been worn down from their original heights by millions of years of weathering and erosion.

▷ **MOLTEN REMNANTS**
Rocks left over from the basalt lava eruptions that formed the Deccan Traps 66 million years ago have been exposed by the erosion of less resilient rocks around them.

◁ **ANCIENT LAND**
The ancient rock beneath the Kazakh steppe was once an entirely separate island continent geologists call Kazakhstania. Now, it is entirely trapped deep inside Asia.

THE SHAPING OF ASIA

Asia is bigger than any other continent and contains the world's highest mountain and largest mountain range. Yet it is also a young continent, formed from a complicated assembly of fragments.

The giant continent

All the continents have joined each other and split apart again and again—except Asia. It has stayed largely intact throughout Earth's history and is a recent creation, an amalgam of many pieces. The vast slab of ancient rock beneath Siberia, known as the Siberian, or Angaran, Platform, forms the continent's core, but it was joined by the other parts relatively recently. Even when the Siberian Platform was part of the supercontinent Pangaea 200 million years ago, south and east Asia were an assortment of separate island continents.

Between 100 and 50 million years ago, these fragments began to come together rapidly. First, north and south China and Southeast Asia fused onto Eurasia's southeast corner. Then, far to the south, India's extraordinary journey began. About 80 million years ago, India was attached to southern Africa, but then it broke away and hurtled north across an ancient ocean and smashed into the southern edge of Eurasia 60–40 million years ago, helping cement all the other fragments in place. Asia as we know it was born.

Volcanic India

India's Deccan Plateau is dominated by one of the largest volcanic features in the world, the Deccan Traps. The traps are what geologists call a large igneous province and are made of layers of volcanic basalt over 6,600 ft (2,000 m) thick. Even now, they spread over an area of 193,000 square miles (500,000 square km), but before being worn away by erosion they covered half of India.

Unlike some lavas, basalt is very runny and can flood onto the surface in huge quantities. The eruptions began 66.5 million years ago, when India had only recently become detached from Africa. One theory suggests that at that time India was above a hotspot in Earth's interior where the island of Réunion is now. At a hotspot, a fountain of hot magma

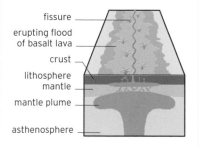

CREATION OF THE DECCAN TRAPS
Basalt lava erupted via a fissure in Earth's crust above a mantle plume, a fountain of hot rock rising in the mantle. The lava hardened and formed the Deccan Traps.

fissure
erupting flood of basalt lava
crust
lithosphere
mantle
mantle plume
asthenosphere

KEY EVENTS

130 million years ago
A rain forest forms in an area that will become Malaysia. A portion of this survives in Taman Negara National Park today.

60–40 million years ago
The Indian Plate crashes into the southern part of the Eurasian Plate, causing the crust to crumple into the Himalayas.

5.5 million years ago An ancient sea that had spread from eastern Europe to western Asia dries out as land lifts and sea levels drop, leaving behind the Caspian, Aral, and Black Seas and Lake Urmia.

252–250 million years ago Intense volcanic activity due to a mantle plume causes eruptions over a large area that will become eastern Siberia, creating a huge plain of volcanic rock called the Siberian Traps that still exists today.

66 million years ago Huge volcanic eruptions on a landmass that will become India occur. Lava spreads and solidifies into the Deccan Traps.

50–40 million years ago The rise of the Himalayas blocks rainfall heading north of the new mountain range, causing the area to dry out and become the Gobi Desert.

11,000 years ago Sumatra, Java, and Borneo become islands as sea levels rise.

◁ ROOF OF
THE WORLD
The vast extent
of the Tibetan
Plateau, with its
fringing ranges of
lofty, snow-capped
Himalayan Mountains,
can be seen from space.

called a mantle plume rises and melts its way through the crust. The giant eruption may have been a contributing factor toward the extinction of the dinosaurs—the main culprit being the meteor strike—and scientists once believed that they might have been killed off by a cooling of the world's climate caused by gases released by this eruption.

▷ SOURCE OF
THE GANGES
The rivers flowing down from the Himalayas, including the Ganges, sweep down 25 percent of all the world's sediment, to create wide, fertile plains.

Mountains rising

Following its separation from Africa, India moved north at a rate of about 8 in (20 cm) a year, faster than any other known plate movement. As it rammed into the southern edge of Asia, it exerted a huge impact.

In collisions between plates, one plate is usually subducted beneath the other. But these two plates were so closely matched in density that neither would sink, so the crust between them crumpled up into the highest and largest mountain range the world

has ever seen—the Himalayas. In just 50 million years, the rocks have been thrust up more than 6 miles (9 km) to create jagged peaks—including Everest, the world's highest mountain at 29,029 ft (8,848 m). In fact, the impact has not stopped the Indian Plate from moving—it still creeps forward at 1¼ in (3 cm) a year, and as a result the Himalayas rise ½ in (1 cm) every year. Geologists think the only reason they are not higher still is that, as it piles up, the rock also flows outward—in effect, the Himalayas are similar to the bow wave in front of a moving boat.

Violent volcanic eruptions in India 66.5 million years ago probably contributed to the extinction of the dinosaurs

BUILDING THE ASIAN CONTINENT

KEY ▬ Convergent boundary

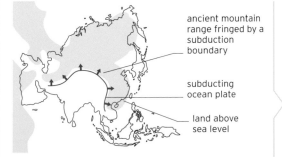

ancient mountain range fringed by a subduction boundary

subducting ocean plate

land above sea level

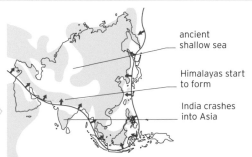

ancient shallow sea

Himalayas start to form

India crashes into Asia

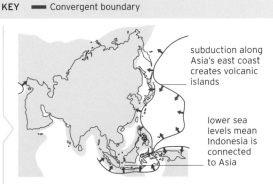

subduction along Asia's east coast creates volcanic islands

lower sea levels mean Indonesia is connected to Asia

94 MILLION YEARS AGO Asia is much smaller than today. China has only recently joined, and India is still attached to Africa, far to the south. An ancient ocean plate to the south subducts beneath Asia.

50–40 MILLION YEARS AGO India crashes into the southern edge of Eurasia to create Asia's current shape and throw up the Himalayan Mountains. A shallow sea in central Asia recedes.

18,000 YEARS AGO The shallow seas that once separated Asia from Africa have receded completely. Lower sea levels due to water being locked away in ice sheets mean Indonesia is connected to Asia.

Pamukkale Hot Springs

Staircaselike terraces of hot springs in southwest Turkey

W. Asia

Pamukkale, which means "cotton castle" in Turkish, evokes a sense of awe due to its white landscape and turquoise pools—arranged in a series of steps, or terraces—that carry water from hot springs.

Travertine steps

Pamukkale's terraces are composed of travertine, a bright white substance made from calcium carbonate deposited by water from the hot springs. There are 17 springs with temperatures ranging from 95 to 207°F (35 to 97°C). For thousands of years, water that has percolated deep into Earth's crust and been heated there has re-emerged near the top of a 330-ft (100-m) high hill. As it flows downhill, it passes through a series of pools lined by travertine. The mineral-rich pools are reputed to have health-giving properties.

▷ **TERRACED POOLS**
This area of stepped pools at Pamukkale is 215 ft (65 m) long and 90 ft (27 m) wide. It has around 20 levels, and its pools contain water at about 97°F (36°C).

The Zagros Mountains

An imposing natural barrier running through the Middle East, which displays evidence of the ongoing effects of a tectonic plate collision

W. Asia

The Zagros is the largest mountain range in Iran, also extending into parts of Turkey and Iraq. Some sections rise above 13,000 ft (4,000 m) and are permanently snow-capped, with the highest peak being Ghash Mastan at 14,550 ft (4,435 m). Wild and mostly semiarid, the Zagros started forming around 30–25 million years ago when two tectonic plates—the Arabian Plate and the Eurasian Plate—collided, resulting in the folding of underlying rock layers. A sizable amount of deformation is still occurring today as a result of this collision, and the region is subject to frequent earthquakes.

◁ **RIPPLED EARTH**
Its many parallel ridges give the Zagros a wavelike appearance.

▷ **NIGHT FEEDER**
A nocturnal predator and scavenger, the striped hyena occurs in small numbers throughout the Zagros. It features prominently in Middle Eastern folklore.

HOW A TRAVERTINE TERRACE FORMS

Travertine terraces form when a hot spring containing calcium carbonate erupts on a hill. The flowing water deposits travertine on the slope below the opening, while a buildup of the mineral in the opening eventually chokes it up. A new opening is then forced to develop a little higher up. In this way, the site of the spring gradually shifts uphill, leading to a travertine staircase.

warm water pools

former opening for spring, now blocked off by travertine deposit

travertine

previous channel, now blocked off

channel containing rising hot water

The **springs flow** at about **106 gallons (400 liters) per second**

The Caucasus Mountains

A wall of mountains stretching across the meeting point of Asia and Europe

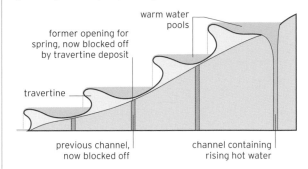

W. Asia

THE ORIGINS OF THE CAUCASUS

The Caucasus started forming when a landmass comprising parts of present-day Iran, Turkey, Armenia, and nearby countries collided with Eurasia. Around the same time, the Paratethys Sea was split into what are now the Black and Caspian Seas.

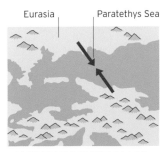

Eurasia   Paratethys Sea

ABOUT 30 MILLION YEARS AGO

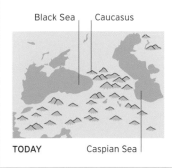

Black Sea   Caucasus

TODAY   Caspian Sea

The formidable Caucasus mountain range is situated mainly in the Asian countries of Georgia, Armenia, and Azerbaijan but also partly in European Russia. It consists of two main subranges: the Greater and Lesser Caucasus. The highest peak is Mount Elbrus (in the Russian section) at 18,510 ft (5,642 m). The mountains feature a remarkable array of vegetation zones, ranging from forests and alpine meadows to highland semideserts, and also the world's deepest-known cave, the Krubera Cave in Georgia.

▷ **MAGNETIC MINERAL**
Dashkesan, a region of the Lesser Caucasus, is a source of the iron-rich mineral magnetite, the most magnetic of all minerals.

▷ **SUNRISE VIEW**
The Caucasus are steep, craggy, and relatively young mountains. Many of the peaks in the range carry glaciers—there are more than 2,000 glaciers in the entire range.

PARTS OF THE HIMALAYAS

Although one continuous arc of mountains, the Himalayas are considered to have two main parts: the Western and Eastern Himalayas. Many of the range's highest peaks, such as Everest, Lhotse, and Kanchenjunga, are clustered together in the eastern section. Neighboring mountain ranges, usually considered separate from the Himalayas, include the Karakoram and Hindu Kush to the northwest and Hengduan Mountains to the northeast. To the north of the Himalayas is the Tibetan Plateau.

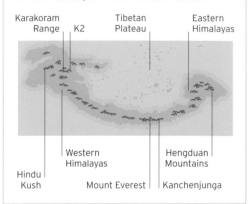

The Himalayas

The world's highest mountain range, with more than 50 peaks reaching a height of 24,000 ft (7,200 m) or more

C. Asia

Running across northern Pakistan and India to Nepal and Bhutan, with some parts in China, the Himalayas are not only Earth's highest mountain range but also relatively young, having formed within the last 50 million years.

Abode of snow

Some 1,400 miles (2,300 km) long and 160–220 miles (250–350 km) wide, the Himalayas form a barrier between the Tibetan Plateau to the north and the plains of the Indian subcontinent to the south. The name of the range comes from the Sanskrit word *Himalaya*, which means "abode of snow." Although extremely inhospitable, the highest peaks of the Himalayas present an irresistible magnet to mountaineers seeking the greatest possible challenges.

△ **HIGH-ALTITUDE HUNTER**
A few thousand snow leopards are believed to live in the Himalayas. In summer, their range extends as high as 20,000 ft (6,000 m).

Parts of the Himalayas are **still rising** at a rate of about ³/₁₆ in (4 mm) a year

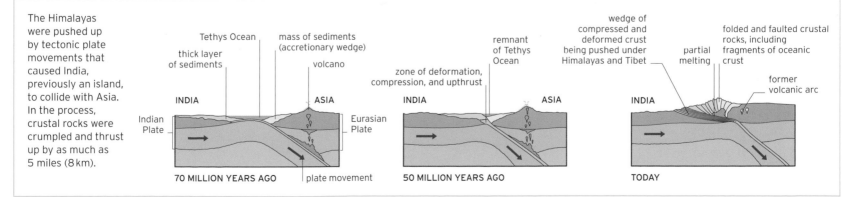

△ **FIRST LIGHT**
At dawn every day, the high snow-clad peaks light up while the valleys remain in darkness.

◁ **FLORAL CARPET**
In July and August, some lower slopes are covered in brightly colored flowers, such as these Himalayan fleeceflowers.

HOW THE HIMALAYAS FORMED

The Himalayas were pushed up by tectonic plate movements that caused India, previously an island, to collide with Asia. In the process, crustal rocks were crumpled and thrust up by as much as 5 miles (8 km).

Tethys Ocean

thick layer of sediments

mass of sediments (accretionary wedge)

volcano

INDIA

Indian Plate

ASIA

Eurasian Plate

70 MILLION YEARS AGO

plate movement

zone of deformation, compression, and upthrust

remnant of Tethys Ocean

INDIA

ASIA

50 MILLION YEARS AGO

wedge of compressed and deformed crust being pushed under Himalayas and Tibet

partial melting

folded and faulted crustal rocks, including fragments of oceanic crust

former volcanic arc

INDIA

TODAY

The Altai Mountains

A beautiful range of high mountains situated almost in the middle of Asia

C. Asia

Straddling an area where Russia, China, Mongolia, and Kazakhstan meet, the Altai stretch for 1,200 miles (2,000 km) across central Asia. Much of the range exceeds 10,000 ft (3,000 m) in height, and there are numerous glaciated peaks over 13,000 ft (4,000 m) in the Mongolian sections. The highest peak, Mount Belukha (14,783 ft/4,506 m), lies in the north at the headwaters of the Ob–Irtysh river system in Russian Siberia. In the east, the Altai merge into the high plateau of the Gobi desert (see pp.266–67) and the cold steppes of Mongolia. Partly because they include a variety of habitats—such as tundra, forest, and alpine vegetation—the Altai are home to a wide range of wildlife, including Siberian ibex and several species of deer, wolves, lynx, and brown bears.

◁ **HEAVY FRAME**
Resembling a heavily built goat, the Siberian ibex is typically found on steep rocky slopes above the tree line. It feeds on small shrubs, grass, and sedges.

▽ **GOLDEN MOUNTAINS**
Altai means "gold peak" in Mongolian. The snow-dusted peaks of the Altai mountains are seen here with yellow steppe grass in the foreground.

The Tian Shan Mountains

One of the world's longest and highest mountain ranges, boasting more than 60 peaks that exceed 20,000 ft (6,000 m) in height

C. Asia

Also known as the Tien Shan, the Tian Shan consists of a series of subranges that extend through Kyrgyzstan, a part of Kazakhstan, and western China. Their overall length is about 1,700 miles (2,800 km), which is longer than the Himalayas. The peaks above 13,000 ft (4,000 m)—including the highest, Victory Peak at 24,406 ft (7,439 m)—are capped by glaciers. The Chinese words *Tian Shan* mean "heavenly mountains." The range has been listed as a World Heritage Site—partly because of its unique and rich biodiversity, as well as its diverse landscapes that contrast with the vast adjacent deserts.

▷ **BLUE MOUNTAINS**
A part of the Borohoros, one of the Tian Shan's subranges, appears vivid blue due to atmospheric scattering of blue-wavelength light.

C. Asia

The Karakoram Range

The greatest concentration of high mountains in the world

Spanning the northern regions of Pakistan and India, with parts in China and Tajikistan, the Karakoram lies to the northwest of the Himalayas and contains nearly all of the world's 100 highest peaks that are not in the Himalayas (see pp.222–23). These include the world's second highest mountain, K2, which rises to a height of 28,251 ft (8,611 m).

Climbing challenges

The Karakoram Range is about 310 miles (500 km) in length and is heavily glaciated. Some individual glaciers, such as the Baltoro (see p.233), are among the world's longest glaciers outside the polar regions. The Karakoram is sparsely populated, and visitors are almost exclusively intent on scaling one or more of its peaks, some of which are extremely difficult to climb. They include K2 and the Trango Towers, a group of granite columns situated close to the Baltoro Glacier.

About **1 in 5 people** who have **tried to climb K2** have **died** in the attempt

△ **NAMELESS TOWER**
Called the Nameless Tower, the granite peak at the center here is part of the Trango Towers. Its near-vertical sides rise some 2,950 ft (900 m) above the ridgeline, and the spire reaches up to 20,469 ft (6,239 m).

▷ **SAVAGE MOUNTAIN**
Seen here with a serac (pointed block of glacial ice) in the foreground, K2, also called Mount Godwin-Austen, is known as the Savage Mountain because of the extreme difficulty of its ascent.

HIGH PEAKS OF THE KARAKORAM

1 K2 28,251 ft (8,611 m)
2 Gasherbrum I 26,509 ft (8,080 m)
3 Broad Peak 26,414 ft (8,051 m)
4 Gasherbrum II 26,362 ft (8,035 m)
5 Gasherbrum III 26,089 ft (7,952 m)

S. Asia

Mount Everest

Earth's highest mountain, which sits in the Eastern Himalayas

Also known in Nepal as Sagamartha (meaning "sky head") and in China as Chomolungma (meaning "Holy Mother"), Mount Everest soars to 29,029 ft (8,848 m) above sea level.

Faces and ridges

Everest is a highly inhospitable place. It is dangerous to climb: fatalities are mainly a result of avalanches, hypothermia, and altitude sickness. The mountain is roughly the shape of a three-sided pyramid, with a southwest face, a north face, and an east face, each very steep. At the intersections of these faces are three ridges—southeast, northeast, and west—leading to the summit. The first confirmed ascent took place in 1953, by Tenzing Norgay and Edmund Hillary, via a route using the southeast ridge. Successful ascents have now been made by at least 18 different routes of varying difficulty using various combinations of ridges and faces.

THE GEOLOGY OF EVEREST

Everest's summit pyramid is composed of limestone that formed on the seafloor some 470 million years ago, during Earth's Ordovician period. Beneath that are other layers of sedimentary rock, and deeper still, metamorphic rocks such as gneiss, with intrusions of igneous rocks.

Ordovician limestone

leucogranite

sedimentary rocks such as shale and limestone

gneiss

igneous rocks intrusion metamorphic rocks

▷ AVALANCHE

Avalanches are frequent on Everest. In April 2015, an earthquake triggered a deadly avalanche that swept into the southern base camp and killed 22 people.

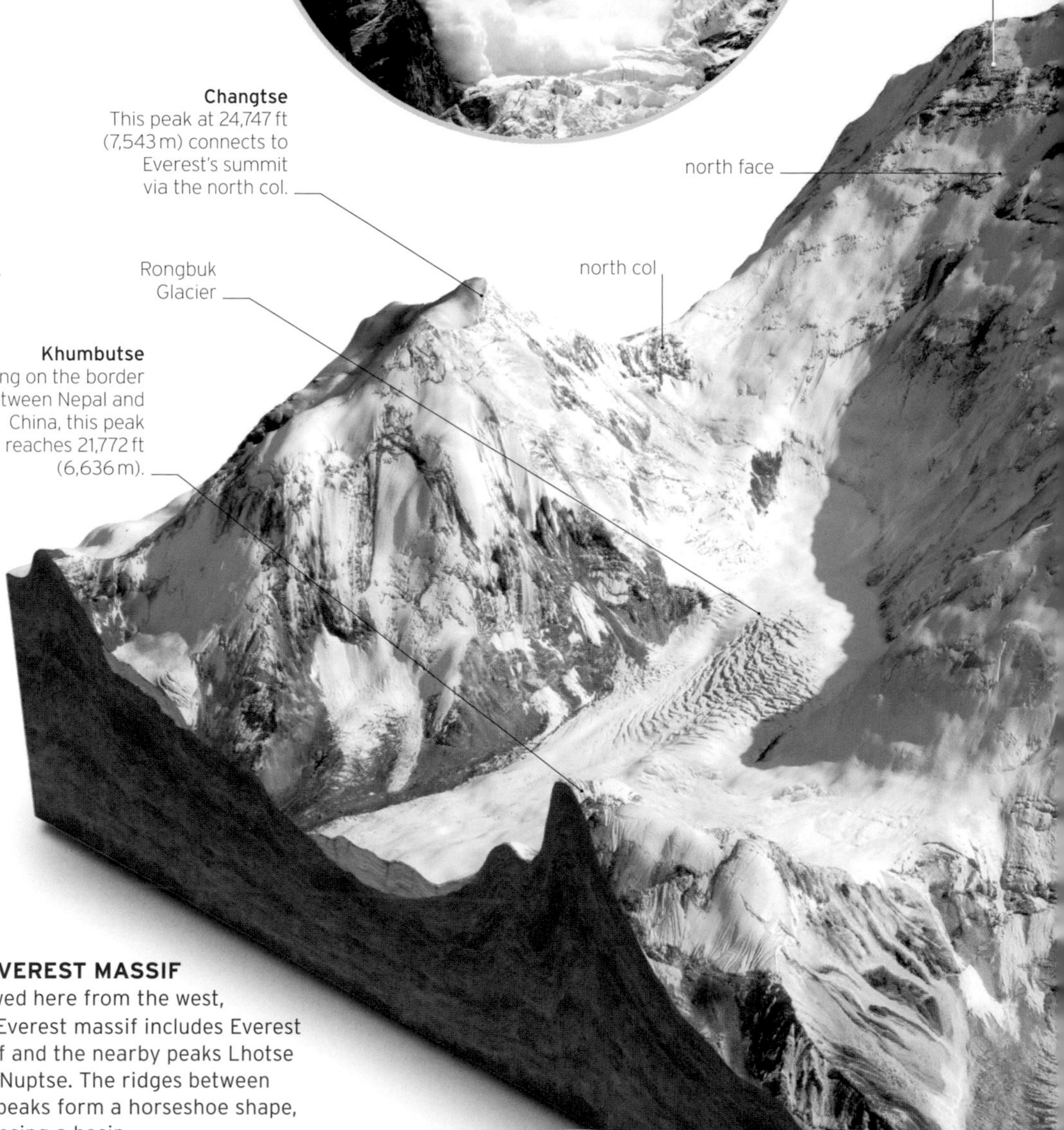

northeast ridge

Changtse
This peak at 24,747 ft (7,543 m) connects to Everest's summit via the north col.

north face

Rongbuk Glacier

north col

Khumbutse
Lying on the border between Nepal and China, this peak reaches 21,772 ft (6,636 m).

▶ EVEREST MASSIF

Viewed here from the west, the Everest massif includes Everest itself and the nearby peaks Lhotse and Nuptse. The ridges between the peaks form a horseshoe shape, enclosing a basin.

▷ BASE CAMP

There are two main base camps on Everest on different sides of the mountain. These campsites are used by climbers before and after ascents and by hikers. The south camp (seen here), in Nepal, is close to the Khumbu Icefall at 17,598 ft (5,364 m).

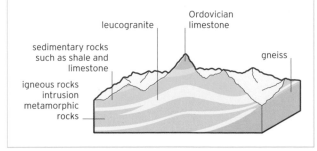

There is **66 percent less oxygen** at **Everest's summit** than at sea level

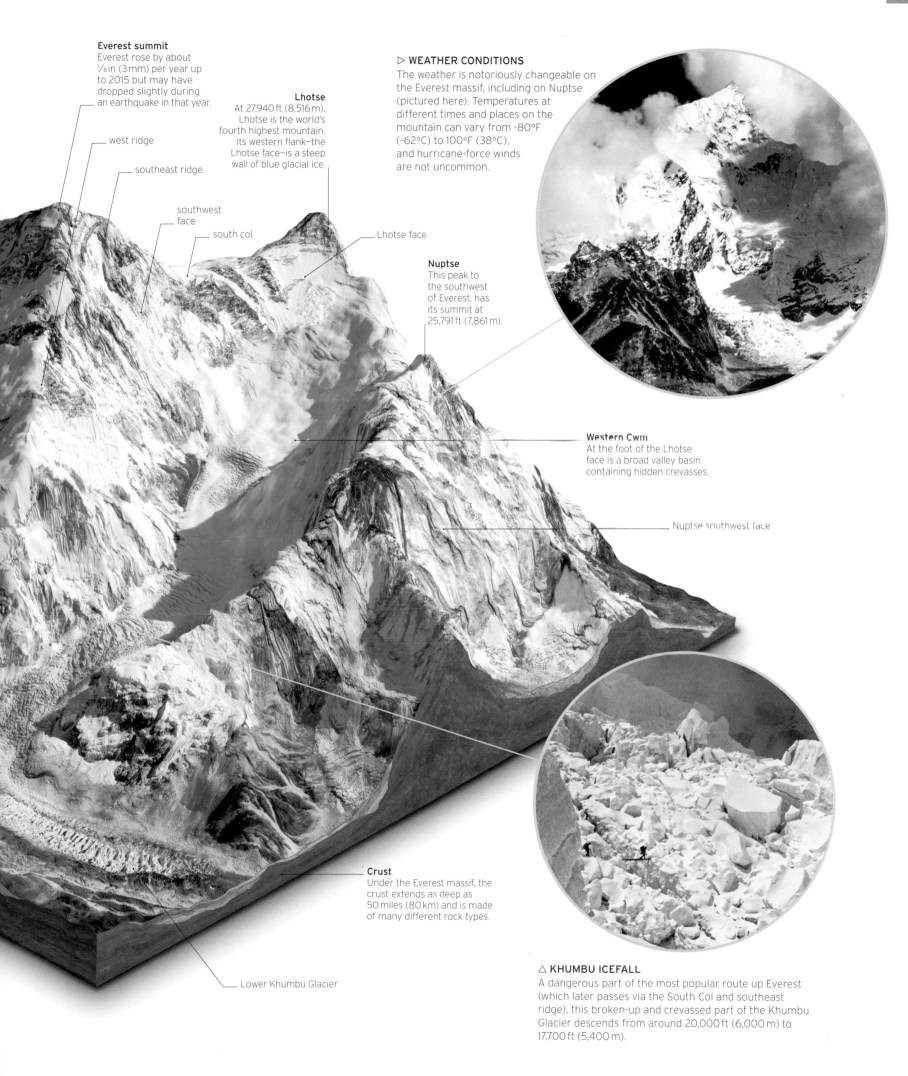

Everest summit
Everest rose by about
1/8 in (3 mm) per year up
to 2015 but may have
dropped slightly during
an earthquake in that year.

west ridge

southeast ridge

southwest
face

south col

Lhotse
At 27,940 ft (8,516 m),
Lhotse is the world's
fourth highest mountain.
Its western flank–the
Lhotse face–is a steep
wall of blue glacial ice.

Lhotse face

▷ **WEATHER CONDITIONS**
The weather is notoriously changeable on
the Everest massif, including on Nuptse
(pictured here). Temperatures at
different times and places on the
mountain can vary from -80°F
(-62°C) to 100°F (38°C),
and hurricane-force winds
are not uncommon.

Nuptse
This peak to
the southwest
of Everest, has
its summit at
25,791 ft (7,861 m).

Western Cwm
At the foot of the Lhotse
face is a broad valley basin
containing hidden crevasses.

Nuptse southwest face

Crust
Under the Everest massif, the
crust extends as deep as
50 miles (80 km) and is made
of many different rock types.

Lower Khumbu Glacier

△ **KHUMBU ICEFALL**
A dangerous part of the most popular route up Everest
(which later passes via the South Col and southeast
ridge), this broken-up and crevassed part of the Khumbu
Glacier descends from around 20,000 ft (6,000 m) to
17,700 ft (5,400 m).

Zhangye Danxia

An extraordinary landscape of brightly colored sedimentary rocks and sculpted landforms in northern China

E. Asia

△ RAINBOW RIDGES
The rock strata of the Zhangye Danxia are predominantly reddish in color, but also include some orange, yellow, pale blue, and green-hued layers.

Stretching across 200 square miles (500 square km) in Gansu Province, China, the Zhangye Danxia Landform Geopark is known for the unusual colors and shapes of its sedimentary rocks. With its undulating, mainly red-hued ridges, parts of it resemble a sea of fire with rolling waves. Danxia is a broad term used to describe the types of geological formations seen in this area, and Zhangye is a city close to the park.

Source of color

From around 80 million years ago, sedimentary rocks began forming in the area through the deposition of sediments in lakes. Variations in the mineral composition of the sediments account for the range of rock colors seen today. Starting from about 20 million years ago, the whole area was uplifted and the sedimentary layers tilted as a result of tectonic stresses. Subsequently, the processes of weathering and erosion have resulted in the spectacular sights visible today.

▷ **ROCK FORTRESS**
To the west of the Zhangye Geopark lies the Binggou Danxia. It features sandstone outcrops in shapes that resemble manmade fortresses.

The Danxia's colored rock layers took **tens of millions** of years to form

Mount Fuji

An iconic, beautifully symmetrical volcano that can be seen from Tokyo on a clear day

Rising to a height of 12,388 ft (3,776 m), and located 55 miles (90 km) southwest of Tokyo, Mount Fuji is the highest mountain in Japan. Mostly formed 11,000–8,000 years ago, its attractive cone is one of the country's most famous symbols, often celebrated in art and literature, as well as visited by sightseers and climbers. Although it is classified as an active volcano, Fuji's last eruption—which covered a huge area to its east in ash—was in December 1707. Since then, there have been no further signs of volcanic activity.

△ **CELEBRATED CONE**
Fuji's cone has a base around 12 miles (20 km) wide and is almost perfectly symmetrical. It is snow capped from October to June.

INSIDE MOUNT FUJI

Fuji is a stratovolcano—a steep-sided volcano composed of layers of solidified lava alternating with ash and other ejected material. The successive layers have built the volcano into a tall cone.

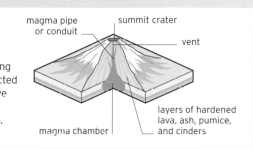

magma pipe or conduit
summit crater
vent
layers of hardened lava, ash, pumice, and cinders
magma chamber

▷ **BLOSSOM SEASON**
The spring flowering of cherry trees is admired so much in the Mount Fuji area that there is an official cherry blossom viewing season, called *Hanami*.

Kliuchevskoi

The highest active volcano outside the Americas, situated on Russia's Kamchatka Peninsula and regarded as sacred by the native population

N. Asia

▽ **LAVA RIVER**
In this photo taken in 2015, a lava flow runs down Kliuchevskoi while plumes of steam, ash, and other volcanic gases escape into the atmosphere.

Also known as Klyuchevskaya Sopka, Kliuchevskoi is one of the most active, as well as the highest of a dense chain of more than 160 volcanoes that run down the Kamchatka Peninsula. The volcanoes are the result of the Pacific and North American plates subducting (sliding under) the small Okhotsk Plate, on which Kamchatka sits. This process provoked the formation of magma at depth, which moved upward to create the volcanoes.

Dangerous beauty

Originally formed around 6,000 years ago, Kliuchevskoi has been erupting almost continually since at least the late 17th century, including significant eruptions in 2007, 2010, 2012, 2013, and 2015. Due to its frequent eruptions, the peak has not been climbed often. A beautiful, symmetrical, snow-capped volcano, Kliuchevskoi is regarded by some of Kamchatka's native people as the place where the world was created.

The Tengger Volcanic Complex

An unearthly looking group of volcanic cones inside an ancient caldera, all set within lush tropical vegetation

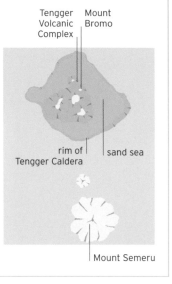

SE. Asia

TENGGER AND SEMERU

The Tengger Caldera is roughly square in shape. Its flat, sand-filled floor is called the Sand Sea. Within the caldera are five overlapping stratovolcanoes. A few kilometers to the south is the large active stratovolcano Semeru.

Tengger Volcanic Complex

Mount Bromo

rim of Tengger Caldera

sand sea

Mount Semeru

The Tengger Volcanic Complex is part of a national park on the Indonesian island of Java. The complex lies within a large caldera, the remains of an enormous volcano that disintegrated in a cataclysmic eruption over 45,000 years ago. Over the past few thousand years, several new volcanic cones have grown from its floor. Of these, the most easily recognized—because its top has been blown off, leaving a wide summit crater—is Mount Bromo.

Recent eruptions

Mount Bromo is the youngest, most active, and most frequently visited of the volcanoes. It has erupted repeatedly since 1590. Outside the caldera is a larger volcano, Semeru, which also regularly belches out huge clouds of steam and smoke. At 12,060 ft (3,676 m), Semeru is Java's highest mountain.

△ TENGGER PANORAMA
In this view, the volcanic cones comprising the Tengger Complex can be seen at the center, with smoke emanating from Mount Bromo. In the distance is Mount Semeru.

▽ DAWN PERSPECTIVE
Here, part of the Tengger Caldera's rim is visible, with a hamlet—a base for exploring the complex—perched on it. On the right is the mist-covered caldera.

The **caldera** within which **the Tengger complex** sits is **10 miles (16 km)** wide

Mount Pinatubo

A brooding volcano in the Philippines with a history of extraordinarily violent eruptions

SE. Asia

△ **SUMMIT LAKE**
Since 1992, Pinatubo's 1½-mile (2.5-km) wide caldera has been filled with a turquoise lake. With a depth of 2,000 ft (600 m), it is the deepest lake in the Philippines.

The volcano Pinatubo on the island of Luzon in the Philippines does not erupt frequently, but when it does, it explodes with extreme ferocity. It has produced some exceptionally large eruptions at intervals of anything from 500 to 8,000 years.

Pinatubo in 1991 and today

The most recent of Pinatubo's eruptions occurred in June 1991, when a series of explosions sent a vast quantity of rock and ash into the atmosphere and produced pyroclastic flows (avalanches of hot gas and ash) that incinerated land up to 11 miles (17 km) away. More than 800 people died, and thousands more were displaced. Part of the top of the volcano collapsed, leaving a summit caldera that is now filled with a tranquil lake. Ash darkened the sky for many days, and fine particles of dust and tiny droplets of sulfuric acid spread all around Earth. These droplets blocked out sunlight, lowering global temperatures by 1°F (0.5°C) for a year.

Even today, some of the effects of the eruption can be seen in the form of vast, deep fields of ash all around the volcano. Pinatubo is closely monitored for any further activity.

THE 1991 ERUPTION OF PINATUBO

Pinatubo's 1991 eruption is the largest, most violent volcanic event of the past 100 years in terms of the volume of ash, lava, pumice, and other materials ejected. Here, it is compared to four other massive eruptions since 1950.

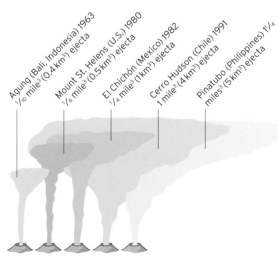

Agung (Bali, Indonesia) 1963
⅒ mile³ (0.4 km³) ejecta

Mount St. Helens (U.S.) 1980
⅙ mile³ (0.5 km³) ejecta

El Chichón (Mexico) 1982
¼ mile³ (1 km³) ejecta

Cerro Hudson (Chile) 1991
1 mile³ (4 km³) ejecta

Pinatubo (Philippines) 1¼ miles³ (5 km³) ejecta

△ **MUSHROOM CLOUD**
During Pinatubo's 1991 eruption, a cloud of volcanic ash was launched 21 miles (34 km) into the air, forming a mushroom shape. This is far higher than most commercial airliners fly.

The Fedchenko Glacier

The longest glacier in the world outside of the polar regions, flowing through valleys in the Pamir Mountains of Tajikistan

C. Asia

The Fedchenko is an exceptionally long, narrow glacier, whose length has been measured from satellite imagery at 48 miles (77 km)—slightly longer than some other very long nonpolar glaciers, such as the Siachen Glacier and the Baltoro Glacier (see opposite) in the Karakoram mountain range. The Fedchenko Glacier covers a total area of about 270 square miles (700 square km) and has dozens of tributaries. It is located in such a remote part of the world that it remained undiscovered until 1878 and was not fully explored until 1928. Subsequently, it was named after Alexei Fedchenko, a Russian explorer well known for his travels in Central Asia.

▷ **NARROW COURSE**
Here, the lower part of the glacier snakes diagonally across the landscape from bottom left to top right. A few tributaries can also be seen.

The Yulong Glacier

One of the best known glaciers in China, located on Jade Dragon Snow Mountain

E. Asia

The Yulong Glacier is one of 19 small glaciers that adorn the flanks of a massif (a compact group of mountains) in Yunan Province, southwestern China. This massif is known as Mount Yulong or Jade Dragon Snow Mountain. The Yulong Glacier is a cirque glacier (one occupying a depression high up on a mountain) with an extra tongue of ice, called a hanging glacier, below it. It is one of the most accessible glaciers in China because it lies near the top of a cable car route. Unfortunately, the glacier is shrinking, and at its current rate of ice loss it is expected to disappear in about 50 years' time.

▷ **CRACKED SURFACE**
The steeply descending "hanging" part of the Yulong Glacier, situated at an altitude of about 12,000 ft (3,650 m), has a heavily crevassed, broken-up surface.

△ **WHITE FLURRY**
A powder snow avalanche falls from one of the peaks in the Gasherbrum group onto the Baltoro Glacier, which is in the foreground.

C. Asia

The Baltoro Glacier

A major glacier running through part of the Karakoram mountain range, with glorious views of many of the world's highest peaks

At 39 miles (63 km) long, the Baltoro Glacier runs roughly west to east through Pakistan-administered Kashmir. It forms part of the route for mountaineers on their way to climb either K2, the world's second highest mountain, or a collection of the world's highest peaks known as the Gasherbrum group.

Peak views

The surface of the glacier is rough and uneven, broken by crevasses and covered by seracs and rock debris. Where the Baltoro comes together with one of its largest tributaries, the Godwin-Austen Glacier, is a broad area known as Concordia. This spot offers mountaineers magnificent views, in different directions, of K2 and three other peaks rising over 26,000 ft (8,000 m).

◁ **ICE CONES**
Huge pinnacles and cones of ice called seracs—some of them as large as houses—can be seen all over the glacier.

HOW CREVASSES AND SERACS FORM

Crevasses are caused by tensional stretching of a glacier as it passes over humped or unevenly sloping bedrock. They can be transverse or longitudinal but sometimes the two types combine, producing pillars or pinnacles of ice—seracs—where they intersect.

direction of stretching

crevasse at right angle to flow

crevasse along direction of flow

serac between crevasses

flow direction

TRANSVERSE CREVASSES

LONGITUDINAL CREVASSES

SERACS

Six of the world's **17 highest peaks can be viewed** on a trek up the glacier

C. Asia

The Biafo Glacier

A long, nearly straight glacier in the Karakoram Mountains of Pakistan

The Biafo Glacier is a 39-mile-(63-km-) long glacier that runs northwest to southeast in a western region of the Karakoram (see p.225) and is heavily covered with rock debris on its lower parts. It is possible to walk along the entire length of the glacier, up to its highest point, at 16,824 ft (5,128 m). At its top end, the Biafo meets another highway of ice, the Hispar Glacier. Nearby is one of the largest basins of snow outside polar regions, known as Snow Lake. Occasionally, evidence of wildlife can also be seen in the vicinity of the glacier, including ibex and markhors (types of mountain goats) or, much more rarely, snow leopards.

SOURCES OF DEBRIS IN GLACIERS

Rocks that have come loose from nearby mountains are the main source of debris in or on a glacier. Dust is also continuously blown onto the glacier surface, and rocks are "plucked" from the bedrock by becoming frozen onto the ice.

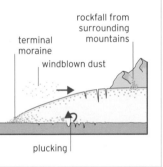

terminal moraine

windblown dust

rockfall from surrounding mountains

plucking

△ **ICY HIGHWAY**
The surface of the lower part of the glacier contains deep crevasses, meltwater streams, rock fragments, and large piles of rocks. Lines of high mountains with tributary glaciers run down both sides.

CROSSING THE CHASM
A member of an Everest expedition traverses a bridge constructed from aluminum ladders above a large crevasse in the Khumbu Icefall.

S. Asia

The Khumbu Glacier

A large, partly rock-covered glacier in Nepal, famous mainly because it lies on some popular routes for approaching and climbing Mount Everest

The Khumbu Glacier has two main parts (see panel, below). Its upper part, on the western flank of Mount Everest (see pp.226–27), is a colossal icefall—a steeply descending, deeply crevassed mass of broken ice. With its source area at 24,900 ft (7,600 m), the ice moves downhill relatively fast, at about 3¼ ft (1 m) per day. As a result, large crevasses can open up with little or no warning. Smaller ones can become buried under the snow, creating treacherous snow bridges through which unwary climbers can fall. Crossing the icefall is a dangerous exercise, but much of the route up and down is along ladders and ropes that are attached each year by experienced guides.

▽ **SPIKY SHARDS**
Seracs (towers of ice) commonly develop on the Khumbu Glacier and can fall at any moment, sending large blocks of ice tumbling down the steep slopes.

▷ **MORNING DESCENT**
A line of climbers can be spotted descending the icefall. Early morning, when the ice is at its most stable, is a favored time for moving on it.

Crevasses in the Khumbu Icefall can be as deep as **165 ft (50 m)** or more

THE KHUMBU GLACIER IN PROFILE

The glacier has two parts: the steeply declining, chaotically crevassed Khumbu Icefall at its top end, which forms part of one of the main routes to Everest; and the gently inclined lower part, which is covered in rock debris and descends down a long, straight-sided approach valley.

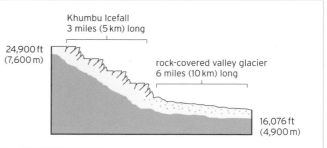

Khumbu Icefall
3 miles (5 km) long

24,900 ft
(7,600 m)

rock-covered valley glacier
6 miles (10 km) long

16,076 ft
(4,900 m)

The Dead Sea

An extremely salty Middle Eastern lake, with rapidly receding shorelines marking Earth's lowest point on dry land

W. Asia

△ **SHORELINES OF THE PAST**
Concentric terraces shows how the shorelines of the Dead Sea have receded over time. As the water levels continue to fall, and the lake shrinks, it becomes saltier.

A landlocked salt lake on the border between Israel and Jordan, the Dead Sea sits in a depression, called a graben, between two tectonic plates in the Jordan Rift Valley. At around 1,410 ft (430 m) below sea level, its shores are the lowest point on Earth's land surface. With a salinity of about 35 percent—up to 10 times that of typical ocean water—the Dead Sea is one of the saltiest bodies of water on Earth and so can support very few life forms, hence its name.

▽ **SALT SUPPLY**
Crystallized salt is found all over the shores of the Dead Sea. Salt from the lake has long been valued for its culinary and medicinal uses.

THE ELEVATION OF THE DEAD SEA

The low elevation of the Dead Sea is due to its position between two tectonic plates. A lake first formed when the land to the west rose up and cut it off from the Mediterranean Sea. As the plates moved apart, the floor of the lake gradually dropped to a level below that of the sea.

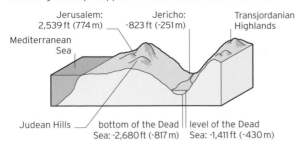

Jerusalem: 2,539 ft (774 m)
Jericho: -823 ft (-251 m)
Transjordanian Highlands
Mediterranean Sea
Judean Hills
bottom of the Dead Sea: -2,680 ft (-817 m)
level of the Dead Sea: -1,411 ft (-430 m)

Shrinking sea

The Dead Sea is mainly fed by the Jordan River, and although it has no outflow, in recent years its surface level has fallen rapidly, largely due to water diversion for commercial use. Receding shorelines have caused the lake to become separated into two basins by the extension of what was previously the Lisan Peninsula. The southern basin has since been divided into a series of evaporation pools for the extraction of salt.

The **level** of the Dead Sea is **dropping** by **more than 3 ft (1 m)** a year

The Ob-Irtysh

*A major river system of northwest Asia, crossing through some
of the most remote and least populated regions on Earth*

NW. Asia

The Ob–Irtysh river system is a combination of two
major Asian rivers that have their sources on different
sides of the Altai Mountains (see p.224) and flow north,
through the Siberian lowlands, to the Arctic Ocean.

Freezing and floods

Measuring a total of 3,460 miles (5,568 km) from the
source of the Irtysh to the mouth of the Ob, the river
system is the longest in Siberia. Its catchment area is
similar in size to the huge Mississippi–Missouri basin
(see p.53). In its lower reaches, the river becomes
braided across a vast, swampy plain that is subject to
seasonal freezing and floods. The Ob–Irtysh empties
through the Gulf of Ob, the world's longest estuary.

▷ **ICE FLOES**
The Ob-Irtysh starts to freeze in the fall from its
mouth, in the north, southward. For more than half
of the year, it is icebound and unnavigable.

N. Asia

The Lena

*An epic trans-Siberian waterway, draining
one-fifth of Russia's territory*

One of the three great Siberian rivers, along with
the Ob–Irtysh and Yenisei, the Lena flows 2,700 miles
(4,400 km) from its source in the Baikal Mountains, just
west of Lake Baikal (see pp.238–39), across Siberia
to its mouth in the Laptev Sea (an arm of the Arctic
Ocean). The Lena has one of the largest basins in
the world, draining one-fifth of Russia's territory. In
its lower reaches, it is free of ice for only 4–5 months
each year, following seasonal flooding.

At its mouth, the river forms the largest delta in
the Arctic and the second largest on Earth. Spanning
170 miles (280 km) and pushing out 70 miles (120 km)
into the ocean, it comprises lakes, channels, sandbars,
and islands, with extensive peat bogs. In winter, it is
a harsh wilderness, but in spring it supports millions
of migratory birds.

◁ **THE LENA PILLARS**
The limestone of these
rock pillars lining the
banks of the Lena in
the Sakha Republic
was formed in a sea
basin around 500
million years ago.

NE. Asia

Lake Baikal

Earth's deepest freshwater lake, containing more water than any other and home to many endemic species

Lake Baikal is a Russian lake situated to the south of the Central Siberian Plateau, near Mongolia. The largest freshwater lake on Earth by volume, it contains around 20 percent of the world's total surface freshwater—similar to that of North America's Great Lakes combined (see pp.50–51). At 395 miles (636 km) long and up to 49 miles (79 km) wide, Baikal has the largest surface area of any freshwater lake in Asia. Its waters have a maximum depth of 5,370 ft (1,637 m), making it the deepest body of freshwater on Earth.

Ancient and dynamic

Also the oldest lake in the world, Lake Baikal formed in a deep rift valley—where the Earth's crust is pulling apart—more than 25 million years ago. The southwest-to-northeast fault along which the lake runs is still active, and the lake is widening by more than ¾ in (2 cm) each year. Lake Baikal's water is renowned for its purity and clarity. Well oxygenated and well mixed throughout its depths, it supports a wide variety of plants and animals, and more than 80 percent of these species are thought to be found only in Lake Baikal.

HALF-GRABEN BASINS

The type of basin in which Lake Baikal sits is known as a half-graben. It was formed through the vertical movement of the Earth's crust along a fault, with the fault bordering just one side. In a full graben, a basin is formed by a depressed block of land bordered on both sides by faults.

sedimentary rock
lake water
bottom of basin
upward movement
footwall
downward movement
border fault
hanging wall

▷ **LATE FREEZE**
Lake Baikal is completely covered with ice for 4–5 months a year. The huge volume of water in its basin means the surface freezes later in the Siberian winter than the region's other lakes. Peschanaya Bay and Baikal's other bays are the first parts of the lake to freeze over.

▶ **THREE BASINS**
Lake Baikal is divided into three basins: North, Central, and South. The deepest parts of the lake are in the Central and South Basins. Both the lake's main source and its single outlet are in the South Basin.

Single outlet
Most lakes have a single outflow into a river system. Lake Baikal drains into the Angara River, a tributary of the Yenisei.

South Basin

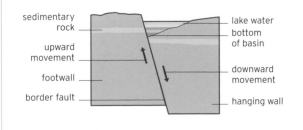

△ **FRESHWATER SEAL**
The endemic Baikal seal is the only exclusively freshwater seal species in the world. One of the smallest seals, it grows up to 4½ ft (1.4 m) in length.

Protected area
Several areas around the lake are protected. The Baikal Nature Reserve by the southeastern shore is part of the World Network of Biosphere Reserves.

Western side
The western side of the lake is flanked by high mountain ranges. The Baikal Mountains rise sharply along the northwestern shore and are the source of the Lena River (see p.237).

Upper Angara River

North Basin

Eastern side
The eastern shoreline slopes more gently than the west. The Barguzin Range is covered in larch forest and features one of the lake's nature reserves on its western slopes.

Svyatov Nos Penisula

Lake Arangatuy

Central Basin

△ **LARGEST ISLAND**
Of Lake Baikal's 27 islands, Olkhon is the largest by far. At 45 miles (72 km) long and 13 miles (21 km) wide, it is the fourth largest lake-bound island in the world.

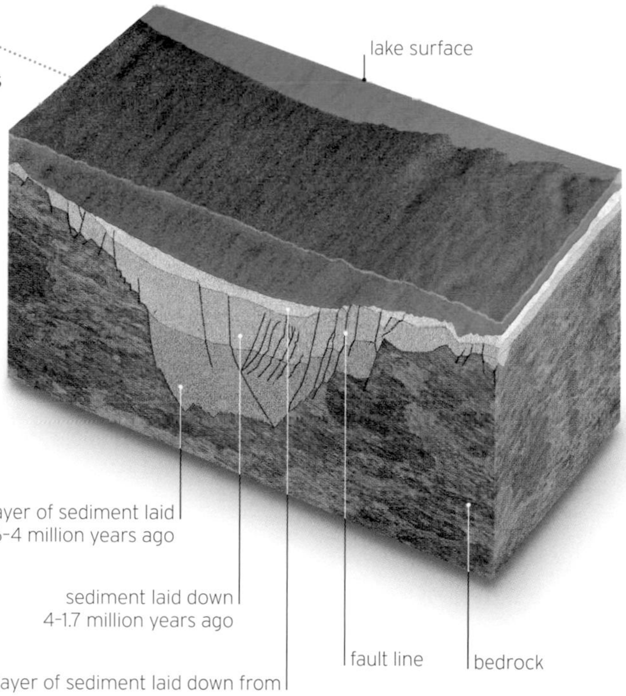

KM MILES

Depth below lake surface

lake surface

bottom layer of sediment laid down 16-4 million years ago

sediment laid down 4-1.7 million years ago

top layer of sediment laid down from 1.7 million years ago to present

fault line

bedrock

△ **DEEP SEDIMENTS**
In parts of Lake Baikal's Central Basin, the sediments on the lake bed are up to 4 miles (7 km) thick. Laid down over millions of years, they contain a record of climatic variation much valued by scientists.

◁ **MAJOR INFLOW**
Most inflow into lakes is surface water that carries a lot of sediment. The Selenga River, Lake Baikal's main inflow, forms one of the world's largest inland deltas as it enters the lake.

Lake Baikal's water is **so clear** that it is possible to see objects **130ft (40m) below the surface**

The Indus

One of Asia's major river systems, crossing the Himalayas and running the entire length of Pakistan

W. Asia

One of the greatest Asian rivers, the Indus is about 1,980 miles (3,180 km) long, and its catchment basin drains an area of more than 425,000 square miles (1.1 million square km).

From its source near Lake Manasarovar on the Tibetan Plateau, the Indus flows northwest through the disputed territory of Jammu and Kashmir, turns southwest to enter Pakistan, and passes through a series of huge canyons near the Nanga Parbat Massif on its journey across the Himalayas. Emerging from the mountains as a fast-flowing river, the Indus descends onto the Punjab Plain, where it is joined by some of its major tributaries and begins to move slowly across a vast floodplain. During the monsoon season, the Indus can grow to a couple of miles wide in places. As it approaches the Arabian Sea, the river branches into the many distributaries that form the Indus River Delta.

▷ **RIVER CONFLUENCE**
In a mountainous region in its upper reaches, the Indus (bottom) meets the Zanskar, a major tributary, near Nimmu in northern India.

The Ganges

One of the world's great waterways, a holy river and a place of pilgrimage

S. Asia

The Ganges is the spiritual and physical lifeblood of India. Considered sacred and worshipped by Hindus, it is thought to support up to one-tenth of the world's population. After the Amazon and the Congo, it is the third largest river on Earth, measured by discharge.

Winding journey

The Ganges flows more than 1,600 miles (2,500 km) across the Indian subcontinent. It forms in the north Indian state of Uttarakhand after the confluence of several ice-fed Himalayan rivers. After passing through narrow mountain valleys, it emerges on the vast Indo-Gangetic Plain to the east of Delhi. From here, it slowly meanders across a wide alluvial plain, passing many centers of holy pilgrimage including Allahabad and Varanasi, toward Bangladesh, where it is joined by the Brahmaputra and Meghna Rivers. The river system subsequently fans out into the largest delta system on Earth, depositing the greatest sediment load of any river into the Bay of Bengal.

△ **BROAD FLOODPLAIN**
The middle course of the Ganges winds sluggishly across a vast floodplain. Over a distance of more than 1,000 miles (1,600 km), the river falls by only about 590 ft (180 m).

△ **COASTAL NETWORK**
Seen here in dark green, the Sundarbans ecoregion stretches inland about 50 miles (80 km). It is largely bordered to the north by farmland.

▷ **EXPOSED ROOTS**
These black mangrove trees have developed aerial roots called pneumatophores, which project into the air during low tide.

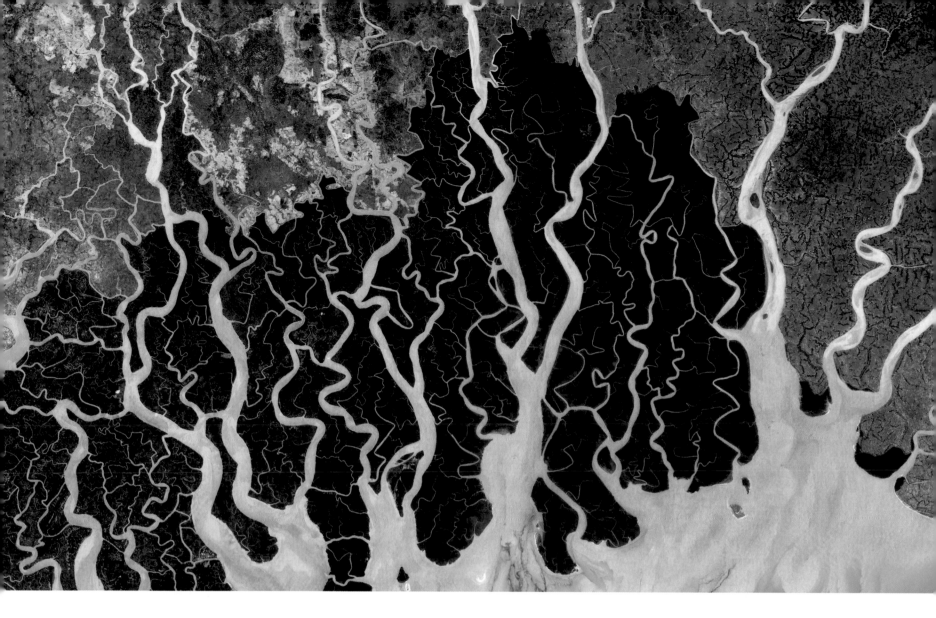

The Sundarbans

A vast stretch of mangrove forest and swampland lying on Earth's largest delta

S. Asia

Stretching more than 160 miles (250 km) across the Bangladesh–India border, where the Ganges Delta merges with the Bay of Bengal, the Sundarbans is a complex network of intertidal waterways, estuaries, mudflats, and islands.

Mangrove forest

The main habitat of the inland parts of the Sundarbans is seasonally flooded freshwater swamp forest consisting of broadleaf trees. The seaward side is dominated by the world's largest continuous tract of mangrove forest. One of the most common mangroves, known as *sundri* or *sundari,* is thought to have given the region its name.

An ecosystem rich in wildlife, the Sundarbans is home to many rare and threatened species, including the Bengal tiger and the Ganges river dolphin.

▽ **VARIED DIET**
As well as eating fish, the black-capped kingfisher varies its diet with shrimp and small crabs.

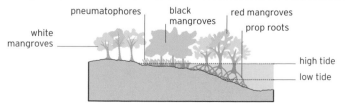

MANGROVE ZONES

There are three main types of mangrove trees, with their own adaptations to particular tidal zones. Red and black mangroves, which live in the tidal zone, have specialized roots to absorb oxygen from the air. White mangroves, common in higher and drier areas, lack these adaptations.

pneumatophores   black mangroves   red mangroves

white mangroves   prop roots

high tide

low tide

The Yellow River

One of Asia's longest rivers, and the world's muddiest, the
"Mother River" of the Chinese people

E. Asia

The most silt-laden river on Earth, the Yellow
River takes its name from its coloration by the fine,
windblown sediment, known as loess, that it carries
along its lower course. It is the second longest river
in China, after the Yangtze, and one of the longest in
the world. After rising high on the Tibetan Plateau,
the Yellow River follows an arching route eastward
across the northern plains of China to the Yellow
Sea, a journey of 3,393 miles (5,460 km).

△ **FIRST BEND**
The Yellow River's upper course in Sichuan
Province forms a huge S-shape, which is
known as the First Bend.

Cradle to grave

Known as both China's Mother River and the Cradle
of Chinese Civilization, the Yellow River historically
sustained one of the country's most fertile and
productive regions. However, the river basin
is extremely prone to flooding, and a series of
devastating floods has also earned the river the
names the Ungovernable and China's Sorrow
(see panel, right). As a result, much of the lower
course of the Yellow River has been ramparted
by levees and many of its tributaries have
been dammed to control its flow.

◁ **WIDE-RANGING TOAD**
The Asiatic toad is common in China, living in a
range of high-humidity habitats that include valleys
and floodplains of the Yellow River. It eats beetles,
bees, ants, and mollusks.

The Yangtze

Asia's longest river, flowing across eastern China from the Tibetan Plateau
to the East China Sea; China's most commercially important river

E. Asia

Known in China as the Chang Jiang—"Long River"—
the Yangtze is the longest river in Asia and the third
longest in the world, after the Nile and the Amazon.
From its glacier-fed source in the Tanggula Mountains
on the Tibetan Plateau, to its mouth in the East China
Sea, it flows 3,900 miles (6,300 km). The catchment area
of the Yangtze covers one-fifth of the area of China and
is home to one-third of the country's population.

Epic journey

On its journey across eastern China, the Yangtze flows
through steep valleys, towering gorges, lake-strewn
plains, and farmland, before forming a delta, where
the megacity of Shanghai lies. Since 2006, the river's
middle course has been regulated by the world's largest
hydroelectric dam, the Three Gorges Dam, which has
created a reservoir 370 miles (600 km) long. The Grand
Canal—the longest human-made waterway on Earth
at 1,100 miles (1,700 km)—links the lower section of
the Yangtze to the Yellow River and Beijing to the north.

△ **GOLD DUST RIVER**
In its upper course, the river passes through the mountainous
Yunnan Province. Here the Yangtze is known as the Jinsha Jiang,
"Gold Dust River," due to the alluvial gold panned from its waters.

◁ **MUDDY WATERS**
The silt-laden Yellow River flows through deep gorges before it descends onto the North China Plain.

CHANGING COURSES

The Yellow River has changed course many times over the centuries, sometimes as the result of human activity. It moved 26 times and flooded nearly 1,600 times between 602 BCE and 1938 CE, resulting in the loss of millions of lives.

Yellow River | Yangtze

KEY

Current channel
Former channel
Former shoreline

SE. Asia

The Mekong

Southeast Asia's greatest river system, passing through six different countries

Like the Yellow River and the Yangtze, the Mekong has its source on the Tibetan Plateau. It runs in a southeast direction through western China, Myanmar, Laos, Thailand, Cambodia, and Vietnam, crossing some national borders and forming others. At its mouth, in the South China Sea, it forms a delta just south of Ho Chi Minh City. With a length of about 2,700 miles (4,350 km), the Mekong is the longest river in Southeast Asia. The very long river basin includes many climatic zones and a diverse range of habitats including upland plateaus, forests, savannas, grasslands, wetlands, and mangroves. As a result, it is an area of extremely high biodiversity.

△ **DELTA FLOODPLAINS**
This aerial view shows part of the huge Mekong Delta in southern Vietnam. The flat floodplains in the south contrast with the hilly areas in other parts of the delta.

▷ **MEKONG GIANT**
The Mekong is home to many large fish species. The giant freshwater stingray can be over 16 ft (5 m) long and 6 ft (2 m) wide, and has the longest stinging spine of any stingray.

stinging spine used in defense

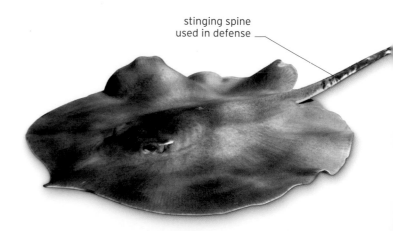

INSIDE A DOLINE
Explorers are dwarfed by the first doline in the cave system, Watch out for Dinosaurs–it is so big a skyscraper could fit inside it.

SE. Asia

Hang Son Doong

A recently discovered limestone cave system of immense scale, situated within jungle-covered mountains in a remote part of Vietnam

One of the largest cave systems discovered on Earth, Hang Son Doong is up to 820 ft (250 m) high, 660 ft (200 m) wide, and stretches a total of 5½ miles (9 km). Located near the Laos border in central Vietnam, it is part of a network of around 150 caves within the Annamite Mountains. First discovered in 1991, Hang Son Doong was unexplored until 2009.

Mountain river cave

Meaning "mountain river cave," Hang Son Doong has been carved out of some of the oldest limestone in Asia by the fast-flowing Rao Thuong River over the last 2–5 million years. Giant openings (dolines), created as massive sections of the ceiling collapsed, have allowed areas of vegetation to develop, including trees up to 100 ft (30 m) tall. These are dwarfed by some of the stalagmites found within the cave, which, at more than 245 ft (75 m) high, are the largest on Earth.

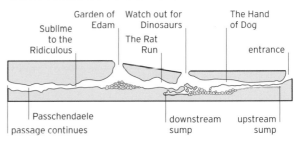

CHAMBERS AND PASSAGES

The Hang Son Doong system is straight, relatively flat, and lacks side passages. This is due to the fact that it has formed along a fault in the limestone. It has many discrete sections and remarkable features, often with evocative names. These include the Hand of Dog, a huge, dog-paw-shaped stalagmite, and the two forested areas, Watch out for Dinosaurs and the Garden of Edam.

Garden of Edam — Watch out for Dinosaurs — The Hand of Dog
Sublime to the Ridiculous — The Rat Run — entrance
Passchendaele passage continues — downstream sump — upstream sump

Hang Son Doong is **so large** that it has its **own microclimate,** including **wind and clouds**

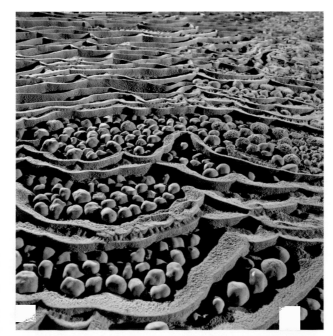

◁ **EARTHLY TREASURES**
Calcite deposits called cave pearls are found in parts of the cave. Usually less than ½ in (1 cm) wide, here they are the size of oranges.

△ **GREEN COVER**
Around the borders of the open sections, where dim light reaches the ground, colonies of ferns and algae cover terraces of limestone.

E. Asia

South China Karst

One of the world's largest and finest examples of a karst landscape, featuring a variety of dramatic rock formations

Covering an area of about 190,000 square miles (500,000 square km), mostly within the provinces of Guizhou, Guangxi, Yunnan, and Chongqing, South China karst is the world's largest karst terrain—a landscape characterized by an array of weathered limestone formations. The diversity, number, and spectacular nature of the geological features make South China karst the greatest example of a humid tropical and subtropical karst landscape on Earth. Since 2007 it has been a UNESCO World Heritage Site.

forests (see pp.248–49). In fact, the region is regarded by geologists as being home to the defining examples of formations such as tower karst (fenglin), cone karst (fengcong), and pinnacle karst (shilin), which creates dense stone forests (see below). It also features many giant sinkholes (or dolines), gorges, natural bridges, and table mountains. Beneath the ground there are extensive cave systems containing subterranean rivers and vast caverns rich in stalactites, stalagmites, and other cave deposits.

Caves, towers, and pinnacles

South China karst is perhaps best known for its isolated stone towers, often rising more than 330 ft (100 m) from a broad mosaic of farmland, and its clusters of cone-shaped hills covered in lush, cloud-draped

▷ **TRAINED TO FISH**
Cormorants are used in a traditional method of fishing on the rivers that cross South China karst. The great cormorant is a favored species.

Mature landscape

The limestone rock in which South China karst is formed was laid down on the ocean floor over a long period before being uplifted. This produced an exceptionally thick, strong, and horizontal bedrock from which the massive and relatively stable landforms have been carved. Although the region's warm and wet climate has enabled the chemical weathering of the limestone to occur at a relatively rapid rate (see pp.248–49), it has been estimated that the fengcong and fenglin karsts of Guizhou and Guangxi have been evolving over a period of 10–20 million years. The result is what is often described as the ultimate mature karst landscape.

One of the region's **vast caves** houses a village of more than **100 people**

▷ **STONE FOREST**
These tightly bundled limestone pinnacles forming stone forests near the village of Shilin, in Yunnan Province, are considered classic examples of what has become known as shilin karst.

△ COLLAPSED CAVE

Moon Hill, a huge stone arch near Yangshuo in Guangxi, is all that is left of what was once a limestone cave. It is now a popular climbing site.

SOUTH CHINA KARST TYPES

Often viewed as a precursor to the development of fenglin karst's high towers, fengcong karst dominates areas like the Yunnan-Guizhou plateau in the north and west of the region, where the bedrock has been more recently uplifted and subjected to a new wave of erosion. Fenglin karst is mainly found in areas that have been stable over long periods, such as the lowlands of Guangxi in the south and east.

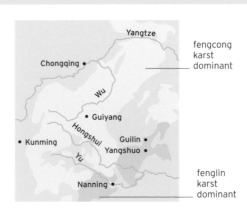

△ BROAD PLAINS

Although the plains between the towering formations of South China karst have been farmed for hundreds of years, the relatively thin soils and lack of water retention in some areas present major challenges to farmers.

The formation of a
KARST LANDSCAPE

Karst landscapes are formed by the weathering of a terrain's underlying carbonate rock, such as limestone or dolomite. Through a process known as dissolution, rain, surface water, and groundwater made acidic by carbon dioxide in the atmosphere and organic matter in soil start to dissolve the rock down cracks and faults, and along its bedding planes. As the rock degrades over time and the cracks become larger, a variety of surface features are produced and an underground drainage system may start to form.

Underground erosion

Percolating water may create subterranean rivers that eventually carve out extensive cave systems. Sinkholes, or dolines, a defining feature of karst landscapes, either develop gradually as surface openings enlarge or are formed suddenly by cave collapses. As they expand and merge, they create sunken regions, and any high ground left between them becomes linked cone-shaped hills, such as the fengcong formations of South China karst. Over time, erosion by surface water can transform these hills into the isolated towers of fenglin karst. The formation of karst landscapes is helped by moist climates and lush vegetation—providing carbon dioxide as it decays—so the most developed karst terrains occur in tropical regions.

The **bedrock** of South China karst **formed on the seabed** more than **250 million years ago**

▶ **SOUTH CHINA KARST**
This cross-section of a karst landscape combines many of the features present in South China karst. It represents a well-developed karst system that has been created under perfect geological and environmental conditions.

△ **CAVE DWELLER**
Caves provide shelter for some animals. Large colonies of great roundleaf bats roost in South China karst caves.

Fenglin tower
Fenglin towers are steep-sided and stand isolated on the flat karst plain.

land converted to agriculture

river meanders on surface

sinkhole

Sediment
A layer of sediment acts as a barrier between the river and the limestone, enabling the river to stay above ground.

Foot cave
Water streams from a cave, known as a foot cave, at the base of a conical hill.

Losing stream
A losing stream first runs along the surface before being captured by an underground drainage system.

Gour
A series of stone dams called gours, formed by mineral deposits, has created a tiered cascade pool.

EVOLUTION OF KARST FEATURES

Karst features can be seen to evolve in a sequence from a flat karst plain to fengcong then fenglin karst. Once a drainage system is established, sinkholes get bigger and join together, producing the hills of fengcong karst. If surface water erodes their bases and the sunken areas between them, they develop into the isolated towers of fenglin karst.

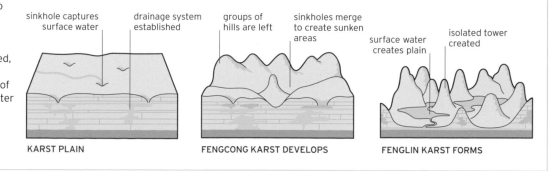

sinkhole captures surface water

drainage system established

KARST PLAIN

groups of hills are left

sinkholes merge to create sunken areas

FENGCONG KARST DEVELOPS

surface water creates plain

isolated tower created

FENGLIN KARST FORMS

◁ **TREE-COVERED PEAK**
Patches of monsoon forest cling to the steep and craggy limestone rock faces of even the tallest fengcong cones. The forests of the South China karst region are key habitats for several species of rare and endangered animals.

Fengcong cone
Fengcong karst is characterized by linked conical hills that share a common bedrock and form a continual terrain. The fengcong cones of South China karst reach more than 980 ft (300 m) high.

karst plain

remnant of old cave

△ **DEEP DEPRESSION**
Sinkholes, or dolines, are depressions in the ground often caused by the rock layer above a cave network suddenly collapsing. Sinkholes vary greatly in size, with some of those in South China karst reaching widths and depths of more than 2,000 ft (600 m).

Limestone rock
The limestone bedrock of South China karst is extremely thick and broken into blocks by bedding planes, joints, and faults.

Fissure
A fissure created by vertical faults in the limestone allows water from the surface to feed into the cave below.

submerged cave

dry cave, now without running water

◁ **CAVE DECORATIONS**
Over time, underground caves formed within karst terrains will often become decorated with calcite deposits, known as speleothems. Many of the ancient caves in South China karst are renowned for their elaborate speleothems, including stalactites and stone pillars.

Underground river
Fed by losing streams and water seeping through sinkholes and fissures, an underground river continues to carve out a passage in the limestone.

rock debris from collapsed roof

debris washed in by stream in past

Ha Long Bay

A large inlet into the Gulf of Tonkin, containing a scattering of hundreds of spectacular karst islands

SE. Asia

Covering an area of about 600 square miles (1,500 square km) off the coast of Vietnam, Ha Long Bay was created about 8,000 years ago when sea-level rise drowned an existing karst landscape. Karst consists of limestone blocks that have been partially dissolved by rainwater over tens of millions of years. In tropical areas, it typically creates conical hills and towers with internal caves and sinkholes (see panel, right). When the sea flooded into Ha Long Bay, these hills and towers were turned into some 1,600 islands, most of which are fringed by vertical cliffs. There are hundreds of caves within the islands, but none is long or deep because their extent is limited by the sizes of the islands. Because of their precipitous nature, most of the islands are uninhabited by humans. However, several communities live on the water in floating houses, sustaining themselves by fishing and aquaculture. The bay's shallow waters are biologically highly productive and support hundreds of species of fish, mollusks, and crustaceans. Animals living on the islands include lizards, monkeys, bats, and many bird species.

◁ **PERILOUSLY LOW**
Cat Ba, a relatively large island on the southwestern edge of the bay, is home to the Cat Ba langur, a critically endangered monkey, with a population estimated at less than 100.

△ **INTRICATE SHORELINES**
Many of the islands in the bay have convoluted shapes. Some have been given distinctive names, such as Man's Head Island, Rocky Dog, and Fighting Cocks Islet.

Some caves have **tunnels** leading into **inland lakes** that have **no other access**

CONES, TOWERS, AND CAVES

Much of the Ha Long Bay karst consists either of large, cone-shaped blocks (known as fengcong), or isolated, towerlike islands (called fenglin). Cave types include wave-eroded notch caves at sea level, horizontal foot caves just above sea level, and larger, much older caves called phreatic caves, which are well above sea level.

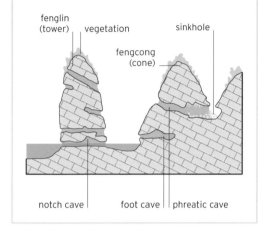

fenglin (tower)   vegetation   sinkhole

fengcong (cone)

notch cave   foot cave   phreatic cave

▽ TRANQUIL WATERS
A group of kayakers ply their way between karst islands that rise up to 660 ft (200 m) above the waters of the bay.

Shiraho Reef

A Japanese coral reef of world-class importance in terms of biodiversity, home to a unique type of coral

SE. Asia

Stretching to 2 miles (3 km), Shiraho Reef first came to notice in the 1980s as an example of biodiversity, with some 120 species of corals and 300 fish species. It also contains the world's largest colony of an unusual species of coral known as blue coral. This belongs to a group known as octocorals, whose colonies mostly have flexible, branching skeletons. However, blue coral has a hard, rigid skeleton. Unfortunately, the reef's coral coverage has greatly decreased in recent years, but efforts are being made to conserve it.

◁ BRANCHING CORAL
The vertical plates of blue coral can form massive colonies. Despite its name, its color varies from blue through turquoise and green to yellow-brown.

Dragon Hole

The world's deepest underwater sinkhole, located on a reef in the South China Sea

SE. Asia

Dragon Hole is just over 980 ft (300 m) deep. Its great depth was realized only when it was measured for the first time in 2016—previously, a sinkhole called Dean's Blue Hole in the Bahamas, which is 663 ft (202 m) deep, held the record for the world's deepest blue hole. Blue holes are sinkholes that form through erosional processes in blocks of limestone—while the limestone is on dry land and before the holes are flooded by sea-level rise. Local fishermen believe this is where the Monkey King found his golden cudgel in the 16th-century Chinese novel *Journey to the West*.

▷ BLUE EYE
Viewed from above, the hole looks dark blue due to its great depth compared to the light blue of the water on top of the reef.

SE. Asia

The Krabi Coastline

An area of southern Thailand notable for its fantastic karst formations

The limestone rock of the Krabi coastline originally formed about 260 million years ago. At that time, a shallow sea covered what is now south Asia and slowly built up deposits of shells and coral fragments. These eventually formed layers of limestone, which were later pushed upward and tipped over when India began to collide with Eurasia some 50 million years ago. Around Krabi and Phang Nga Bay to its north, chemical erosion of these limestone layers by rainwater (which combines with dissolved carbon dioxide to form carbonic acid), followed by a rising sea level, created a landscape of thousands of craggy karst islands and hills. These include a number of isolated towers that rise to heights of up to 690 ft (210 m).

HOW THE COASTLINE FORMED

The Krabi coastline is a drowned karst landscape—one in which an existing landscape of limestone hills and towers became inundated by the sea (see panel, p.248).

group of karst hills

seawater

cave

isolated karst tower

△ **HIDDEN PARADISE**
At the island of Phi Phi Le, the drowning of the coastline has created a hidden lagoon and secret beach.

▷ **HANGING ROCK**
Near the entrance to a small cave, stalactites covered in tropical vegetation hang from an eroded mass of limestone.

Andaman Sea Reefs

Some of the largest continuous areas of coral reefs in south Asia, with an enormous variety of fish, corals, and other invertebrates

SE. Asia

Situated in a northeastern corner of the Indian Ocean, the Andaman Sea bathes the coasts and coastal islands of Thailand and Myanmar on its eastern side and the Andaman and Nicobar Islands on its western side. Most of the coral reefs here are fringing reefs, and altogether they cover an area of some 2,000 square miles (5,000 square km). The reefs contain a wide diversity of marine life, with over 500 fish species and 200 coral species recorded. The reefs and nearby islands are also important feeding and breeding grounds for endangered sea turtles.

△ **COMB CORAL**
This soft coral is known for its distinctive comblike shape. It has long, unbranched stems and can grow up to 5 ft (1.5 m) wide.

Nusa Tenggara

A chain of coral islands in southern Indonesia, fringed by reefs that contain a high diversity of marine life

SE. Asia

Stretching eastward from Bali, Nusa Tenggara includes several large islands such as Lombok, Sumbawa, and Flores, as well as 500 smaller islands. Many of the fringing reefs around these islands have barely been explored, but as a whole the area is known to include around 500 species of reef-building corals. Common animals seen on the reefs include eagle rays, humphead parrotfish, and various species of nudibranch (sea slug) and octopus.

△ **DENSE HABITAT**
Just beneath the surface, a mass of stony corals covers a fringing reef a few hundred feet off the coast of Flores in eastern Nusa Tenggara.

Siberian Taiga

A massive forest on the edge of the Arctic, ruled by conifers, where winter lasts most of the year

N. Asia

The vast boreal forest known as Siberian taiga covers around 2.6 million square miles (6.7 million square km) from the Ural Mountains east to the Pacific Ocean. It stretches from northern Russia's Arctic fringes, where the taiga merges into the tundra, as far south as Mongolia. Like all boreal forests, Siberian taiga is primarily a conifer habitat, beginning as a mixture of spruce, pine, and larch in the west before giving way to unbroken swathes of larch in its eastern Russian and Mongolian reaches.

Changing canopy

In more southerly areas, dense tree growth results in "closed canopy" forests, with little light penetrating to the moss-covered forest floor beneath. However, with increasing distance northward, the forest cover becomes sparser, and lichen replaces moss on the ground. In more temperate parts, dominant conifers become mixed with broadleaf species such as birch, aspen, and willow.

▷ PINPOINTING PREY
The great gray owl mainly eats tiny rodents, which it locates with exceptional hearing—pinpointing them even under snow.

In the east, marsh and bogland support understory shrubs such as cranberry and bilberry, while in the west, poor drainage and permafrost mean such areas are mainly given over to spongy, shallow bogs with no tree cover at all.

Cold climate

Winter in the Siberian taiga lasts six to seven months, with the climate dominated by cold Arctic air. The average temperature range is broad—from 65°F (-54°C) in winter to 70°F (21°C) in summer—but extremes of -76°F (-60°C) and 104°F (40°C) are not uncommon, and the yearly average is below freezing.

The taiga contains **as many trees as all the rain forests** combined

◁ SNOW FOREST
Snow lasts into May in the taiga, returning in September or earlier. However, snow-covered trees serve as an insulator, keeping the ground warmer than the air above the forest.

△ CHANGING TERRAIN
In many parts of the taiga, conifers cluster around rivers, where richer soils offer better growing conditions. Forested sections are divided by mountains at higher elevations.

FROM TAIGA TO TUNDRA

Taiga tree growth varies greatly according to latitude. In southern areas, trees are taller and more closely packed together, but density and height both diminish toward the north. Eventually, only a few small trees survive before the landscape changes entirely from taiga to tundra.

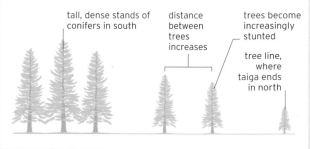

tall, dense stands of conifers in south

distance between trees increases

trees become increasingly stunted

tree line, where taiga ends in north

Eastern Himalayan Forest

A wealth of trees growing in the world's highest mountain range, supporting a wide variety of plant and animal life

S. Asia

The eastern Himalayas region is characterized chiefly by temperate broadleaf evergreen and deciduous forests. These cover an area of about 32,000 square miles (83,000 square km) at altitudes of 6,600–10,000 ft (2,000–3,000 m) and stretch from central Nepal eastward through Bhutan into northeast India.

Plant treasury

This is an ecoregion rich in plant life, where tree type varies according to altitude and geography. In the temperate evergreen forests, oaks and other trees such as magnolia and cinnamon grow alongside thickets of rhododendrons—some areas, such as Bhutan, may contain up to 60 species of rhododendrons alone. Temperate deciduous forest areas are dominated by maples, birches, and walnuts, which give way to magnolias and maples in the wetter eastern reaches of Nepal, together with more tropical shrubs such as schefflera and pockets of bamboo. At least 125 mammal species are found in the ecoregion, including some, such as the Namdapha flying squirrel, that live nowhere else in the world. Threatened species include the stump-tailed macaque and clouded leopard.

▷ **FLORAL DISPLAY**
Rhododendron blooms dot south-facing slopes, where they flourish alongside oaks, epiphytic orchids, ferns, and mosses.

▽ **RAINBOW PLUMAGE**
The Himalayan monal pheasant is one of around 500 bird species that live in the forest. The iridescent male is the national bird of Nepal.

Taiheiyo Montane Forest

High and hilly, one of Japan's seven forest ecoregions, covering parts of three islands with both broadleaf and fir trees

E. Asia

Hardy deciduous trees mixed with firs and bamboo characterize Taiheiyo montane forest. Running along the Pacific side of Japan's main island of Honshu, as well as Shikoku and Kyushu islands, this temperate ecoregion covers an area of about 16,200 square miles (42,000 square km).

In this forest, the climate is humid all year round, but trees must withstand sharp seasonal temperature changes. Winters are cold and snowy, with average temperatures below freezing, but during summer, they rise to 77°F (25°C) or more. Beech and fir are dominant species, growing alongside maple and oak, with an understory of sasa, a type of dwarf bamboo. The forest's seeds, nuts, and bark provide food for mammals such as the Asiatic black bear and sika deer, and birds such as the Oriental white-eye, all of which help replenish the forest by dispersing seeds.

◁ **PINK TRAILS**
The Taiheiyo's beech and maple trees are joined in some areas by other hardwoods such as oaks and Japanese cherries, which stand out against pockets of firs.

Upper Yangtze Forests

A landscape of evergreen and deciduous forests between the Yangtze and Yellow rivers, home to China's iconic giant panda

E. Asia

The forests of the Upper Yangtze extend eastward from the Hengduan Mountains and across the provinces of Sichuan and Shaanxi in south-central China. Covering an area of 150,600 square miles (390,000 square km), this ecoregion comprises three subregions: the evergreen broadleaf forests of the Sichuan Basin, the Daba Mountains evergreen forests, and the deciduous forests of Qinling.

The forests in the north of the region are made up of deciduous and mixed conifer species that thrive in the area's cooler, more temperate climate. The forests of the lower-lying Qinling Mountains support a dense bamboo understory, providing food and shelter for species such as the rare giant panda. In the warmer Sichuan Basin, subtropical evergreen broadleaf trees flourish, and it is here that the dawn redwood, a deciduous conifer previously known only from fossil remains, was discovered growing in the 1940s.

▽ **AUTUMN LEAVES**
In autumn, the oak, walnut, and maple trees of the Qinling Mountain foothills create a colorful patchwork landscape. The area is also home to the primitive ginkgo, whose leaves turn golden in autumn.

CLIMATE

Due to its higher altitude, the climate in the Upper Yangtze is generally temperate, with mild winters, but temperatures in the lower-lying Sichuan Basin can reach 84°F (29°C). Humidity in the entire region is generally high.

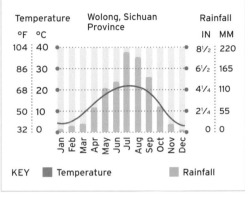

Temperature — Wolong, Sichuan Province — Rainfall

KEY ■ Temperature ■ Rainfall

Together, the Upper Yangtze forests contain one-fifth of all China's mammal species

▽ **BAMBOO DIET**
Once common in lowland areas, the giant panda has moved to the mountains due to human expansion. As bamboo makes up 99 percent of its diet, it can only live where bamboo forests thrive.

SE. Asia

Bornean Rain Forest

A threatened habitat governed by three nations, home to more than 15,000 plant species and one of the oldest and most biodiverse rain forests on Earth

Shared by Malaysia, Brunei, and Indonesia, the island of Borneo is home to Asia's largest rain forest, which is also one of the world's most ancient at 130 million years old—around 70 million years older than the Amazon Rain Forest (see pp.120–21). Borneo's rain forest is hugely biodiverse: while the island amounts to just 1 percent of Earth's landmass, its rain forest contains around 6 percent of the planet's plant and animal species.

A wealth of hardwood

The plants include a family of tropical hardwood trees called dipterocarps, many of which reach heights of 200 ft (60 m). Most of the 600-plus species of dipterocarps grow in Southeast Asia at elevations up to 3,300 ft (1,000 m), and Borneo's lowland rain forests contain more of these trees than any other location. The island's 270 species of dipterocarps include the highly prized Borneo ironwood, whose wood is so dense it never needs treating. Borneo's rain forests, including those covering its more mountainous interior, are also rich in many other types of life. More than 360 new plant species have been discovered here

since 1995, and the island is home to more than 1,400 species of amphibians, mammals, birds, reptiles, and fish—many found nowhere else.

Rain forest under threat

However, this species treasury is in peril. Until the early 1970s, more than three-quarters of Borneo's 287,000 square miles (743,330 square km) were covered in thick tropical rain forest that was densest in the lowlands. Since then, at least one-third has been destroyed. Fires and the planting of oil palm plantations are partly to blame, but it was demand for the highly valuable dipterocarps that resulted in logging on an industrial scale, particularly in the Malaysian states of Sabah and Sarawak in the north of the island, where an estimated 80 percent of rain forest has been lost. To preserve this habitat, in 2007, 85,000 square miles (220,000 square km) of rain forest in the center of the island were designated as a protected area, known as the Heart of Borneo.

◁ **CALLING CARD**
The nocturnal tree-dwelling tokay gecko is just one of many reptiles in Borneo and is named for its "to-kay!" call. It grows up to 16 in (40 cm) long.

△ **PROTECTED TREES**
The crowns of the tallest trees, known as emergents, rise above the morning mist that covers the rest of the rain forest in the Danum Valley, a conservation area of lowland dipterocarp forest in Sabah.

Up to **240 tree species** can be found in just **2½ acres (1 hectare)** of **Borneo's rain forest**

HOW A STRANGLER FIG GROWS

In most cases, a sticky strangler fig seed germinates high on a branch of a rain forest tree, where it was left by a monkey, bird, or bat that had eaten the fruit. The seedling sends long roots down the trunk of its host to enter the soil. Eventually, a scaffolding of roots encases the trunk, and the host's root system must compete with the strangler fig's. The fig's thick foliage also shades the host tree's crown. Ultimately the host dies, leaving the fig tree behind.

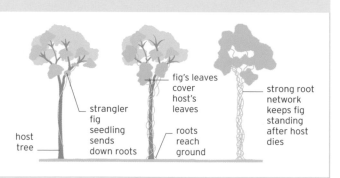

host tree

strangler fig seedling sends down roots

roots reach ground

fig's leaves cover host's leaves

strong root network keeps fig standing after host dies

▷ **LIFE SUPPORT**
Borneo's rain forests support many highly endangered animals, including the critically endangered Bornean orangutan, which depends on dipterocarp trees for food and shelter.

◁ **GIANT ORCHID**
The queen of orchids, believed to be the world's largest orchid, may be found growing in a fork of tall trees. This giant orchid is one of Borneo's more than 1,700 orchid species.

The Terai-Duar Savannas

A fertile band at the base of the Himalayas, where Earth's tallest grasses grow

S. Asia

Forming a narrow belt of savanna and marshy grassland dotted with remnants of old forests, the Terai-Duar stretches from the southern Himalayan foothills in Nepal eastward to Bhutan and India. Several rivers, including the Ganges, flow into the region, creating vast alluvial fans of sand, silt, and gravel—ideal for grass and reed growth.

A forest of grass

The Terai's grasses are the tallest in the world, with some species, collectively known as elephant grass, reaching heights of 22 ft (7 m) or more. These grass forests, as they are known, provide perfect cover for hoofed mammals such as swamp deer, pygmy hogs, and buffalo, and the region's national parks are a sanctuary for the Indian rhinoceros and Bengal tiger. The ecoregion also supports mugger crocodiles and rare gharials, as well as many species of birds, including three species found nowhere else on Earth.

△ **FIRST FLOWERS**
Kans grass is the first grass to germinate after the floodwaters retreat in the Terai-Duar.

The Eastern Steppe

The largest intact temperate grassland on Earth, encompassing a vast, unbroken, windblown plain and prone to seasonal temperature extremes

E. Asia

Part of the huge Eurasian Steppe, the Eastern Steppe is dominated by grasslands across most of its 342,600 square miles (887,330 square km). It extends from southern Siberia to the coastal hills of northeast China and from the Altai Mountains east to the Greater Khingan range.

Harsh land of extremes

This is a much harsher region than its counterpart, the Western Steppe (see p.175), so it has not been subjected to as much agricultural activity. However, the grazing of sheep and goats remains a constant threat to what is a fragile and easily degraded habitat. The region receives just 10–20 in (250–500 mm) of precipitation a year, but hardy grass species manage to grow, despite a temperature range of -4°F (-20°C) in winter to 104°F (40°C) in summer. At higher elevations, rainfall increases, and snow accumulated on mountain peaks descends as streams into the dry lands below.

◁ **STEPPE COUNTRY**
The Eastern Steppe's vast grasslands in Mongolia look empty, yet are rich enough to feed a diverse range of wildlife, from marmots and grazing animals to many birds of prey.

CLIMATE

Heat and humidity characterize the Terai-Duar all year round, but temperatures often rise to 104°F (40°C) in the late dry season. Annual monsoon rains result in floods throughout the region, renewing vital deposits of fertile silt.

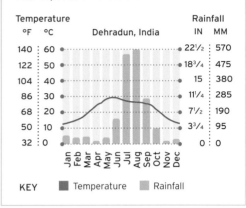

Temperature			Rainfall	
°F	°C	Dehradun, India	IN	MM
140	60		22½	570
122	50		18¾	475
104	40		15	380
86	30		11¼	285
68	20		7½	190
50	10		3¾	95
32	0		0	0

Jan Feb Mar Apr May Jun Jul Aug Sep Oct Nov Dec

KEY ■ Temperature ■ Rainfall

◁ **RIVER CROSSING**
A rare Indian rhinoceros cools off in a river in Nepal's Royal Chitwan National Park. Of all five rhino species, this one is most at home in water.

Siberian Tundra

A cold and often frozen landscape lying mostly above the Arctic Circle, where life persists even in the most inhospitable of conditions

N. Asia

Northeastern Russia's Siberian tundra ranges from the edge of the taiga in the south (see pp.254–55) through the central region of the north coast and eastward as far as the Chukchi Peninsula.

Long winters where temperatures plunge to -40°F (-40°C) are the norm in Siberian tundra, while summers are short and cool, warming to just 54°F (12°C). Winds blow at speeds of up to 60 mph (100 kph), and the frozen soil layer known as permafrost (see p.175) can be as thick as 2,000 ft (600 m). Yet clumps of grasses, fungi, and low-growing shrubs manage to survive, providing a vital source of nourishment for swarms of insects and migrating bird and mammal species.

▷ **FIELDS OF COTTON**
Arctic cotton grass grows in wet ground and blooms during the short summers, when it provides food for reindeer calves.

▷ **SNOW SURVIVOR**
Insulated by two coats of hair, reindeers survive on sedges, mosses, and lichens, which they find even beneath the snow.

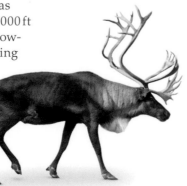

The Arabian Desert

Asia's largest desert, shared among nine nations, contained within a ring of mountains, and home to one of the largest continuous expanses of sand on Earth

W. Asia

Covering 900,000 square miles (2.3 million square km), the Arabian Desert is so large that it spans almost the entire Arabian Peninsula. The bulk of it lies within Saudi Arabia, but it also extends southwest into Yemen, southeast into Oman, east along the Persian Gulf into the United Arab Emirates and Qatar, northeast into Kuwait and Iraq, and northwest into Jordan, just nudging into the tip of Egypt.

Region of extremes

With vast stretches of sand in the south, gravel or salt-encrusted plains (*sabkhahs*) in the east, volcanic fields to the west, and temperatures ranging from 122°F (50°C) on summer days to below freezing on winter nights, this environment at first seems too harsh for survival. However, a surprising number of creatures live here, particularly insects such as locusts and dung beetles, as well as spiders and scorpions. These serve as food for many snakes and lizards, while mammals, including jerboas, goats, and gazelles, find enough vegetation in oases and fringe areas to sustain them.

Beneath the desert surface lies a huge groundwater reserve that was trapped here during the Pleistocene Epoch, 2.6 million–11,700 years ago. This, together with the area's vast oil reserves, has been tapped in recent years: the former used for irrigation, the latter for extraction of petroleum products.

both sexes have long, ringed horns

wide, shovel-shaped hooves aid walking on loose sand

▷ **DESERT SURVIVOR**
Once extinct in the wild, the rare Arabian oryx has been reintroduced and now roams the desert, foraging on grasses and roots. Its brilliant white coat reflects the worst of the desert Sun's heat.

◁ **SHIFTING SEA OF SAND**
In the south of the desert lies Ar Rub 'al Khali, or the Empty Quarter, an area of sand roughly the size of France. Strong winds called *shamals* move huge amounts of sand twice a year, reshaping the dunes.

HOW A STAR DUNE FORMS

While most dunes are crescent-shaped, almost 10 percent of the planet's sand dunes are star-shaped. Winds blowing from several directions push piles of sand together, forming three or more "arms" of sand radiating from a high central point. The star dunes of the Arabian Desert are found in the eastern part of the Empty Quarter.

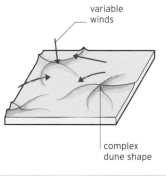

variable winds

complex dune shape

W. Asia

The Dasht-e-Lut

A salt desert with spectacular dunes and one of the hottest places on the planet

Southeastern Iran's Dasht-e-Lut, or Lut Desert (*lut* means "completely barren"), is a torrid salt desert. In a satellite study of global land surface temperatures, the Lut's readings were the hottest for five out of seven years. In 2005, a temperature of 159.3°F (70.7°C) was recorded.

Black sand, high heat

The extreme heat is due partly to the desert's geology. Much of the Lut's surface is made up of black sand—mainly magnetite, a type of iron ore. Magnetite absorbs more radiant energy than light-colored sand, causing the surface temperatures to rise. Other distinguishing features of the Lut include yardangs. These huge, high rock ridges created by seasonal wind erosion give the desert a corrugated appearance, particularly on its western edge.

△ **LUNAR LANDSCAPE**
Southern Jordan's Wadi Rum, or the Valley of the Moon, reveals a volcanic past that created granite and sandstone mesas rising 2,620 ft (800 m) above the desert's surface.

Average rainfall in the Arabian Desert is less than 4 in (100 mm) a year

△ **MOBILE DUNES**
In the eastern part of the Lut's estimated 20,000 square miles (51,000 square km), desert winds shape the more sandy areas into massive mobile dune fields.

The Karakum Desert

East of the Caspian Sea and south of the Aral Sea, a sparsely vegetated desert of three terrains—and a crater that has burned for decades

C. Asia

The Central Asian Karakum Desert covers about 135,000 square miles (350,000 square km), occupying around 70 percent of Turkmenistan.

Three deserts in one

Southern salt marshes give way to a mineral-rich central plain, then to barren, higher northern areas, constantly eroded by winds. The center contains sand ridges that are 245–295 ft (75–90 m) high, as well as barchans (crescent-shaped dunes) and clay-lined depressions where the average annual rainfall of 3–6 in (70–150 mm) accumulates to form temporary lakes. Grasses, wormwood, and saxual trees provide food and shelter for wildlife.

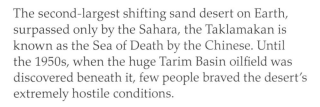

yellow-brown coat provides camouflage

blunt claws aid digging

▷ **DOOR TO HELL**
In 1971, a crater created by a Soviet drilling accident released toxic methane gas near Darvaza village. The fire that was lit to burn off the gas has never gone out.

◁ **PROTECTED FEET**
The sand cat's foot pads are covered in thick fur, helping it walk on the hot, rocky terrain of the Karakum.

The Taklamakan Desert

China's largest desert, composed of fields of shifting dunes that are sporadically convulsed by tornado-like sandstorms

C. Asia

The second-largest shifting sand desert on Earth, surpassed only by the Sahara, the Taklamakan is known as the Sea of Death by the Chinese. Until the 1950s, when the huge Tarim Basin oilfield was discovered beneath it, few people braved the desert's extremely hostile conditions.

Black hurricanes

Constantly shifting dunes (see panel, right) make the 123,550 square miles (320,000 square km) of the Taklamakan difficult to navigate, despite a series of isolated oases scattered at the base of the Tian Shan foothills on its northern border. Mobility aside, the dunes themselves are formidable: some dune chains are up to 1,640 ft (500 m) wide and 490 ft (150 m) high, and the distance between the chains varies from ½ to 3 miles (1 to 5 km). Fierce winds also cause *karaburans*, or "black hurricanes": sudden violent storms of sand and gravel so dense that they block out sunlight, resulting in dust clouds that rise as high as 13,000 ft (4,000 m).

Although several rivers feed into the desert, bringing snowmelt from the mountains that surround it on three sides, actual precipitation is minimal—only ½–1½ in (10–38 mm) a year. What groundwater there is lies just 10–16 ft (3–5 m) below the surface, and even that is increasingly being compromised by the transportation of oil and gas out of the Tarim Basin via a highway constructed in 1995.

Although classed as a **cold desert**, Taklamakan **summer temperatures** can soar to **100°F (38°C)**

The Thar Desert

The most densely populated desert in the world, named for its iconic sand ridges accumulated over the past 1.8 million years

C. Asia

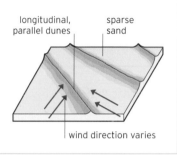

▷ **AT RISK**
The Thar Desert provides a haven for the Egyptian vulture, endangered elsewhere partly due to habitat destruction.

The Thar, or Great Indian Desert, belongs to two nations. Pakistan lays claim to around 15 percent, but the greater part lies in India, mainly in Rajasthan.

Undulating dunes give the Thar Desert its name, derived from *thul*, the local word for "sand ridges." The region is made up of metamorphic gneiss, sedimentary rocks, and alluvial deposits that, over about 1.8 million years, have been covered by windborne sand. Parts of the Thar consist of sandy plains, others of continually shifting dunes. Barren hills and vast saline lakebeds are also scattered throughout the desert.

Seasonal dry southwesterly winds blow from March to July, when temperatures rise to 122°F (50°C). From July to September, monsoons bring an average 4 in (100 mm) of rainfall to the western Thar and as much as 20 in (500 mm) to the east. This is enough to support sparse scrub and grasses, as well as drought-tolerant tree species such as acacia. Grasses provide food for mammals and birds, including the critically endangered great bustard.

△ **SEA OF SAND**
Huge crescent-shaped dunes resembling rolling ocean waves cover the landscape of the arid western Thar.

HOW SEIF DUNES FORM

Most dunes are formed by multidirectional winds. When winds vary between two different, usually acutely angled, directions, longitudinal or "seif" dunes are created. These may extend for 60 miles (100 km) or more.

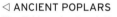

longitudinal, parallel dunes

sparse sand

wind direction varies

◁ **ANCIENT POPLARS**
The Taklamakan's relatively temperate climate does support some vegetation, including Euphrates poplars, which grow near the Tarim River and stabilize the sand dunes.

DUNE MIGRATION

Taklamakan dunes move up to 30 ft (9 m) per year due to prevailing northwesterly winds. Sand blown to the top of a dune falls to the other side, gradually shifting the position of the dune.

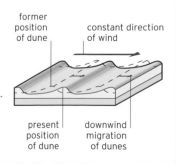

former position of dune

constant direction of wind

present position of dune

downwind migration of dunes

The Gobi Desert

High, harsh, and dry, a desert of climatic extremes, rugged beauty—and very little sand

The Gobi Desert stretches a staggering 500,000 square miles (1.3 million square km) across Asia's remote heartland, making it the continent's second largest desert. Once part of the Mongol Empire, today it straddles northern China and southern Mongolia.

The waterless place

Bordered by Siberia to the north and the Tibetan Plateau to the south, the Gobi lies in the shadow of the Himalayas, which block most clouds laden with life-giving rain. Indeed, the Gobi derives its name from the Mongolian word *gēbi*—the "waterless place." Seasonal monsoons occasionally moisten its southeast corner, but the desert is mainly arid—average annual rainfall is just 4–6 in (100–150 mm), although snow sweeps into the north from the Siberian Steppes. Summers are brutally hot, with temperatures soaring to 122–151°F (50–66°C) during the day; in winter, night temperatures can plummet to -36°F (-38°C). Within just 24 hours, temperatures may shift by as much as 60°F (33°C).

Most of the Gobi's terrain is rocky, forming a pavementlike surface (see panel, below), which tempted merchants to cross it as part of the ancient Silk Road. Rugged valleys, chalk or gravel plains, and rocky massifs abound, but sand, for the most part, is thin on the ground. Exceptions include the massive dunes at the base of the Gobi Altai Mountains in southern Mongolia.

◁ **BORN SURVIVOR**
Wildlife adapts to the Gobi Desert in remarkable ways. The long-eared hedgehog, which shelters in burrows during the day, can go without food or water for up to 10 weeks.

Only **five percent** of the Gobi Desert is made up of **sand**

HOW A DESERT PAVEMENT FORMS

Water and wind together form hard desert pavements. Over millennia, silt particles on extremely dry ground are either blown away by wind or filtered beneath the surface during the rare occasions when moisture reaches the desert floor. This leaves gravel behind, which fuses into a conglomerate whenever the minerals in it, such as gypsum, absorb enough soil moisture to liquefy, then set. The result is a "pavement."

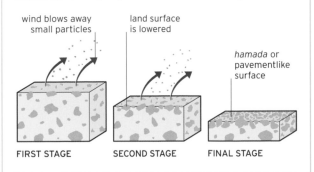

wind blows away small particles

land surface is lowered

hamada or pavementlike surface

FIRST STAGE

SECOND STAGE

FINAL STAGE

△ **RED BEDS**
The color red dominates the Gobi due to the large amounts of iron oxide present in its rocks. "Red bed" formations are typically made up of sandstone, siltstone, and shale.

▷ **SINGING SAND**
At Khongoryn Els, Mongolia, "Singing Dunes" at the foot of the Altai Mountains rise as high as 980 ft (300 m). Their name refers to the sound made when wind sweeps the dunes.

Ancient life
South Australia's Kati Thanda-Lake Eyre shrinks and grows
according to the continent's erratic rainfall. The pink color
comes from pigment in the cell membranes of salt-loving
bacteria, some of the oldest forms of life on Earth.

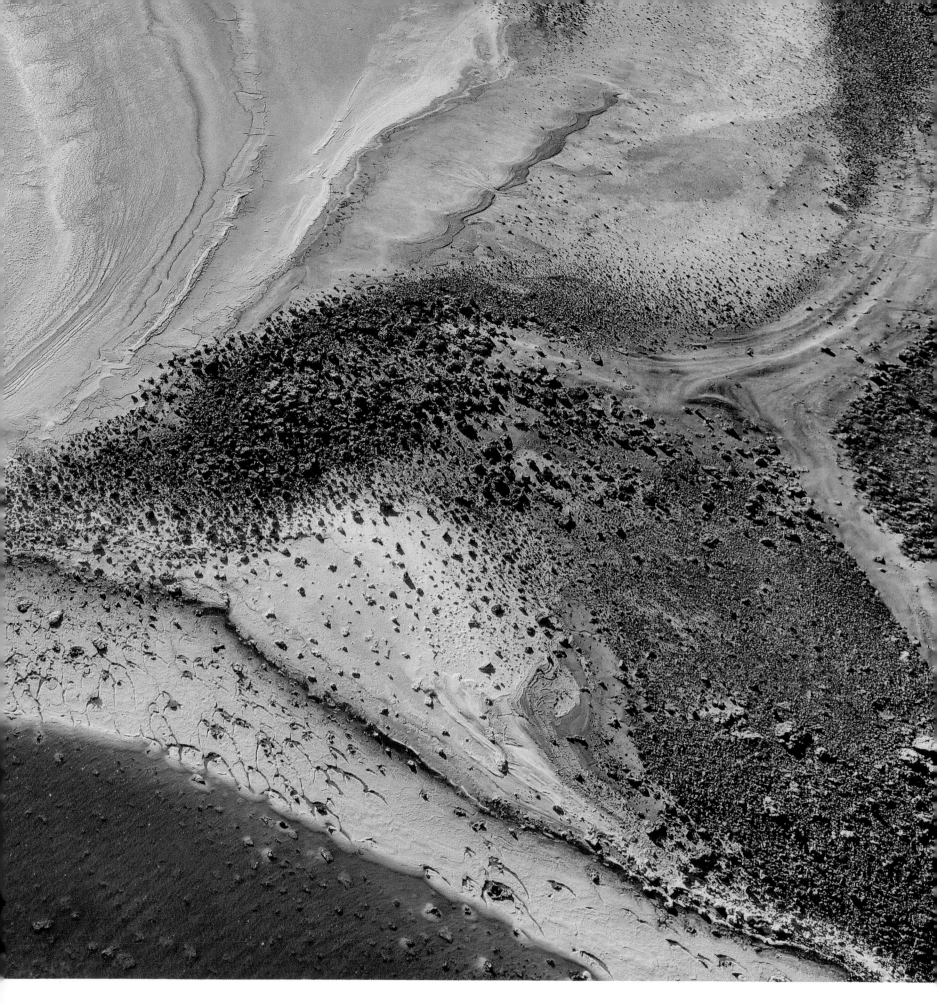

Australia & New Zealand

AN ANCIENT LAND
Australia and New Zealand

The smallest continent, Australia is surrounded by water and sits in the middle of the huge Australian Plate. This plate is largely inundated by the Pacific and Indian Oceans but is dotted with islands—some large like New Guinea, others tiny like the Fijian islands. New Zealand straddles the boundary between the Australian and adjacent Pacific Plates.

Australia is dominated by the vast and flat outback, scorched by the tropical sun. In the east, it rises to the Great Dividing Range of mountains, and beyond lies the moist and mild coastal plain. Off the northeast coast stretches the world's biggest coral reef, the Great Barrier Reef. Temperate New Zealand's two islands lie far to the east—a mix of lofty, snowcapped mountains, rolling plains, and active volcanic features.

INDIAN OCEAN

KEY

- ■ Precambrian (pre-541 million years ago)
- ■ Paleozoic (541-252 million years ago)
- ■ Mesozoic (252-66 million years ago)
- ■ Cenozoic (66 million years ago to present day)

New

Arafura Sea

Torres

Timor Sea

Melville Island

Arnhem Land

Kimberley Plateau

Barkly Tableland

Tanami Desert

Great Sandy Desert

Macdonnell Ranges

A U S T R A L I A

Hamersley Range

Gibson Desert

△ Uluru (Ayers Rock) 2,831 ft

Simpson Desert

▼ Lake Eyre shore -53 ft

Lake Eyre Basin

Great Victoria Desert

Lake Torrens

Flinders Ranges

Nullarbor Plain

Darling Range

Kangaroo Island

GEOLOGY

Australia is the lowest, oldest, and flattest continent. The ancient eastern mountains rise to only 7,310 ft (2,228 m). Far from any plate boundaries, the continent is geologically stable and has no volcanoes.

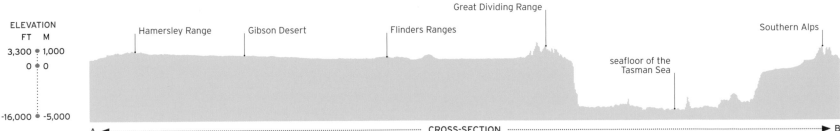

ELEVATION FT M					
	Hamersley Range	Gibson Desert	Flinders Ranges	Great Dividing Range	Southern Alps
3,300 ● 1,000					
0 ● 0				seafloor of the Tasman Sea	
-16,000 ● -5,000					

A ◄———————————————— CROSS-SECTION ————————————————► B

Guinea

Solomon Islands

△ *Mount Wilhelm*
14,794 ft

Strait

Cape York Peninsula

Great Barrier Reef

Great Dividing Range

Coral Sea

Fraser Island

Bourke
125°F

Darling

Great Dividing Range

Murray

Charlotte Pass
-9°F
△ *Mount Kosciuszko*
7,310 ft

Bass Strait

Flinders Island

Tasmania

Tasman Sea

Km
0   150   300

0   150   300
Miles

North Island

New Zealand

Aoraki (Mt Cook)
12,218 ft
△ *Southern Alps*

South Island

KEY DATA

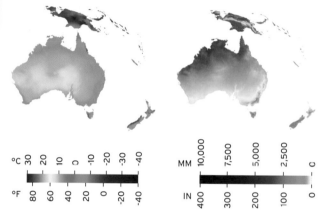

▲ **Highest point** Mount Wilhelm, Papua New Guinea: 14,794 ft (4,509 m)

▼ **Lowest point** Lake Eyre shore, Australia: -53 ft (-16 m)

● **Hottest record** Bourke, Australia: 125°F (52°C)

● **Coldest record** Charlotte Pass, Australia: -9°F (-23°C)

CLIMATE

Most of Australia is very dry and often scorching. Only along the east coast is there much rainfall, with a Mediterranean climate in the south and a tropical monsoon climate in the north. New Zealand's climate is cooler and more moderate.

AVERAGE TEMPERATURE

AVERAGE RAINFALL

°C 30 20 10 0 -10 -20 -30 -40

°F 80 60 40 20 0 -20 -40

MM 10,000 7,500 5,000 2,500 0

IN 400 300 200 100 0

ECOSYSTEMS

Most of Australia's interior is desert, dry grassland, or scrub. Tropical rain forests grow in the north; while temperate forests grow in the south. New Zealand is mostly temperate forest.

KEY

■ Tropical broadleaf forest
■ Temperate broadleaf forest
■ Mediterranean woodland and scrub
■ Tropical and subtropical grassland
■ Temperate grassland
■ Desert and scrub
■ Montane grassland

▷ **OLD IRON**
The banded iron formations (alternate layers of iron-rich deposits), exposed here in the Hamersley Range on one of Australia's cratons, are 2.5 billion years old.

▷ **ANCIENT RIVER**
The Finke River first started to flow over 350 million years ago. At that time, it flowed freely over the plain above its present-day gorge.

◁ **SMOKING ISLAND**
Mount Tavurvur is one of the most active vents on the highly volatile Rabaul caldera on New Britain, Papua New Guinea.

THE SHAPING OF AUSTRALIA AND NEW ZEALAND

Australia is a flat, time-worn continent. It holds some of Earth's oldest rocks, and its remarkable landscapes have been carved out over hundreds of millions of years.

Ancient crust

The western half of the continent is the Australian Shield of old crustal rock, built around three ancient sections of stable crust called cratons. Zircon crystals in the Jack Hills, discovered in rocks on one of the cratons, are 4.4 billion years old—the oldest datable material on Earth. Another craton is, along with South Africa's Kaapvaal Craton, one of only two unaltered chunks of Archean crust on Earth. The Australian Alps formed over 360 million years ago—as a result, they have been worn lower than other ranges and have no jagged peaks. Instead, they are made of plateaus and gorges.

Australia has the **most ancient landscape** and holds **Earth's oldest rocks**

Volcanic islands

Along its northern edge, Australia has crunched into a complex melee of small tectonic plates to the north of Papua New Guinea. There are more than five small plates here, including the South and North Bismarck Plates under New Britain. These small plates jostle to and fro as they are squeezed between the giant Australian and Pacific Plates converging from either side. Some of the small plates will not last long because they are subducting beneath other plates and melting into Earth's mantle. Molten magma from the mantle bursts through the overriding plates in a series of volcanoes to make the islands of Melanesia—the most volcanically active region in the world. There are 22 volcanoes here that have erupted to form islands, while others erupt under the sea. The Rabaul caldera on the northern tip of New Britain contains several violently explosive vents.

Solo journey

Australia was attached to India and Antarctica as part of the great southern continent of Gondwana 150 million years ago, but about 130 million years ago a rift opened inside Gondwana and the three landmasses split away. About 100 million years ago, India broke

KEY EVENTS

300-280 million years ago
The Great Dividing Range forms when Australia collides with landmasses that will become South America and Zealandia.

85 million years ago A landmass that will eventually fall below sea level and become Zealandia splits away from Australia while it is still attached to Gondwana.

50 million years ago Australia and Papua New Guinea break away from Antarctica and drift north, while Antarctica drifts southward to the South Pole.

500,000 years ago
The Great Barrier Reef first forms, but it has receded and expanded since then.

450 million years ago An ancient ocean forms where Australia will eventually be. The sandstone seafloor will later be pushed upward by tectonic movement–both Uluru and Mount Olga are remnants of this seafloor.

180 million years ago
Rain forests arise that will eventually survive to become the Gondwanan rain forests in southeastern Australia.

80 million years ago
A landmass that will become Australia and Papua New Guinea breaks away from the rest of Gondwana.

25 million years ago
What was once an ancient seafloor thrusts upward due to tectonic activity and becomes the Nullarbor Plain.

6,000 years ago
New Guinea becomes an island as sea levels rise after the ice age.

▷ UNLIKELY PEAKS
There has been enough vertical movement along the usually horizontal fault that cuts through New Zealand to throw up the Southern Alps.

△ UNIQUE WILDLIFE
As a result of Australia's isolation from other continents, much of its wildlife has evolved independently and is unique to the continent.

away, too, and headed north—but Australia was hauled to the chilly South Pole in Antarctica's embrace. From 85 million years ago, Australia began to crack away from Antarctica until, about 45 million years ago, it finally broke free and started creeping northward—and it is still heading in that direction.

Although the Australian Plate is now pressing up against the Eurasian Plate, unlike the Indian Plate it has never come into direct contact by land—Australia and the nearby islands remain separated by deep water. Australia's wildlife has evolved independently, and some species are completely unique to the continent.

Hidden continent

The landmass that was to become New Zealand split away from Antarctica and Australia 80 million years ago. A great rip formed along Australia's western edge and along the edge of Marie Byrd Land in Antarctica. New Zealand drifted north, and the Tasman Sea opened to leave the two islands isolated.

In 2017, scientists confirmed what had long been suspected—that the split moved not just present-day New Zealand but an entire continent, over 90 percent of which is now submerged.

This submerged continent, known as Zealandia, is now considered the world's eighth continent due to the type of crustal rock that it is made of. New Zealand is, in fact, half on the Australian Plate and half on the Pacific Plate, and for 25 million years has been torn apart along the Alpine Fault (which has thrown up the Southern Alps) as the two plates move in opposite directions.

HIDDEN CONTINENT
The islands of New Zealand and New Caledonia, although now separated by sea, are just the highest points of the now submerged continent of Zealandia.

extent of Zealandia

BUILDING THE AUSTRALIAN AND ZEALANDIAN CONTINENTS

KEY — Convergent boundary — Divergent boundary — Transform boundary

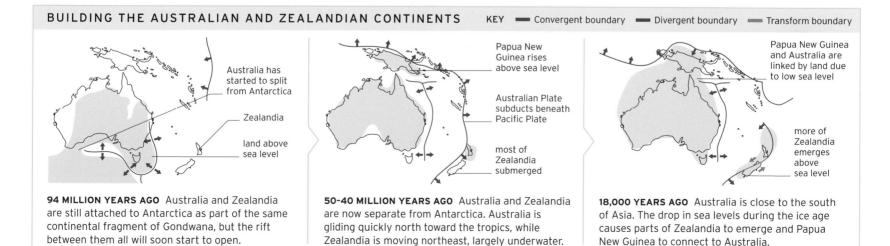

Australia has started to split from Antarctica

Zealandia

land above sea level

Papua New Guinea rises above sea level

Australian Plate subducts beneath Pacific Plate

most of Zealandia submerged

Papua New Guinea and Australia are linked by land due to low sea level

more of Zealandia emerges above sea level

94 MILLION YEARS AGO Australia and Zealandia are still attached to Antarctica as part of the same continental fragment of Gondwana, but the rift between them all will soon start to open.

50-40 MILLION YEARS AGO Australia and Zealandia are now separate from Antarctica. Australia is gliding quickly north toward the tropics, while Zealandia is moving northeast, largely underwater.

18,000 YEARS AGO Australia is close to the south of Asia. The drop in sea levels during the ice age causes parts of Zealandia to emerge and Papua New Guinea to connect to Australia.

The Great Dividing Range

The main mountain range of Australia, stretching for more than 2,200 miles (3,500 km) down the country's eastern side

E. Australia

The third longest land-based mountain range in the world, the Great Dividing Range has many parts, including, for example, the Australian Alps in the south, the Blue Mountains and Warrumbungles in New South Wales, and the Clarke Range in Queensland. Its origins are complex and partly connected to the breakup of the ancient supercontinent of Gondwana that started more than 100 million years ago.

south. The mountains are the source of many of the country's major rivers, including the Murray–Darling (see p.282). The highest peak, Mount Kosciuszko (7,310 ft/2,228 m), is in the Australian Alps. The region's wildlife is as varied as its landscapes and includes kangaroos and platypuses. The area also boasts rich agricultural, timber, and mining resources.

Diverse vistas

Landscapes across the range vary from hills covered with lush tropical rain forest in Queensland to snow-covered peaks farther

△ **RAZOR SHARP**
This rock pinnacle, known as the Breadknife, and similar outcrops in the Warrumbungles subrange are remnants of ancient volcanoes.

◁ **INDUSTRIAL TREASURE**
The valuable metalloid element antimony, which has a variety of industrial uses, is mined at a few different sites in the Great Dividing Range.

RUGGED WILDERNESS
Located in the Blue Mountains, the Grose Valley is a wilderness area featuring a combination of dry open forest with dramatic gorges, cliffs, and canyons.

HOW THE AUSTRALIAN ALPS FORMED

The history of the Great Dividing Range is very complex, but one part of it, the Australian Alps, is theorized to have originated when a microcontinent called Zealandia (see pp.272-73) rifted away from southeastern Australia. Today, Zealandia is mostly submerged (the parts above land are New Zealand), while the rift margin on the Australian side has been eroded down to form the Australian Alps.

faults — rift valley
tectonic plate movement — upwelling hot mantle rock

120 MILLION YEARS AGO

sediment — mid-ocean ridge — narrow sea
southeastern Australia — Zealandia

80 MILLION YEARS AGO

Australian Alps — Tasman Sea — extinct mid-ocean ridge
southeastern Australia

TODAY
oceanic lithosphere — submerged continent of Zealandia

N. New Zealand

Tongariro Volcanic Park

A volcanic wonderland in the center of New Zealand's North Island

Tongariro is a large national park in New Zealand centered on three active stratovolcanoes. These are Mount Tongariro, which has a complicated structure consisting of at least 12 overlapping volcanic cones and numerous craters; Ngauruhoe, a single, highly symmetrical volcanic cone; and Ruapehu, the largest active volcano in New Zealand. The summits of Ngauruhoe and Ruapehu—which at 9,177 ft (2,797 m) is the highest point on the North Island—are considered sacred by the local Maori people. The whole park, which is dominated by volcanic features such as ancient craters and solidified lava flows, has a stark beauty.

TONGARIRO'S VOLCANOES

Much of the park is volcanic terrain (gray) surrounded by forested areas (green). The main concentration of small volcanic features lies at the northern end, around Mount Tongariro, while Ruapehu, which tends to produce a large eruption every 50 years, looms in the south.

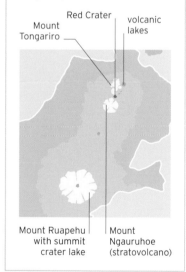

Mount Tongariro — Red Crater — volcanic lakes
Mount Ruapehu with summit crater lake — Mount Ngauruhoe (stratovolcano)

▽ **CRATERS AND LAKES**
In this view of Tongariro volcano, several craters and lakes are visible. The lake colors come from minerals leached out of surrounding rocks.

N. New Zealand

Rotorua

A region of New Zealand's North Island that boasts an array of geothermal features, such as geysers, colorful hot springs, and bubbling mud pools

A large part of New Zealand's North Island has a volcanic nature and history. Near its center is Lake Taupo, the site of a supervolcano that last erupted about 1,800 years ago, and to the southwest of this is the Tongariro Volcanic Park (see p.275). To the northeast of Lake Taupo is a hotbed of geothermal activity often referred to as Rotorua, although only some of the geothermal features it contains are close to the city of Rotorua itself. In addition to geysers, hot springs, fumaroles, and mud pools, more unusual sights in the region include a hot-water waterfall (the Kakahi Falls), acidic boiling lakes, sinter and marble terraces, a mud volcano, and a mud spa.

▽ **BRIGHT DEPOSITS**
The orange coloration at the edges of this hot spring, one of several in an area called Wai-o-tapu, are caused by deposits of arsenic- and antimony-containing minerals.

colored hot spring

water percolates downward through cracks in rock

spouting geyser

terraces of limestonelike material

▷ **SULFUR TERRACES**
Some flowing hot springs create stepped terraces by depositing limestonelike minerals and, in this case, yellow sulfur.

Mud pool
A mud pool is created by a combination of hot water, steam, and acidic gases from a fumarole dissolving nearby rocks.

Dormant fountain geyser
This is currently inactive due to a fracture lower down in the geyser system.

Fumarole
A hole in the ground that emits steam and gas is called a fumarole.

Rising hot water
Heated water expands and moves upward after contact with hot rocks.

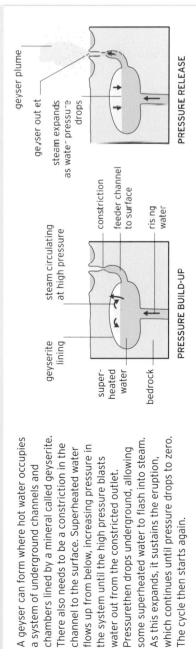

△ **SPOUTING GEYSER**
On the outskirts of Rotorua City is a geyser field containing seven active geysers. The Pohutu geyser, shown here, produces the biggest and loudest eruptions.

heated water saturating rock

Reservoir of superheated water
When high pressure allows the water temperature to rise above boiling point, the water is called superheated.

The Pohutu geyser erupts up to 20 times a day, and eruptions can reach heights of up to 100 ft (30 m)

porous rock can hold large quantities of heated water

Geyserite
This watertight and airtight mineral lines all the plumbing of a geyser.

Hot rock
Underlying magma provides the energy that heats the overlying rock and powers the geothermal features.

Feeder channel
This channel is dry because the rock fracture has blocked the water supply.

rising steam and gases, rather than hot water, occupy some channels

Rock fracture
Caused by an earthquake, this fracture has broken the channel supplying hot water to a geyser above.

△ **MUD SWIRLS**
Bubbling mud pools occur at several sites around Rotorua. As the bubbles burst, they create intricate swirling patterns on the mud surface.

▲ **GEOTHERMAL FEATURES**
This illustration shows a typical geothermal landscape that includes many of the features seen at Rotorua. The geothermal areas lie over volcanic calderas. These still have magma under them. The magma heats overlying rock, which in turn heats water that has percolated downward. This heated water supplies the geothermal features.

HOW A GEYSER ERUPTS

A geyser can form where hot water occupies a system of underground channels and chambers lined by a mineral called geyserite. There also needs to be a constriction in the channel to the surface. Superheated water flows up from below, increasing pressure in the system until the high pressure blasts water out from the constricted outlet. Pressure then drops underground, allowing some superheated water to flash into steam. As this expands, it sustains the eruption, which continues until pressure drops to zero. The cycle then starts again.

geyser plume

geyser outlet

steam expands as water pressure drops

PRESSURE RELEASE

constriction

feeder channel to surface

steam circulating at high pressure

rising water

geyserite lining

super-heated water

bedrock

PRESSURE BUILD-UP

W. Australia

Hyden Rock

An ancient granite mound in Western Australia,
with a wavelike cliff on one side of its base

Hyden Rock is famous for an unusual rock formation, which looks like a tall, breaking ocean wave, along a section of its base. Known as Wave Rock, it is thought to have been created by weathering and erosion of the base of the mound. The initial stage of weathering took place underground through slightly acidic groundwater breaking down the granite. This produced a concave pocket of fragmented granite within the formerly solid base. When the land surface around Hyden Rock was later lowered by erosion, the pocket of fragmented granite was also removed, leaving behind Wave Rock. Crystals from it have been dated at 2.7 billion years old, making it one of the oldest formations in Australia.

◁ **WAVE ROCK**
The formation is around 360 ft (110 m) long and 46 ft (14 m) high. The gray and amber stripes are the result of rainwater dissolving out minerals.

▽ **SMOOTH FORMATION**
The wave illusion is largely the result of the rounded overhang. Above the overhang, the rock's top surface is smooth and domed.

◁ **ICY HEIGHTS**
Aoraki, or Mount Cook, is exceptionally challenging to climb. The surrounding area is an extensive, high crevassed ice-field from which several major glaciers descend.

THE ALPINE FAULT

Along the Alpine Fault, the Pacific Plate is sliding past and also slightly toward the Australian Plate, moving 10-12 ft (3-4 m) per century. This has caused the crust of the Pacific Plate to be pushed up along the fault, leading to the creation of the Southern Alps.

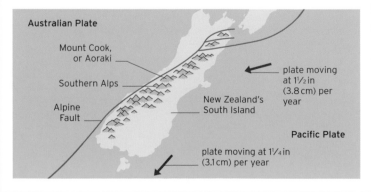

Australian Plate

Mount Cook, or Aoraki

Southern Alps

Alpine Fault

New Zealand's South Island

plate moving at 1½ in (3.8 cm) per year

Pacific Plate

plate moving at 1¼ in (3.1 cm) per year

FALL LANDSCAPE
In the fall, the eastern flanks of the Southern Alps are relatively dry and sparsely vegetated, in contrast to the wet, forested side.

The Southern Alps

A high, steep mountain range running almost the whole length of New Zealand's South Island, with a pristine environment of glaciers, lakes, and forests

S. New Zealand

The Southern Alps are the most significant mountain range in New Zealand, extending for 310 miles (500 km) across the country's South Island, from southwest to northeast. Aoraki, or Mount Cook, the highest point in New Zealand, rises from the center of the range, peaking at 12,218 ft (3,724 m). Another 16 peaks exceed 10,000 ft (3,048 m) in height. Most of the rivers of South Island, including the longest, the Clutha River, flow from the Southern Alps.

Inching skyward

A boundary between two tectonic plates, called the Alpine Fault, runs to the west of the main line of the Southern Alps. It is of great significance to the mountains' origins and continuing uplift (see panel, left). The Southern Alps are continuing to rise, at a rate of up to fractions of an inch annually, but are also being eroded at an equally fast rate, aided by the sheer steepness of the mountains, high rainfall on the western slopes, and the presence of several glaciers. Due to the high rainfall—which can be up to 390 in (10,000 mm) a year—the western slopes of the Alps are covered in temperate rain forest up to a height of about 3,300 ft (1,000 m). This leads to the unusual appearance of glaciers terminating a short distance away from lush rain forest. The eastern slopes are much drier and contain several large glacial lakes, such as Lake Tekapo and Lake Pukaki, which are both a brilliant powder blue color. Endemic birds in the region include the kea, the rare and flightless southern brown kiwi, and the New Zealand rock wren.

▽ **CLEVER BIRD**
The kea, the world's only alpine parrot, is well known for its curiosity and intelligence, including an ability to use tools and solve logical puzzles. It is found only in the Southern Alps.

The **Southern Alps** were **so-named** because of their **resemblance** to the **European Alps**

W. New
Zealand

The Franz Josef Glacier

One of two glaciers that descend steeply from the Southern Alps toward the west coast of New Zealand's South Island

Named in 1865 after the Emperor of Austria, the Franz Josef Glacier is about 7 miles (12 km) long. Its lower parts used to flow through a temperate rain forest near the coast. However, since 2008 its terminus has retreated rapidly and now hardly reaches the tree line.

The glacier is known to advance and retreat cyclically according to changes in regional climate. It is currently shrinking, but between 1983 and 2008 it advanced 1 mile (1.5 km). The driving force behind the glacier's whole existence is a high rate of precipitation in the upper area of the Southern Alps—up to 100 ft (30 m) of snow falls onto the glacier's ice accumulation area each year—and annual changes in this precipitation rate can affect the glacier's behavior for years to come.

△ **AQUA TUNNEL**
One part of the glacier contains tunnels carved out by meltwater. This one, dubbed the Aqua Tunnel, is large enough to walk through.

△ **BROKEN FLOES**
The surface of the glacier's meltwater lake sometimes freezes over in winter. The floes (flat expanses of ice) later break up into angular slabs.

▷ **MELTWATER LAKE**
At the foot of the glacier stands a 4-mile- (7-km-) long meltwater lake. A century ago, the valley was still filled with the glacier itself.

W. New
Zealand

The Tasman Glacier

*New Zealand's largest glacier, with a skiing
arena at its top and a silty lake at its foot*

The Tasman Glacier flows down a long valley to
the east of New Zealand's highest mountain, Aoraki,
or Mount Cook, originating in a large ice accumulation
area at an altitude of 9,200 ft (2,800 m). In total, it is
15 miles (24 km) long.

Ski runs, crevasses, and lakes

The upper part of the glacier is smooth and supplies
one of the world's longest ski runs, about 7 miles
(11 km) in length. Farther down, it breaks up into
a maze of crevasses and ice tunnels. The lower third
is covered in a thick layer of rock debris—the result of
ice melting. At the foot of the glacier is a large meltwater
lake that has enlarged considerably in recent decades. It
is up to 820 ft (250 m) deep and is gray colored due to
suspended silt. The lake has a steady temperature close
to freezing because icebergs are continually discharged
into it from the glacier's terminus.

In 2011, millions of **tons of ice** fell from the **glacier** into its **lake** after an **earthquake**

△ BRAIDED FLOW
An intricate braided river (a network of
branching channels with intervening
sand and gravel bars) carries water
away from the melting glacier.

FEATURES OF A RETREATING GLACIER

A retreating glacier such
as the Tasman leaves behind
some characteristic landscape
features below its terminus
(or snout). These may include
drumlins (elongated hills lined
up in the direction of retreat),
kettle lakes (water-filled
depressions left behind
when partially buried ice
blocks melt), and eskers
(long, winding ridges of
sand and gravel).

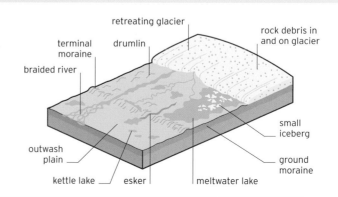

retreating glacier

rock debris in
and on glacier

terminal
moraine

drumlin

braided river

small
iceberg

outwash
plain

ground
moraine

kettle lake

esker

meltwater lake

The Murray-Darling

The largest river system in Australasia with one of the most extensive catchment areas on Earth, responsible for irrigating the "food basket" of Australia

SE. Australia

Consisting of the Murray, Australia's longest river, and its main tributary, the Darling, the Murray–Darling river system is the largest in Australasia. Draining the southeast of Australia, its basin covers an area of more than 386,000 square miles (1 million square km)— around 14 percent of the country's land area. The Murray–Darling basin is the most important agricultural region of a largely arid country, growing three-quarters of Australia's irrigated crops and producing one-third of its total food supply.

From mountains to mouth

Rising in the Australian Alps, at the southern end of the Great Dividing Range, the Murray River flows across flat inland plains, forming the border between New South Wales and Victoria for much of its route. The river receives most of its main tributaries from the north along this section, including the Darling, which flows southwest across almost the whole length of New South

The Murray-Darling is thought to be the **driest major river system** in the world

Wales from its source near the Queensland border. Both the Murray and Darling have very low gradients over most of their lengths, resulting in slow, meandering waterways.

After the Murray–Darling confluence, the river enters South Australia and flows through ancient gorges with towering sandstone cliffs separated by a series of shallow lakes. Having flowed a distance of 1,570 miles (2,530 km) from the source of the Murray, the river discharges into the Southern Ocean through a dynamic, sand-dune-flanked point known as Murray Mouth, at the eastern end of the Great Australian Bight.

△ **FRESHWATER TURTLE**
The eastern long-necked turtle has webbed feet and is able to stay underwater for long periods. Unlike other turtles and tortoises, it pulls its neck sideways into its shell.

Jenolan Caves

SE. Australia

A complex Australian cave system containing elaborate formations, the oldest caves ever discovered

Situated on the western flank of the Blue Mountains in New South Wales, the Jenolan Caves are the best-known limestone cave system in Australasia. They sit within a narrow, uplifted ridge of limestone formed 500 million years ago, when the area was submerged beneath the ocean. The Jenolan Caves comprise a series of water-carved passages and caverns that stretch for more than 25 miles (40km), over different levels, with more than 300 entrances. One of the largest caves, the Lucas Cave, has several vast sections including the 177-ft- (54-m-) high Cathedral Chamber, which is famous for its excellent acoustics. Scientists have estimated that the Jenolan Caves date back more than 340 million years, making them by far the oldest caves ever discovered.

▷ **LARGE CRYSTALS**
Deposits of a form of calcite called dogtooth spar, consisting of large crystals formed over long periods, are found in the pools of the Jenolan Caves.

△ **RICH DECORATIONS**
The caves are renowned for their rich and varied deposits, including stalactites, stalagmites, columns, arches, and other intricate calcite formations.

△ **STILL WATERS**
An Australian pelican glides along the Murray River in the Riverland region of South Australia. The Murray-Darling basin is home to almost 100 species of waterbirds.

▽ **RIVER'S END**
Just before emptying into the ocean, the Murray-Darling connects with the Coorong, a narrow channel of lagoons lying behind the Younghusband Peninsula.

Waitomo Caves

N. New Zealand

An underground network of New Zealand caves, many decorated by the luminescence of glowworms

In the area of Waitomo, in the west of New Zealand's North Island, runs a 28-mile- (45-km-) long network of caves, passages, and grottoes beneath fields of open farmland. The Waitomo Caves formed within a section of 30-million-year-old limestone rock up to (330 ft 100 m) thick. Fractures and fault lines in the rock allowed water to percolate through, creating an underground drainage system that gradually increased in size and complexity. Many of the caves and passages have streams with waterfalls running through them, and most feature a combination of calcite formations including stalactites, stalagmites, and columns in shades of brown, pink, and white. The best known of the 300 caves in the region are the Glowworm Caves, Ruakuri Cave, Aranui Cave, and Gardner's Gut.

▷ **GLOWWORM GROTTO**
The ceilings of many of the Waitomo Caves are lit up by thousands of tiny glowworms, the larvae of a species of gnat endemic to New Zealand.

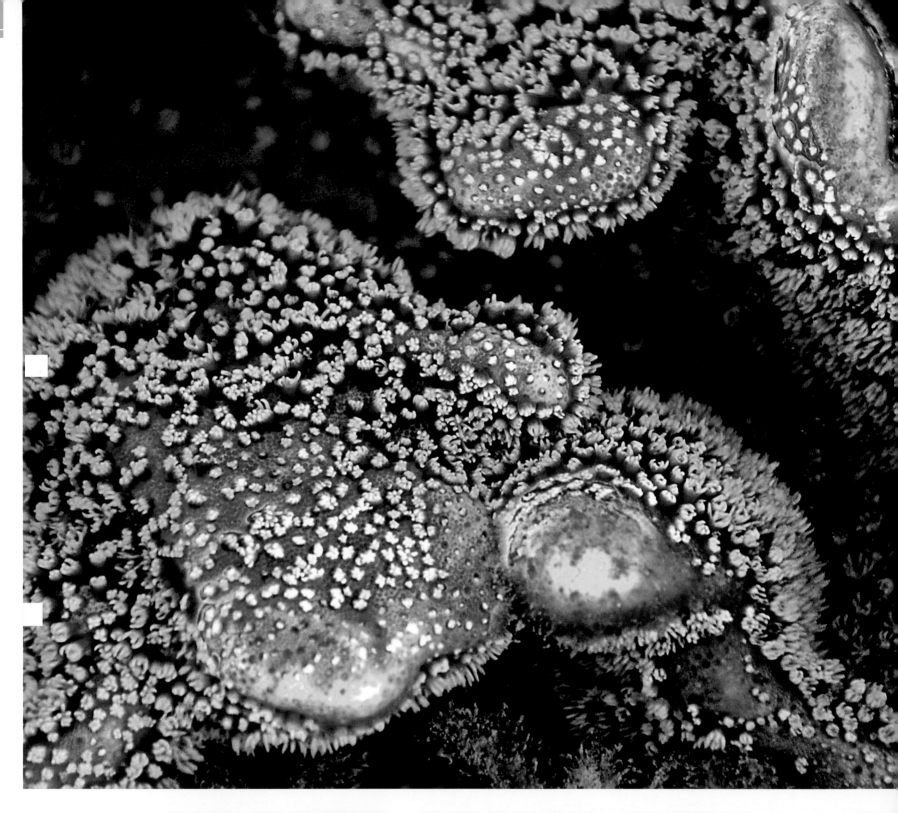

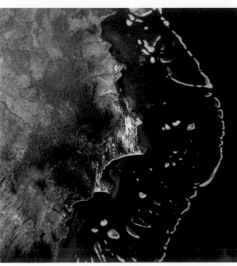

▷ **OCEAN BARRIER**
This view shows about one-tenth of the reef. Part of Queensland is visible on the left, while the reef edge appears as a broken line on the right.

The Great Barrier Reef

The world's most extensive coral reef system, often described as the largest structure ever built by living organisms

NE. Australia

The Great Barrier Reef stretches for 1,400 miles (2,300 km) over a part of the Coral Sea, which lies off the coast of Queensland, Australia. Rather than a single unbroken reef, it is composed of more than 2,900 individual small reefs of various types and 900 coral islands, covering a total area of some 133,000 square miles (344,400 square km). The whole structure has been built through the activity of animals called coral polyps (see panel, below). The reef gets its name from the fact that it acts as a barrier between the coast of Australia and the large waves of the Pacific Ocean.

Biological diversity

Teeming with marine life, the Great Barrier Reef boasts an astonishingly high biological diversity, including about 350 species of stony corals and several hundred species of soft corals, as well as 1,500 species of fish, 17 species of sea snakes, 30 species of whales, dolphins, and porpoises, and 6 species of sea turtles. Over 200 bird species visit the reef, or nest or roost on the islands.

Unfortunately, in recent years the Great Barrier Reef has become severely threatened, the biggest cause of concern being rising ocean temperature as a result of global warming. This leads to coral bleaching (see p.286), a process that can ultimately kill corals permanently.

◁ **LIVING FOSSIL**
The chambered nautilus is one of over 4,000 species of mollusks that inhabit the reef. It has evolved little over many millions of years.

The Great Barrier Reef is the **size of** about **48 million international football pitches**

△ **BUDDING LIFE**
A coral forms when a polyp attaches itself to a rock underwater. It then divides to form a colony—this one contains hundreds of polyps.

◁ **STRUCTURAL COMPLEXITY**
A variety of forms exist in even a small reef section, with patches and platforms of limestone, as well as ribbonlike structures.

STONY AND SOFT CORALS

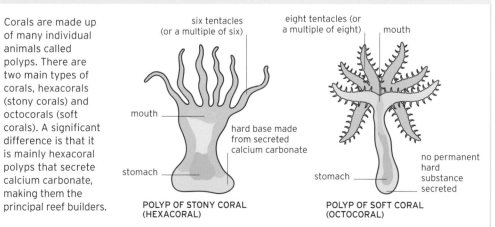

Corals are made up of many individual animals called polyps. There are two main types of corals, hexacorals (stony corals) and octocorals (soft corals). A significant difference is that it is mainly hexacoral polyps that secrete calcium carbonate, making them the principal reef builders.

six tentacles (or a multiple of six)

mouth

stomach

hard base made from secreted calcium carbonate

POLYP OF STONY CORAL (HEXACORAL)

eight tentacles (or a multiple of eight)

mouth

stomach

no permanent hard substance secreted

POLYP OF SOFT CORAL (OCTOCORAL)

The Structure of a
BARRIER REEF

A barrier reef is a type of coral reef widely separated from a continental mainland or island by a deep channel or lagoon. One circumstance in which such a reef can develop is when a volcanic island in the tropics begins to sink (as typically occurs when its volcanic activity ceases). As the island subsides, any fringing reef hugging its shores continues to grow upward. A gradually widening lagoon then develops between the reef and the island. A second way a barrier reef can develop is on and around the edge of a continental shelf when sea level rises. This is how the world's largest barrier reef—the Great Barrier Reef off Queensland, Australia—formed.

Development of the Great Barrier Reef

The Great Barrier Reef has taken some 20 million years overall to develop, although much of the structure visible today has grown since the last ice age on top of an older platform of continental shelf. Around 18,000 years ago, the sea was about 390 ft (120 m) lower and the outer edge of today's barrier reef was the edge of continental Australia. As sea level rose, ocean water and then corals encroached on what had been hills of the Australian coastal plain. By 13,000 years ago, many of these hills had become islands on the continental shelf. Eventually, they were submerged and corals grew over them. Sea level on the Great Barrier Reef has not changed significantly in the last 6,000 years.

CORAL BLEACHING

Bleaching occurs when zooxanthellae (tiny algal organisms), which give corals their colors, leave the coral polyps due to a change in seawater temperature. This is reversible if the temperature change is not too prolonged or severe. But bleached coral becomes more vulnerable to disease. If it becomes colonized by a mat of other types of algae, called "turfing" algae, this signifies irreversible damage.

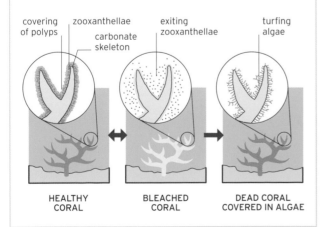

covering of polyps
zooxanthellae
carbonate skeleton
exiting zooxanthellae
turfing algae

HEALTHY CORAL — BLEACHED CORAL — DEAD CORAL COVERED IN ALGAE

▽ **REEF BUILDERS**
Some polyps, called stony corals or hexacorals, secrete small amounts of calcium carbonate (limestone) as part of their body structures, building on the substrate underneath. When polyps die, their limestone skeletons remain, building up the reef.

▶ **THE GREAT BARRIER REEF**
A 155-mile- (250- km-) long section of the reef, off northern Queensland, shows the overall structure. There are many individual reefs in a variety of forms ranging from ribbonlike structures making up the outer reef edge to rounded and crescent-shaped patch and bank reefs.

Mainland of Queensland, Australia

spur of the Great Dividing Range

Normanby River

Cooktown

Bank reef
A reef of this sort is larger than a patch reef and often linear or semicircular in outline. It may support a cay, or small island.

Williamson Reef

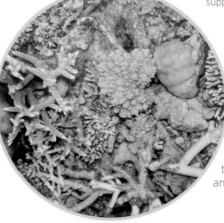

◁ **BLEACHED CORAL**
A bleached coral appears white due to loss of algal organisms that supply the color. The algae provide most of the coral's energy: after they leave, the coral begins to starve and may eventually die.

Fringing reefs
These reefs develop near coasts, around the edges of what were once hills on the mainland. Polluted or sediment-laden water flowing off the mainland is a threat to them.

Patch reef
Reefs of this type are rounded patches that grow up from the continental shelf. Some are topped by forested or sandy islands, and others develop lagoons.

Combe Island National Park

Flinders Group National Park

◁ **CORAL SPAWNING**
Once a year, male and female coral polyps across the reef release a blizzard of tiny sperm and eggs, called gametes. This synchronized release is crucial to chances of fertilization, as the gametes can survive in the ocean for only a few hours.

seamounts

Ribbon reefs
Reefs of this type follow the edge of the continental shelf, tracking the coastline of the last ice age. Some are up to 15 miles (25 km) long.

Cay
This is a small, low-elevation, sandy island formed on the surface of a reef from eroded material that piles up.

Lizard Island

◁ **DROP-OFF**
Ribbon reefs form the outer edge of the barrier reef, bearing the brunt of Pacific storms. They are populated by robust corals and have steep drop-offs on their oceanic sides. These merge with the continental slope below, which descends to 6,600 ft (2,000 m).

continental shelf

continental slope

continental rise

Three Islands National Park

About **one-third** of the reefs making up the **Great Barrier Reef** are **fringing reefs**

Shark Bay

A coastline with exceptional natural features, including seagrass beds, living stromatolites, and rich animal life

W. Australia

Shark Bay lies 500 miles (800 km) north of Perth, on the west coast of Australia, and has a coastline that is more than 900 miles (1,500 km) long, including about 190 miles (300 km) of spectacular limestone cliffs. Another feature of Shark Bay is a dazzling white beach consisting almost entirely of the shells of one species of cockle. The bay contains one of the world's largest seagrass beds, which provides food for about

▷ **BLUE STINGER**
A sting from this local *Mastigias* jellyfish can be painful but is not dangerous, unlike that from other smaller species called irukandji.

10,000 dugongs—large marine mammals related to manatees (or sea cows). In turn, these are preyed on by various species of sharks, including the great white and tiger sharks, which give the bay its name. The area is also home to more than 100 species of reptiles and amphibians, 240 species of birds, 820 species of fish, and more than 80 types of corals.

But Shark Bay is probably most famous for its living stromatolites— modern equivalents of some of the earliest types of life on Earth. These are colonies of microbes that form hard domelike accretions (see panel, below).

◁ **LAYERED LIFE**
In an area called Hamelin Pool lies the world's most abundant collection of living stromatolites. They are thought to have started growing here about 1,000 years ago.

▷ **SHORELINE SPECTRUM**
Mangrove-lined creeks, emerald waters, dunes, and mineral-stained pools come together in this aerial view of a part of the bay.

HOW STROMATOLITES GROW

The surface of a living stromatolite is covered in a mat of microbes called cyanobacteria, which in some cases develop into long, sticky filaments. These trap and bind together sediment particles into a hard layer. The stromatolite grows through the buildup of alternating layers of this hard substance and dead filaments.

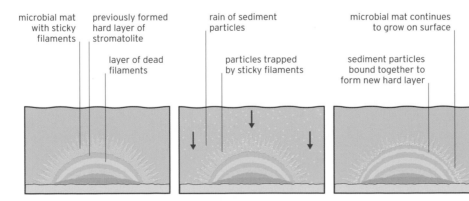

microbial mat with sticky filaments

previously formed hard layer of stromatolite

layer of dead filaments

rain of sediment particles

particles trapped by sticky filaments

microbial mat continues to grow on surface

sediment particles bound together to form new hard layer

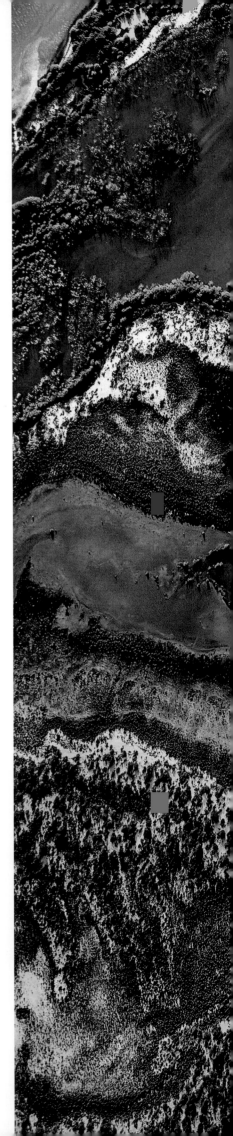

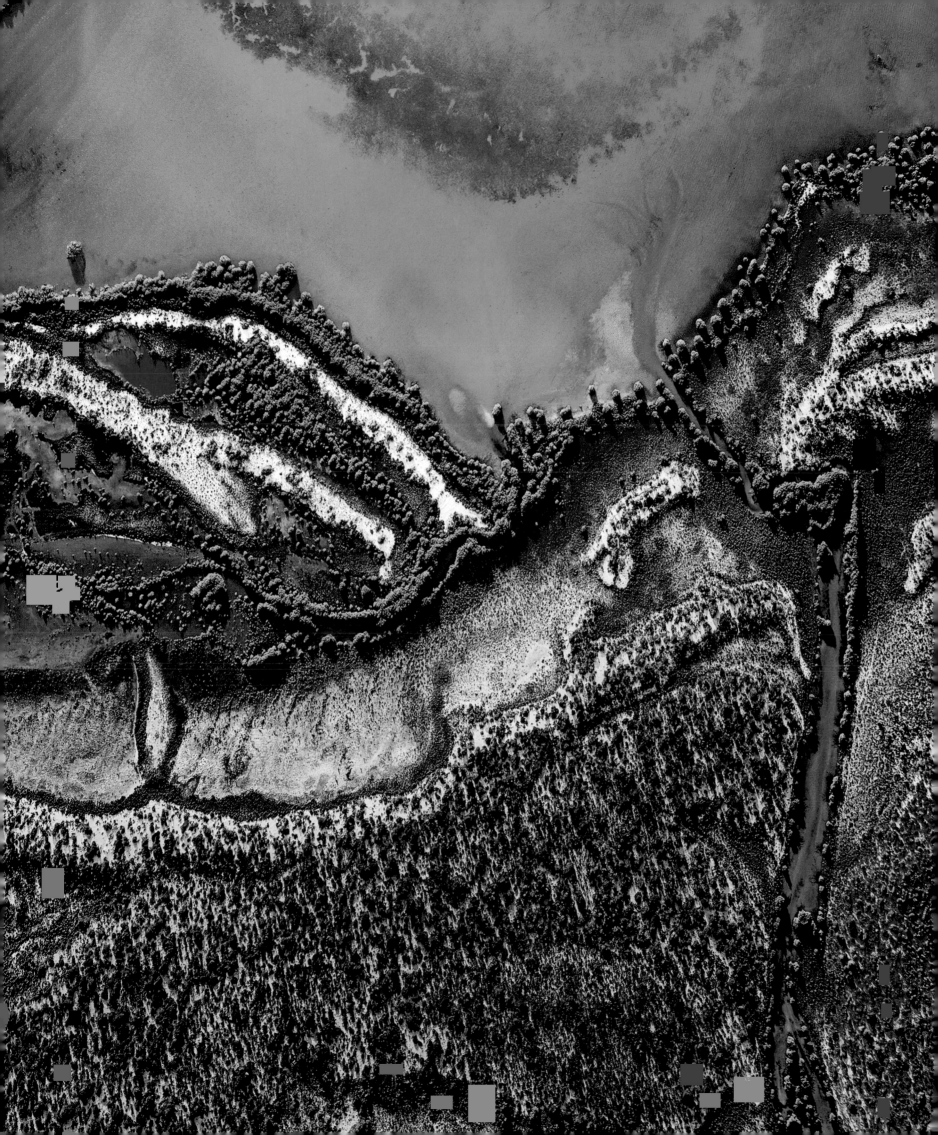

The Twelve Apostles

An iconic group of large sea stacks, formed through the erosion of 20-million-year old limestone cliffs

SE. Australia

HOW SEA STACKS FORM

Wave action creates caves in the sides of a headland, eventually tunneling through it. The arch over the tunnel collapses to leave an island, which is then eroded to form stacks.

The Twelve Apostles are located in Port Campbell National Park, a coastal area of Victoria. The sea stacks are strung out along about 3 mile (5 km) of coastline, with some reaching up to 165 ft (50 m) in height. The Twelve Apostles are one of Australia's best-known geological landmarks, but their name is something of a misnomer, as when they were originally named only nine large stacks existed and one has since collapsed, leaving only eight. However, several prominent headlands and islands nearby are continuing to erode through wave action (see panel, right), so new sea stacks are expected to form in the future. Like all coasts, the shoreline around the stacks is constantly changing.

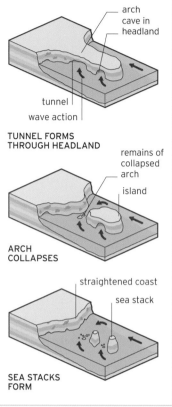

arch
cave in headland
tunnel
wave action

TUNNEL FORMS THROUGH HEADLAND

remains of collapsed arch
island

ARCH COLLAPSES

straightened coast
sea stack

SEA STACKS FORM

Until the **1920s**, the stacks were known as the **Sow and Piglets**

▷ **NEW STACKS IN THE MAKING**
This aerial view shows one existing isolated stack, at left, and other headlands and islands that, as they erode, are expected to form new stacks.

Moeraki Beach

A beach in southeastern New Zealand strewn with some unusually large, almost spherical boulders

E. New Zealand

HOW THE BOULDERS FORMED

The boulders originally grew within marine sediments, which hardened into rock. This was later lifted above sea level and now forms a coastal cliff. From time to time, a boulder is freed up and rolls onto the beach.

Some 45 miles (70 km) northeast of Dunedin, Moeraki Beach appears at first sight as though it might have been the location for a giants' game of bowls or petanque. Almost perfectly spherical gray boulders lie scattered across a 165-ft (50-m) stretch of the beach. The scientific view is that the huge boulders actually formed within clay and silt sediments on the seafloor around 60 million years ago by a process called concretion, in which a cementing mineral binds together other mineral particles into a hard, erosion-resistant mass. This process can take millions of years. Later, the concretions, now embedded within mudstone rock, were lifted up, and today most of them are buried within a cliff at the back of Moeraki Beach. There the boulders are gradually exposed and released onto the beach by erosion of the mudstone (see panel, right).

△ **SUPERSIZED BOULDERS**
The boulders come in two main sizes—around 3 ft (0.9 m) and 6 ft (1.8 m) in diameter—and weigh several tons. Many are half-buried in the sand.

spherical mineral concretions forming around seed nuclei

FORMATION UNDER THE SEA

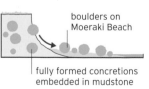

boulders on Moeraki Beach

fully formed concretions embedded in mudstone

TODAY

New Zealand Fjords

A group of fjords in the southwestern corner of New Zealand, featuring abundant marine life and spectacular cliffs and waterfalls

SW. New Zealand

New Zealand's fiords formed some 15,000 years ago when glaciers retreated at the end of the last ice age and the sea flooded in to fill the deep, U-shaped coastal valleys carved out by the glaciers.

Dangerous discovery

Of some 14 major fjords, two of the best known are Doubtful Sound—one of the largest at 20 miles (30 km) in length—and Milford Sound. The former was named in 1770 by Captain James Cook, who refrained from sailing into it because he doubted that he would ever be able to sail out again. The smaller but more famous Milford Sound, which runs 9 miles (15 km) inland, is notable for the spectacular waterfalls that rain down into it, and an enormous mountain—the 5,551-ft-(1,692-m-) high Mitre Peak—that soars above it. Both of these fjords, and others nearby, are home to bottlenose and dusky dolphins, New Zealand fur seals, little penguins, and the rare Fiordland crested penguin. Prevailing westerly winds blowing moist air in from the Tasman Sea produce exceedingly high rainfall across the region, and this supports lush temperate rain forests close by.

◁ **CURIOUS MINERAL**
Small deposits of arsenopyrite are found in the region. If heated, the mineral gives off toxic fumes and becomes magnetic. If struck with a hammer, it emits a garliclike odor.

▽ **MISTY CASCADE**
The 495-ft (-151-m-) high Stirling Falls is one of two permanent waterfalls in Milford Sound. Its Maori name is Wai Manu, meaning "cloud on the water."

The Daintree Rain Forest

Australia's largest tropical rain forest, one of the oldest and most biodiverse in the world, with a striking array of plants and animals

NE. Australia

Covering 460 square miles (1,200 square km) in northeastern Queensland, the Daintree is the largest tropical rain forest in Australia and one of the oldest on Earth. Estimates of its age vary from 135 to 180 million years. Part of the Wet Tropics World Heritage Site, it contains the closest living relatives of prehistoric rain forest plant species that covered the continent millions of years ago. These include the ribbonwood tree, also known as the idiot fruit or green dinosaur. A primitive

evergreen dating back 120–170 million years, it was believed to be extinct until 1971. Half of Australia's bird species, a third of its frogs and mammals, a quarter of its reptiles, and more than 12,000 insect species have been catalogued in the Daintree, many of them living nowhere else.

◁ **RAIN FOREST WARDEN**
The endangered flightless southern cassowary is among the Daintree's residents and is an important disperser of tree seeds.

The Daintree contains the **greatest concentration of birds** in all of **Australia**

Eastern Australian Temperate Forest

Home to more than one-fifth of Australia's eucalyptus trees, the oils of which give nearby mountains their blue haze

E. Australia

▽ **TOWERING TREES**
Near the Great Dividing Range, which forms the ecoregion's inland boundary, eucalyptus trees rise above smaller trees such as acacia.

Stretching from the New South Wales central coast into southeast Queensland, the Eastern Australian Temperate Forest ecoregion covers 85,750 square miles (222,100 square km). Varying altitudes and microclimates support a wide range of vegetation, but the eucalyptus is the region's dominant plant—more than 100 species of eucalyptus, also known as gum trees, grow here. Airborne oil droplets from eucalyptus leaves are believed to refract blue light, lending the area's Blue Mountains their colorful haze. More importantly, these gum tree forests are home to many endangered plant and animal species, including the iconic koala.

▷ **HOLDING TIGHT**
The koala initially carries its young, or joey, in a pouch. Eventually, the joey clings to its mother's back as she feeds on eucalyptus leaves.

△ **STANDING GUARD**
The granite summit of Mount Pieter Botte, or Ngalba-bulal, keeps watch over the Daintree National Park's rare black kauri pines.

◁ **PREHISTORIC LANDSCAPE**
Because of its isolated location and climate, the Daintree has remained largely unchanged since its origin. Today, it is home to 12 of the world's 19 families of primitive flowering plants.

Waipoua Forest

An ancient native New Zealand forest, home to the world's largest stand of kauri trees

N. New Zealand

The subtropical Waipoua Forest lies on the west coast of the Northland Region of New Zealand's North Island. Declared a sanctuary in 1952, Waipoua contains Earth's largest collection of kauri trees—giant, cone-bearing, hardwood conifers that are among the oldest trees in the world and are also some of its most endangered. While ancestors of the kauri were growing during the Jurassic Period, Waipoua's most ancient kauri, called Te Matua Ngahere, or Father of the Forest, is merely 2,000 years old. However, with a girth of 52 ft (16 m) and a height of 121 ft (37 m), it has the largest diameter of any living kauri—and it is still growing.

▽ **SEED-BEARERS**
A single kauri produces both cylindrical male and round female cones. Not until a tree is 25–30 years old will its female cones release fertile winged seeds.

◁ **IMPOSING HEIGHT**
Kauris can reach heights of up to 165 ft (50 m), making them dominant emergent trees that rise above other conifers, such as toatoas and monoaos, in the surrounding forest.

The North Australian Savannas

A diverse tropical grassland of two seasonal extremes, home to hundreds of native plant and animal species

N. Australia

△ **PERFECTLY ADAPTED PALM**
Highly drought-resistant Livistona palms can cope with seasonal extremes because they are able to survive frequent fires, as well as heavy downpours.

North Australia's grassland is one of the world's largest remaining tropical savannas and shares much in common with those of Africa. However, where Africa's savannas have large grazing animals, insects such as termites eat much of the plant material in Australia's savannas, alongside marsupials ranging from the small short-beaked echidna to large red kangaroos. Waterfowl, wading birds, and reptiles live in the wetter parts of the region, and the Cape York Peninsula alone contains around 60 percent Australia's butterfly species.

Due to its midlatitude location, the grassland is subject to the El Niño Southern Oscillation climate cycle. This produces a pattern of high rainfall and drought that gives the area its two contrasting seasons.

◁ **DESIGNED TO RUN**
At up to 7 ft (2.1 m) high, the emu is the savanna's—and Australia's—largest bird. It cannot fly, but long, powerful legs enable it to run at speeds up to 30 mph (48 kph).

CLIMATE

Australia's savannas have two distinct seasons. During the hot, humid wet season (December–March), northwesterly winds blow in heavy rainstorms. From May to October, the dry season sets in, with low temperatures and humidity.

Temperature
°F °C

Katherine, Northern Territory

Rainfall
IN MM

Temperature		Rainfall
122 : 50		10 : 260
104 : 40		8 : 210
86 : 30		6 : 150
68 : 20		4 : 100
50 : 10		2 : 50
32 : 0		0 : 0

Jan Feb Mar Apr May Jun Jul Aug Sep Oct Nov Dec

KEY ■ Temperature ■ Rainfall

▽ **MAGNETIC MOUNDS**
The magnetic or compass termite builds impressive mounds 10–12 ft (3–4 m) high. The wedge-shaped structures are constructed so that their longest sides are aligned north-south, keeping them cool.

Giant grazing animals roamed the savannas about **40,000 years ago**

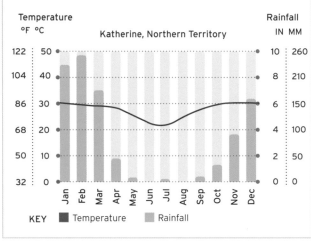

The Great Sandy Desert

An active sandy desert, wetter than the continent's central region, but heat and evaporation steal rain almost as fast as it falls

NW. Australia

△ **LIFE SUPPORT**
An escarpment just above the Oakover River marks the Great Sandy Desert's western edge. The river provides a lifeline for a range of animals.

▽ **EVER-CHANGING DUNES**
The dunes of this desert are thought to have formed about 10,000 years ago. However, constant winds continue to reshape their crests.

Northwest Australia's Great Sandy Desert is bordered to the east by the rocky hills of the Tanami Desert and to the south by the gravelly Gibson. Besides its sand, what makes the Great Sandy different from its neighbors is its climate: the region gets a surprisingly high amount of rain. The average annual rainfall is 10 in (250 mm), even higher in northern sections, and this would normally result in a reasonable amount of plant growth. In the Great Sandy, however, summer daytime temperatures of around 104°F (40°C) mean that evaporation rates are so high that most moisture never has a chance to penetrate the soil. The prevailing winds blow mainly east to west, creating a landscape of constantly shifting red linear sand dunes.

◁ **SMALL WONDER**
The tiny spinifex hopping mouse has the most efficient kidneys of any mammal, so it can survive without water for long periods.

AUSTRALIA'S DESERTS

Deserts make up around one-fifth of Australia's landmass, making it the driest inhabited continent on Earth, but due to their challenging conditions, they are home to less than 3 percent of Australians. The five largest are shown here.

Great Victoria Desert
134,650 square miles
(348,750 square km)

Great Sandy Desert
103,200 square miles
(267,250 square km)

Tanami Desert
71,250 square miles
(184,500 square km)

Simpson Desert
68,150 square miles
(176,500 square km)

Gibson Desert 60,250 square miles
(156,000 square km)

C. Australia

The Simpson Desert

Australia's most centrally located desert, where sand ranges from pink to deep red and forms the largest parallel dune system in the world

The Simpson Desert forms part of Australia's red center: the colloquial name for the Northern Territory's southern desert region that reflects the dominant color of its sands. Stretching south from the Northern Territory's famous outback town of Alice Springs to cross the border of South Australia, the Simpson covers an area of around 68,150 square miles (176,500 square km). It contains the largest parallel sand dune system on Earth. These dunes vary in height from 10 ft (3 m) in the western desert to 100 ft (30 m) on its eastern side, and can extend for 120 miles (200 km) or more.

A desert in bloom

Rainfall in the Simpson is low and irregular, averaging just 5 in (125 mm) annually. However, spinifex grass, shrubs, and acacia trees grow year round, joined by colorful wildflowers that bloom suddenly after a rare rain shower. Rainfall in nearby areas can also cause some parts of the Simpson to flood as inflowing rivers burst their banks. The resulting vegetation provides a haven for various animals, including the rare fat-tailed marsupial mouse. National parks and conservation areas have been set up in parts of the desert to ensure both flora and fauna are protected.

△ **PLAINS DWELLER**
The gibberbird's name reflects its main habitat—the gibber plains of Australian deserts, a type of desert pavement. It rarely flies, preferring to feed, nest, and roost on the ground.

Uluru is higher than the Statue of Liberty, the Eiffel Tower, and the Great Pyramid

△ **PURPLE BLOSSOM**
A member of the nightshade family, bush tomato shrubs thrive in the Simpson Desert's arid conditions.

▷ **PARALLEL DUNES**
The true size and extent of the Simpson Desert's sandy expanses, as well as the vegetation that grows between the dunes, can be seen from the air.

BENEATH ULURU

The rocks that make up Uluru began in the Amadeus Basin, a depression that formed around 900 million years ago. Collected sediment was compressed into a sandstone layer that folded, fractured, and rotated as tectonic activity pushed it upward. Uluru is just the tip of a formation thought to extend as much as 4 miles (6 km) underground.

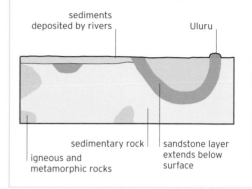

sediments deposited by rivers   Uluru

sedimentary rock   sandstone layer extends below surface

igneous and metamorphic rocks

▽ **ULURU**
Uluru rises 1,142 ft (348 m) above the surrounding desert surface. Around 2¼ miles (3.6 km) long and 1¼ miles (1.9 km) wide, it takes about 3½ hours to walk around the base.

The Gibson Desert

A hostile, largely untouched sandy desert in the center of Western Australia

W. Australia

△ **BIG RED**
Australia's largest living marsupial, the red kangaroo, is common in the Gibson Desert. Adaptations such as hopping conserve energy and minimize contact with hot sand.

Bordered by three other deserts—the Great Sandy (to the north), the Little Sandy (west), and the Great Victoria (south)—the Gibson is one of Australia's most pristine deserts, protected from human impact by its hot, harsh conditions. Few people live here—not surprisingly, given that it is named for an explorer, Alfred Gibson, who perished trying to cross it in 1874. The desert is 60,250 square miles (156,000 square km) of mostly low, undulating dune fields and sandy plains. There are also gravel ridges, formed of pebbles coated in iron oxide, and small saline lakes. Despite summer temperatures of 104°F (40°C) and little water, animals such as kangaroos and emus manage to survive.

▷ **YELLOW BLOOM**
When rain does fall, the Gibson Desert's sand supports thousands of tufts of yellow spinifex grass, one of the few plants able to germinate here.

The Pinnacles

A jagged landscape, with thousands of stone pillars rising out of yellow desert sand, continually covered and uncovered, and all born of rainwater and plant roots

W. Australia

In the center of Western Australia's Nambung National Park lies a coastal desert with a difference. The Pinnacles is the name given to thousands of limestone pillars, rising out of a landscape otherwise covered by sand dunes. How the pillars formed, or why they formed here and not in any other coastal dune environment, is still a subject of debate. However, what is known is that about 30,000–25,000 years ago, they began as crushed seashells, which eventually turned into limestone beneath a dune system.

Root cause

As vegetation developed in the area, rainwater leached through enriched acidic soil, causing a hardened layer to form on top of the softer limestone. Plant roots eventually forced their way down through cracks in this hardpan, setting off a vertical erosion process in the limestone. As winds shifted the sand dunes in a drying climate, the first pinnacles were revealed.

◁ **ROCK SPIRES**
Most of the pinnacles are 3–6 ft (1–2 m) high, but some reach a height of 16 ft (5 m). Winds continually shift the dunes, covering some pinnacles and revealing others.

The Great Victoria Desert

Australia's largest sand-dune desert, shared between two states, dotted with rock pavement, and transected by shrubland

S. Australia

Stretching more than 440 miles (700 km) from west to east across parts of Western Australia and South Australia, the Great Victoria is the continent's largest, yet possibly least desertlike, desert. Although its average annual rainfall is just 6⅜ in (162 mm) and daytime temperatures are as high as 86–113°F (30–45°C), a surprising amount of plant life still manages to thrive here.

Under protection

Eucalyptus trees grow in parts of the region, while acacias and shrubland form a narrow, continuous strip known as the Giles Corridor that stretches

across the desert's entire width. Clumps of spinifex, aristida, and other dry-adapted grasses also break up its dune fields and gibber plains (tightly packed layers of pebbles often covered with a layer of iron oxide). The desert has enough vegetation to attract many animals. It is famous for its reptiles, which number over 100 species, with geckos and skinks being particularly diverse. The region is also home to threatened mammals such as the marsupial mole and sandhill dunnart, and predators such as the dingo and Gould's goanna, a large monitor lizard that averages 3 ft (1 m) in length. The Great Victoria Desert's biodiversity has led to large sections being protected from development.

LUNETTE DUNES

Lunettes are fixed, crescent-shaped dunes found along the edges of temporary saline lakes, as in the Great Victoria Desert. Some are thought to occur due to wind-driven waves, which carry sediment into vegetation, where it becomes trapped.

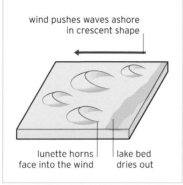

wind pushes waves ashore in crescent shape

lunette horns face into the wind | lake bed dries out

▷ **POWERFUL PREDATOR**
The perentie is Australia's largest lizard and the largest goanna species. Growing up to 6 ft (2 m) long, this formidable predator charges at its prey at a speed of up to 20 mph (32 kph)

▽ **MORE THAN SAND**
Near Kati Thanda-Lake Eyre, Australia's largest salt lake, dunes and spinifex give way to mesas topped by silcrete, a hard layer of gravel and sand cemented together by silica.

Crossing the gap
As the Matusevich Glacier flows toward the coast of East Antarctica, it passes through a narrow gap between hills, causing cracks, or crevasses, to form in the ice. After traversing the gap, the glacier ice floats out onto the sea.

Antarctica

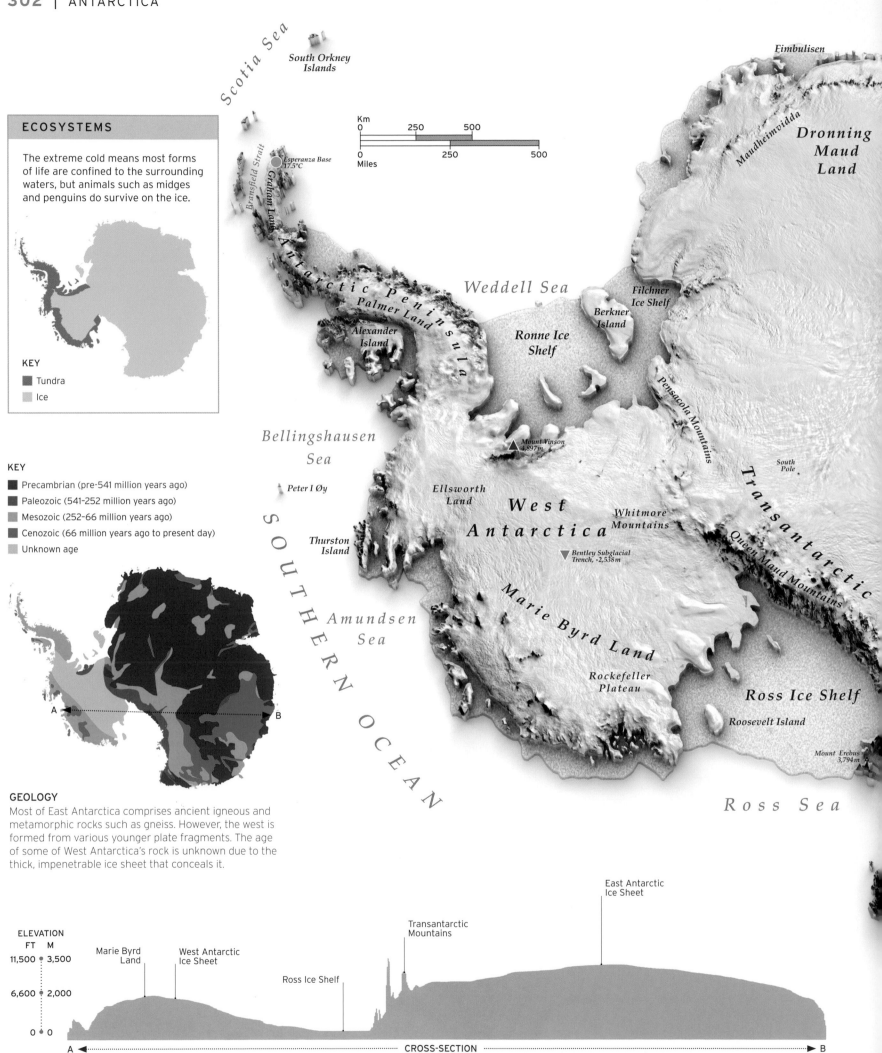

ECOSYSTEMS

The extreme cold means most forms of life are confined to the surrounding waters, but animals such as midges and penguins do survive on the ice.

KEY
- Tundra
- Ice

KEY
- Precambrian (pre-541 million years ago)
- Paleozoic (541-252 million years ago)
- Mesozoic (252-66 million years ago)
- Cenozoic (66 million years ago to present day)
- Unknown age

GEOLOGY

Most of East Antarctica comprises ancient igneous and metamorphic rocks such as gneiss. However, the west is formed from various younger plate fragments. The age of some of West Antarctica's rock is unknown due to the thick, impenetrable ice sheet that conceals it.

Scotia Sea

South Orkney Islands

Fimbulisen

Km
0   250   500
0   250   500
Miles

Esperanza Base
17.5°C

Bransfield Strait

Graham Land

Antarctic peninsula

Palmer Land

Alexander Island

Maudheimvidda

Dronning Maud Land

Weddell Sea

Filchner Ice Shelf

Berkner Island

Ronne Ice Shelf

Pensacola Mountains

Bellingshausen Sea

Mount Vinson 4,897m

South Pole

Ellsworth Land

West Antarctica

Whitmore Mountains

Transantarctic

Peter I Øy

Bentley Subglacial Trench, -2,538m

Queen Maud Mountains

Thurston Island

SOUTHERN OCEAN

Amundsen Sea

Marie Byrd Land

Rockefeller Plateau

Ross Ice Shelf

Roosevelt Island

Mount Erebus 3,794m

Ross Sea

CROSS-SECTION

ELEVATION
FT   M

11,500 • 3,500

6,600 • 2,000

0 • 0

East Antarctic Ice Sheet

Transantarctic Mountains

Marie Byrd Land

West Antarctic Ice Sheet

Ross Ice Shelf

A ◄──── CROSS-SECTION ────► B

THE FROZEN CONTINENT
Antarctica

Entirely surrounded by the Southern Ocean, Antarctica is the most remote continent. South America and Australia are 600 miles (1,000 km) and 1,600 miles (2,500 km) away respectively—but it wasn't always so isolated. Antarctica was part of the Gondwana supercontinent 200 million years ago, before it broke away to drift and settle at the South Pole some 35 million years ago.

The Antarctic Ice Sheet is the world's largest, averaging over 1 mile (1.6 km) thick—and weighing so heavily on the rock beneath that some parts have been pushed 1½ miles (2.5 km) below sea level. Antarctica is divided into two parts, East and West, separated by the Transantarctic Mountains. To the east, an enormous, flat plateau of ice overlies the ancient continental shield. To the west, the icescape and geology are more varied and have much in common with the South American Andes.

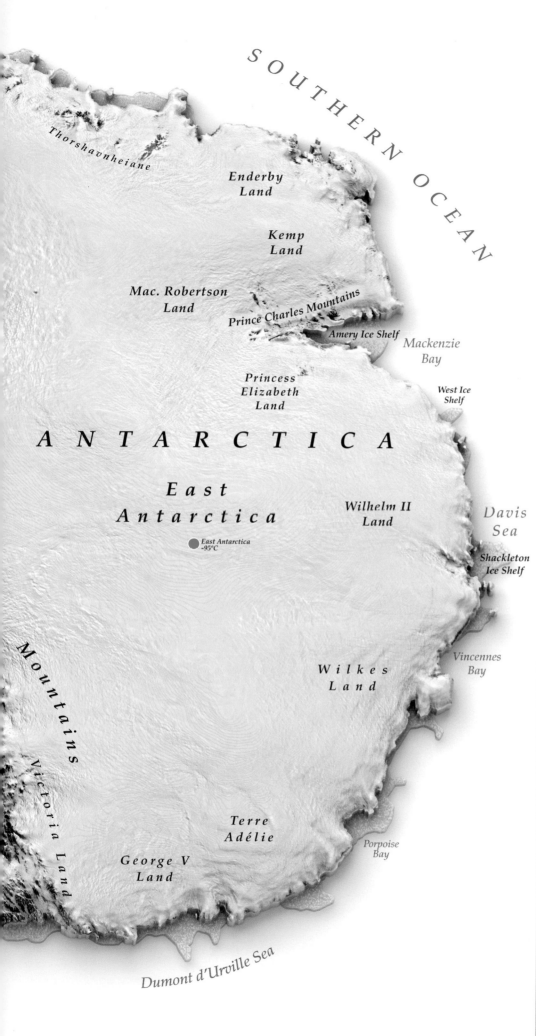

KEY DATA

▲ **Highest point** Mount Vinson: 16,066 ft (4,897 m)

▼ **Lowest point** Bentley Subglacial Trench: -8,327 ft (-2,538 m)

● **Hottest record** Esperanza Base: 63.5°F (17.5°C)

● **Coldest record** East Antarctica: -139°F (-95°C)

CLIMATE

Antarctica is the coldest continent; 99 percent of its land is covered by ice, and freezing temperatures persist all year. High winds can cause violent snowstorms that may last for days.

AVERAGE TEMPERATURE

AVERAGE RAINFALL

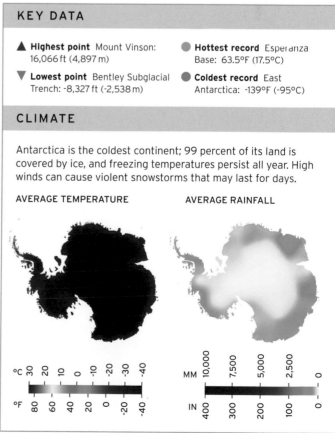

°C 30 20 10 0 -10 -20 -30 -40
°F 80 60 40 20 0 -20 -40

MM 10,000 7,500 5,000 2,500 0
IN 400 300 200 100 0

C. Antarctica

The Transantarctic Mountains

Antarctica's longest mountain range, which was first sighted by people only in 1841

The Transantarctic Mountains are a curved belt of mountains, about 2,200 miles (3,500 km) long, that cross Antarctica. One section divides the continent into its two main parts: East and West Antarctica. Much of the rest of the belt runs along a part of the coast of East Antarctica, including a long stretch where it borders the Ross Ice Shelf. Some large glaciers flow through gaps in the range in this area.

The mountain belt has many subranges, including the Queen Maud Mountains and the Royal Society Range. Eight out of its 10 highest peaks are in just one subrange, the Queen Alexandra Range. Parts of the Transantarctic Mountains were first sighted by the British naval explorer Captain James Ross in 1841, but the Queen Alexandra Range was not discovered until 1908 (by a British Antarctic expedition). Life in the range's interior is limited to microorganisms—such as bacteria and algae. However, in the ancient past a wide variety of animals and plants thrived there, as evidenced by fossils of primitive amphibians and reptiles in rocks laid down 400–180 million years ago.

HIGH PEAKS OF THE TRANSANTARCTIC MOUNTAINS

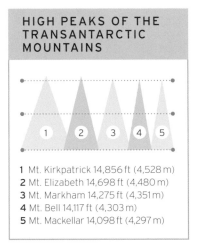

1 Mt. Kirkpatrick 14,856 ft (4,528 m)
2 Mt. Elizabeth 14,698 ft (4,480 m)
3 Mt. Markham 14,275 ft (4,351 m)
4 Mt. Bell 14,117 ft (4,303 m)
5 Mt. Mackellar 14,098 ft (4,297 m)

▽ **BREAKING THROUGH THE ICE**
Among the few features not buried under ice, the tops of the mountains consist of sedimentary rock layers lying on top of granites and gneisses.

△ **WHITE DOME**
This view of Mount Erebus, which is spouting steam at its summit, is from the ice-covered Ross Sea. Erebus has a moderately steep, irregular dome shape.

C. Antarctica

Mount Erebus

Earth's most southerly volcano known to have erupted, with a lava lake and ice caves

Antarctica's most active volcano, Mount Erebus is a 12,448-ft- (3,794-m-) high stratovolcano, located on Ross Island, just off the coast of East Antarctica. Ross Island also contains three other, apparently inactive, volcanoes.

Fire and ice

Erebus was erupting when it was first sighted by British explorer Captain James Ross in 1841, and it is still erupting frequently, if not particularly explosively, today. The volcano is one of just a handful on Earth to contain a long-lasting lava lake. This occupies a vent in the main summit crater and produces regular small explosions of lava. The temperature of the swirling pool of fiery magma is 1,700°F (900°C), and it may be several hundred feet deep. Riddling the sides of the volcano are several ice caves, carved out of its thick blanket of ice and snow by steam and other warm gases escaping from fumaroles (gas vents) on its rocky surface. Above some of the caves are ice chimneys, formed by escaping steam turning to trickles of liquid water, and then ice, on encountering the cold air outside.

△ **HOUSE OF ICE**
This ice cave on Mount Erebus is about 39 ft (12 m) high. It is continuously being reshaped as warm gases melt the ice in some parts, while condensing steam deposits new ice in other parts.

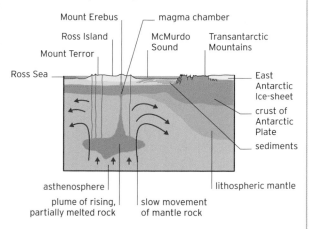

◁ **GAPING HOLE**
Erebus' summit crater is 1,300 ft (400 m) wide and 400 ft (120 m) deep. At its base are several pitlike openings, one of which contains the volcano's lava lake.

UNDERNEATH MOUNT EREBUS

Mount Erebus, as well as Mount Terror and two other currently inactive volcanoes on Ross Island, are thought to sit over plumes of hot material rising up and forming magma under the Antarctic Plate. These plumes account for the past and present volcanism.

Mount Erebus — magma chamber
Ross Island — McMurdo Sound — Transantarctic Mountains
Mount Terror
Ross Sea — East Antarctic Ice-sheet
crust of Antarctic Plate
sediments
asthenosphere — lithospheric mantle
plume of rising, partially melted rock — slow movement of mantle rock

Antarctica

The Antarctic Ice Sheet

The world's largest glacier, having a volume of some 7.2 million cubic miles (30 million cubic km) and holding over 60 percent of Earth's freshwater

The Antarctic Ice Sheet is by far the most extensive continuous mass of ice on Earth, covering an area of 5.4 million square miles (14 million square km). It consists of two main, neighboring, parts. The larger of these, the East Antarctic Ice Sheet, is over 2¾ miles (4.5 km) thick in places and lies on top of a large landmass. It has an area of about 4.6 million square miles (12 million square km). In contrast, the smaller West Antarctic Ice Sheet has an area of about 0.8 million square miles (2 million square km) and a maximum thickness of 2¼ miles (3.5 km), and rests on bedrock that is mostly below sea level. Both parts are slightly dome-shaped, with ice flowing from their highest parts toward the coasts. The rate of ice movement varies from less than 3 feet (1 m) per year in some of the higher regions to a thousand feet (several hundred meters) per year in outlet glaciers and ice streams that drain ice down to the coasts. In some places, the outlet glaciers merge to form vast platforms of floating ice called ice shelves, or extend over the sea as smaller ice tongues.

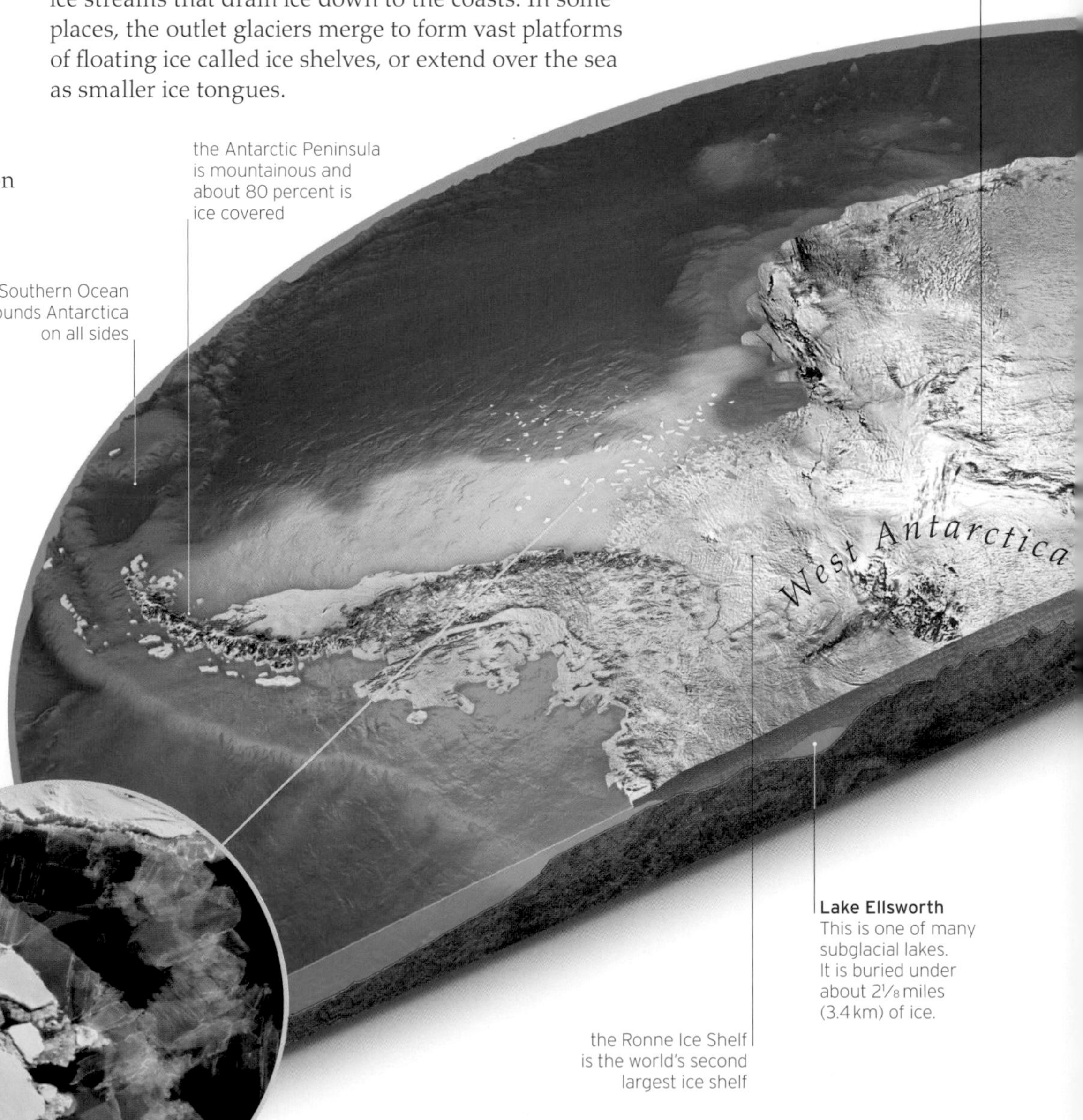

▶ **THE ANTARCTIC ICE SHEET**
The ice sheet, shown here with a cross-section through the Ross Ice Shelf, covers about 99 percent of the Antarctic landmass. The only ice-free areas are the peaks of the highest mountains and a few coastal areas of limited extent.

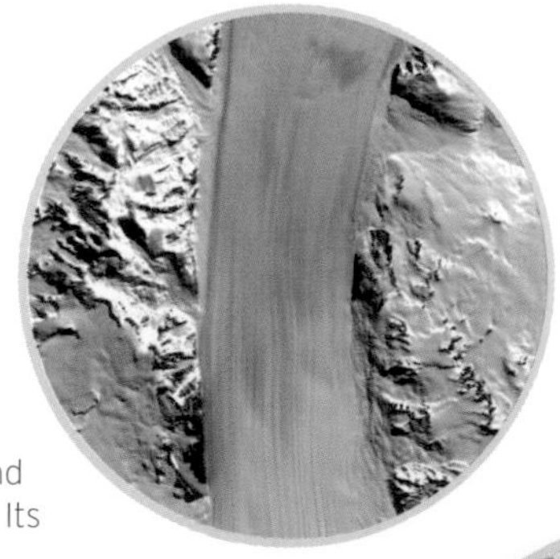

▷ **SPEEDY GLACIER**
The massive Byrd Glacier, which drains part of the East Antarctic Ice Sheet into the Ross Ice Shelf, is about 84 miles (136 km) long and 15 miles (24 km) wide. Its flow rate is high for a glacier—about ½ mile (750 m) per year.

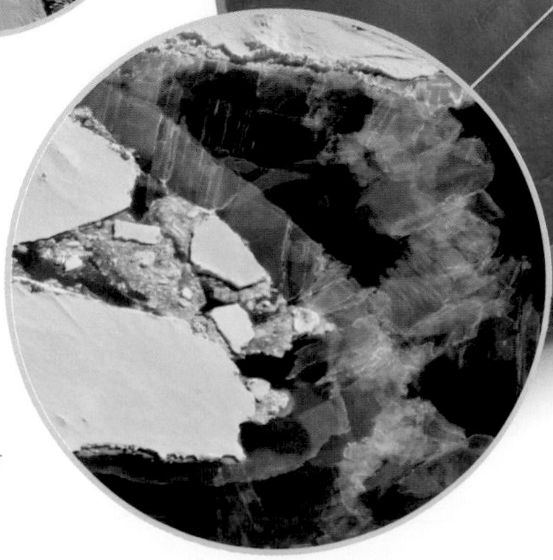

▷ **ICE FLOES**
Sea ice occurs all around Antarctica, although its extent varies seasonally. Unlike ice shelves and icebergs, which originate on land, sea ice is frozen seawater and contains some salt. Flat, freely moving sheets of sea ice are called floes.

Transantarctic Mountains
Along a section of their length, the mountains form a boundary between the two main parts of the ice sheet.

the Antarctic Peninsula is mountainous and about 80 percent is ice covered

Southern Ocean surrounds Antarctica on all sides

West Antarctica

Lake Ellsworth
This is one of many subglacial lakes. It is buried under about 2⅛ miles (3.4 km) of ice.

the Ronne Ice Shelf is the world's second largest ice shelf

Several hundred lakes lie under the ice sheet and on top of the continental crust beneath it

◁ **ABOVE THE ICE**
In the upper Taylor Valley, which lies in the Transantarctic Mountains, the spectacular Taylor Glacier–an outlet glacier from the East Antarctic Ice Sheet–is channeled by the mountains down toward the Ross Sea.

ICE FLOW LINES

Snow falling on the Antarctic Ice Sheet compacts to form ice. Under the force of its own weight, this ice slowly deforms and moves toward the coasts along the flow lines. Imaginary lines called ice divides separate regions of ice moving toward different coasts.

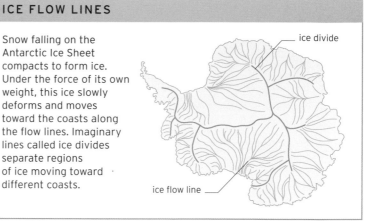

ice divide

ice flow line

High point
The highest area on the ice sheet, called Dome Argus, is more than 13,000 ft (4,000 m) above sea level.

East Antarctica

△ **CLIFFS OF ICE**
The Mertz Glacier ends in an ice tongue that protrudes about 12–15 miles (20–25 km) over the Southern Ocean. At its terminus are cliffs that tower as much as 165 ft (50 m) above the sea surface.

Bedrock
Continental crust underlies the ice sheet. In some areas, its surface is below sea level.

the Ross Ice Shelf floats with around 90 percent of its bulk underwater

Grounding line
This marks the boundary between the floating part of an ice shelf and grounded ice feeding into it.

Icebergs calving from the Ross Ice Shelf

Continental shelf
This occupies a strip about 50–200 miles (80–320 km) wide around the whole of Antarctica.

△ **ROSS ICE SHELF**
The world's largest ice shelf, with an ice cliff along its edge that is more than 370 miles (600 km) long, the Ross Ice Shelf (see p.308) has a total area of about 188,000 square miles (487,000 square km).

The Ross Ice Shelf

The world's largest floating body of ice, approximately the same size as mainland France

S. Antarctica

△ THE BARRIER
The near-vertical ice cliff at the front end of the Ross Ice Shelf is up to 165 ft (50 m) high. Underwater, it plunges down another 980 ft (300 m) or so.

The Ross Ice Shelf is a roughly triangular area of floating ice that extends up to 280 miles (450 km) from the shore of Antarctica, over the southern part of the Ross Sea. It ranges in thickness from about 1,150 ft (350 m) at its front end to around 2,460 ft (750 m) farther back, close to where its base is in contact with land. Named after Captain James Ross, who discovered it in 1841, the ice shelf was originally called the Barrier. The ice comprising the shelf flows seaward at a rate of about 2,950 ft (900 m) per year and is fed by several outlet glaciers and ice streams that flow down into it from the East Antarctic Ice Sheet (see p.306–307).

These include the Byrd, Nimrod, Beardmore, Scott, Shackleton, and Amundsen glaciers. The upper surface of the Ross Ice Shelf is an inhospitable place, subject to strong cold winds that shape it into a series of ridges and troughs called sastrugi.

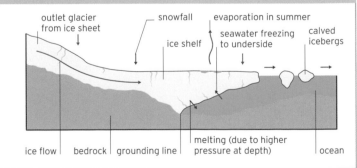

◁ SHELF SHELTERERS
Some emperor penguin colonies occupy the shores of the Ross Sea a short distance from the base of the ice shelf, where the steep cliffs provide some protection against wind.

ICE SHELF GAINS AND LOSSES

An ice shelf gains ice from glaciers flowing into its landward end, from water freezing to its underside, and from new snowfall. It loses ice mainly from icebergs calving (breaking off) from its front edge, and also as a result of some melting and evaporation.

outlet glacier from ice sheet — snowfall — evaporation in summer — seawater freezing to underside — ice shelf — calved icebergs

ice flow | bedrock | grounding line | melting (due to higher pressure at depth) | ocean

Antarctic Tundra

The world's most inhospitable continent for vegetation, where only 1 percent is covered in plant life

W. Antarctica

Given that around 99 percent of Antarctica is covered in ice, it is surprising that any plants survive on Earth's southernmost, most elevated continent. Ice sheets, glaciers, and mountains make up its interior, yet small patches of tundra exist, mainly on the Antarctic Peninsula and several sub-Antarctic islands, as well as on a few exposed inland rocks known as nunataks. Only two flowering plant species grow here—Antarctic hair grass and Antarctic pearlwort—joined by around 100 moss species, 25 liverwort species, and 300–400 lichen species. Due to the region's extreme climate, where icy winds blow year round and average annual temperatures range from 14°F to -76°F (-10°C to -60°C), some algae and lichen species survive inside rocks, growing within tiny pores.

▽ **FLORAL CUSHION**
Antarctic pearlwort grows in cushionlike clumps, reaching a height of around 2 in (5 cm).

The McMurdo Dry Valleys

One of Earth's most extreme desert areas, where organisms survive severe cold, arid conditions

S. Antarctica

The McMurdo Dry Valleys lie west of the McMurdo Sound, bordering the Ross Sea. Covering just 0.03 percent of the continental landmass, they contain the largest ice-free region in Antarctica and constitute an extreme cold desert ecosystem, where temperatures can plunge to -90°F (-68°C) and winds rip through at up to 200 mph (322 kph). The surrounding mountains block the valleys from the East Antarctic Ice Sheet, and humidity levels are extremely low, yet microbial life survives here in all but the driest areas.

△ **STAINED LANDSCAPE**
The outflow of a saline lake, the Blood Falls gets its red color from the water's high iron content.

◁ **SUPER-SALTY**
The shallow Don Juan Pond is so salty it does not freeze even in winter, unlike other water bodies in the dry valleys.

△ **BIGGEST ICEBERG**
In 2000, the largest iceberg ever recorded, called B-15 (to left of center), broke off the Ross Ice Shelf (at bottom in this image).

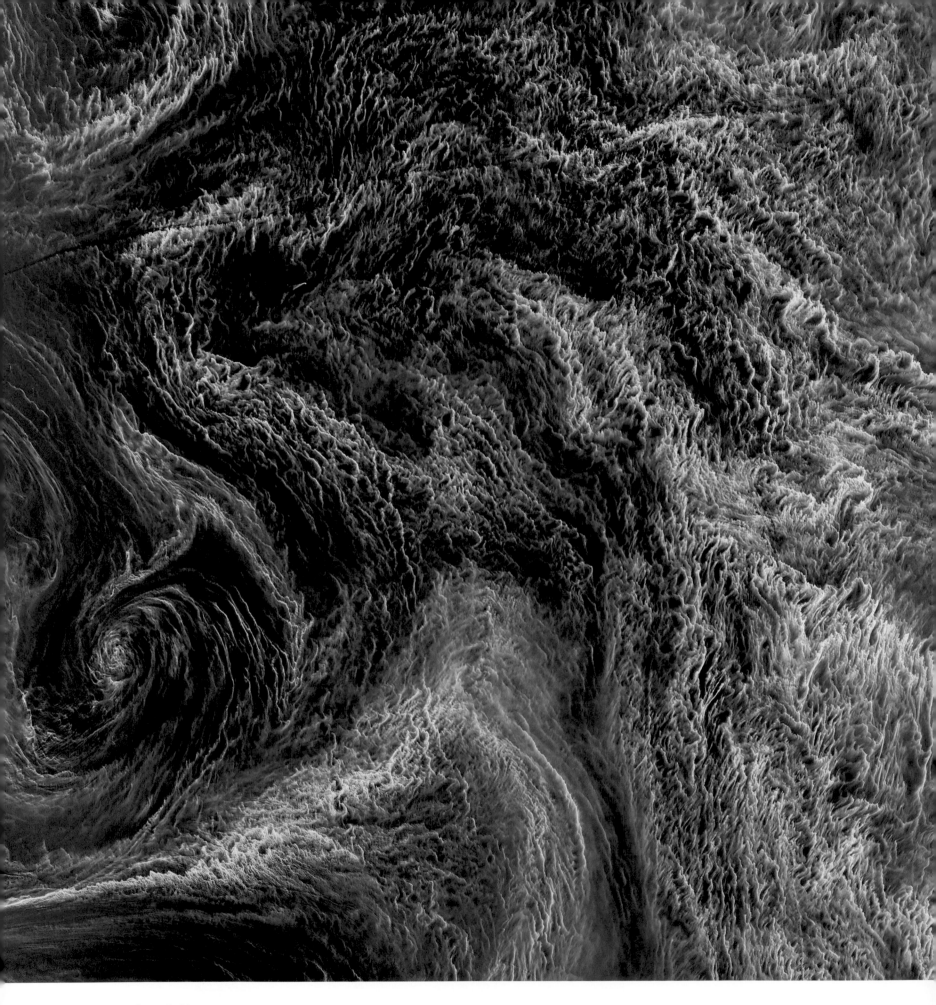

Ocean in bloom
Under the perpetual sunshine of an Arctic summer, a
huge bloom of cyanobacteria—an ancient type of marine
bacteria—spreads out across the waters of the Baltic Sea
between Sweden and Latvia.

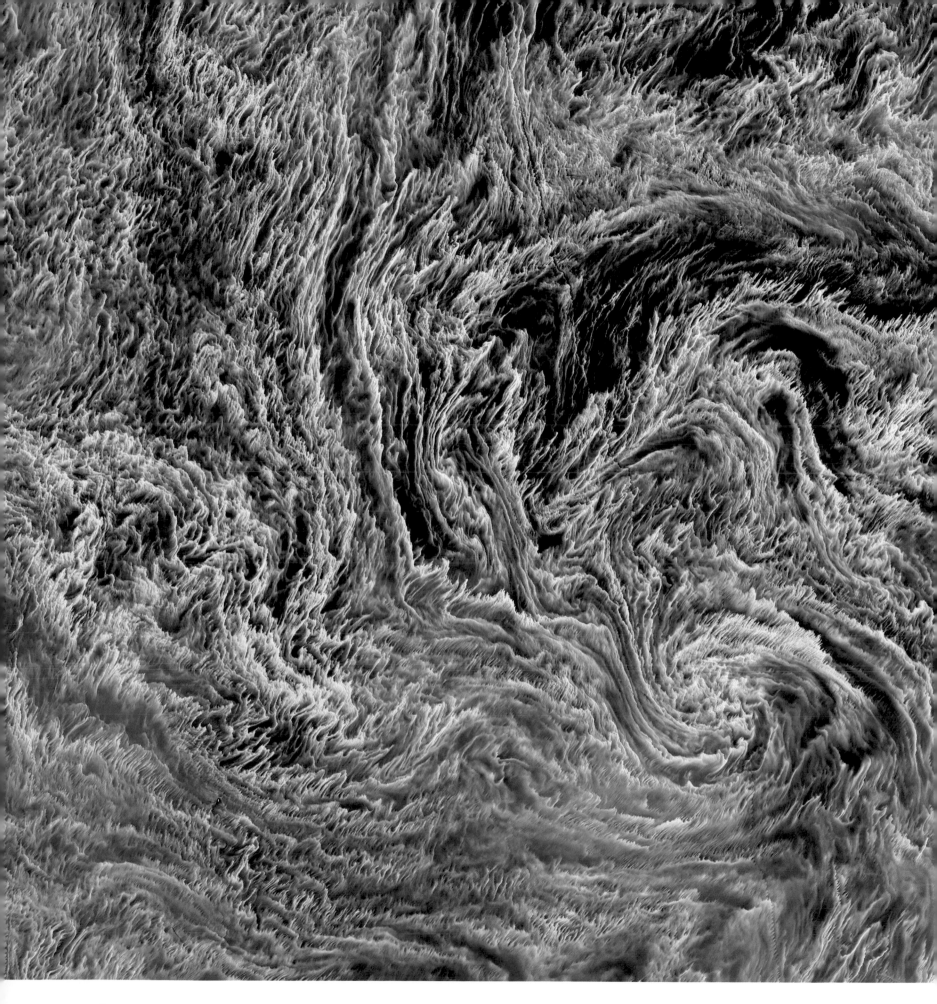

The Oceans

Atlantic
Ocean

The Mid-Atlantic Ridge

*Part of a global oceanic ridge system, a huge underwater
mountain chain that is pushing continents apart*

Running down the middle of the Atlantic Ocean, the
Mid-Atlantic Ridge is a slow-spreading submarine
ridge system that extends around 10,000 miles
(16,000 km) from the Arctic to beyond the southern tip
of Africa. The ridge marks a boundary where the
North and South American tectonic plates are
diverging from the Eurasian and African Plates.

Mountains and valleys

The path of the ridge mirrors the contours of the
continents it not only divides but is also pushing apart
at a rate of ¾–2 in (2–5 cm) per year—much slower
than the 6¼ in (16 cm) per year of the East Pacific Rise
(see pp.322–23). The Mid-Atlantic Ridge rises 1¼ –2
miles (2–3 km) above the seafloor in a long chain of
underwater mountains, and it has a deep central rift
valley along its crest. Where the ridge has pushed
above sea level, it has created landmasses such as
the volcanic islands of Iceland and the Azores.

MAGNETIC BANDS AT SPREADING RIDGES

Earth's magnetic field periodically reverses, changing the
north magnetic pole to the south magnetic pole, and vice
versa. Why this occurs is unknown, but it is recorded in
bands of basaltic rock on either side of mid-ocean ridges.
Metallic elements within new crust created at a spreading
ridge are aligned with Earth's polarity at the time the
basaltic lava emerges. The study of these bands has been
a crucial line of evidence in support of the theory of plate
tectonics because it helped prove that Earth's plates move.

band of reversed
polarity

bands of equal age sit
symmetrically either
side of ridge

band of normal
polarity

magma rises
in mid-ocean
ridge, forcing
plates apart

▽ **CARBONATE TOWERS**
During 2000, an undersea research
team filmed a cluster of ghostly spires
90-200 ft (27-61 m) high. This is the Lost
City, the only known hydrothermal field
made solely of white
carbonate chimneys.

North America

▶ **DIVERGENT BOUNDARY**
Mid-oceanic ridges such as the Mid-Atlantic
Ridge mark areas where tectonic plates
diverge as new crust forms. Deep below
the ocean surface, the upwelling of magma
from the mantle causes volcanic eruptions,
creating mountain ranges, rift valleys, and
hydrothermal vents along the ridge's length.

Transform fault
Instead of forming a straight line,
the ridge follows a stepped path,
with sections offset along faults
called transform faults. Movement
along these faults often sets off
shallow earthquakes.

South
America

continental
crust

△ **SEABED PILLOWS**
Lava erupted along the ridge cools
rapidly in the cold seawater to form
mounds 20 in-3 ft (0.5-1 m) across
called pillow lavas.

westward movement of
South American Plate

oceanic crust

Asthenosphere
Below the lithosphere lies the
asthenosphere, part of Earth's
mantle. Because it remains hot
and fluid, it acts like a lubricant
on the underside of tectonic
plates, allowing them to move.

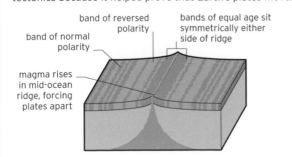

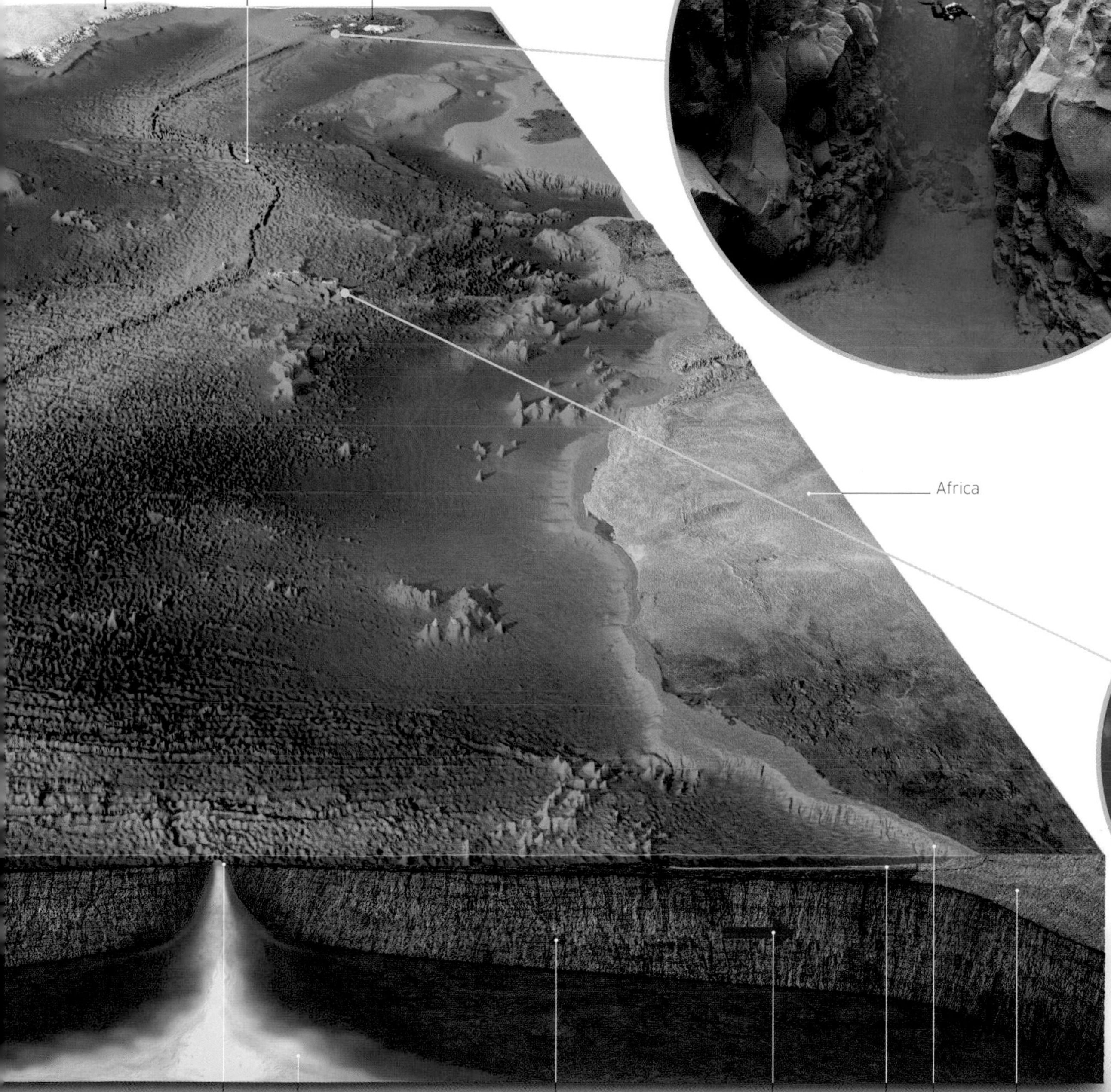

Mid-oceanic ridge
The Mid-Atlantic Ridge is part of a vast global network of mid-oceanic ridge systems—the longest mountain chain on Earth.

Greenland

Iceland

◁ **FRESHWATER FISSURE**
Iceland was formed by hotspot activity (see p.318) merging with the Mid-Atlantic Ridge. Further ridge spreading created a deep fissure, known as Silfra, at the bottom of Thingvellir Valley. Silfra is spring-fed with pure glacial meltwater, providing clear visibility for divers as they swim between the North American and Eurasian tectonic plates.

Africa

△ **ISLAND CHAIN**
The Azores are an example of an island chain formed by a combination of hotspot (see p.318) and ridge activity. When a hotspot occurs near a mid-oceanic ridge, the ridge segments nearest the hotspot are exposed to larger volumes of magma, leading to greater volcanic activity in the area.

continental crust

Surface eruption
Once magma erupts through the seabed, it emerges as lava and cools, creating new oceanic crust and causing seafloor spreading.

Mantle upwelling
Originating in the upper mantle and lower crust, magma is a mixture of molten and semimolten rock. It rises at points of weakness in the oceanic crust.

eastward movement of Eurasian plate

continental shelf

sediment on ocean floor

lithospheric mantle

Most of the Mid-Atlantic Ridge's **central rift valley** is **deeper and wider** than the **Grand Canyon**

The Seychelles

More than 100 tropical islands of remarkable beauty, home to the world's rarest palm and one of the largest raised coral atolls on Earth

W. Indian Ocean

△ **TIDAL CHANNELS**
On Aldabra Atoll, channeling tides have created many habitats, including a saline lagoon, mangrove swamps, and seagrass beds that support numerous aquatic species.

The Seychelles' 115 tropical islands lie off the east coast of Africa in the western Indian Ocean and are divided into two groups. Most of the 41 mountainous granite Inner Islands are located 4° south of the equator and are projections of the large undersea Mascarene Plateau that extends south to Réunion. Among the Inner Islands is Praslin, whose

Vallée de Mai World Heritage Site contains Earth's largest forest of endangered coco-de-mer palms. In contrast, the low-lying, coralline Outer Islands are situated beyond the plateau, 10° south of the equator, and include the Aldabra group, a raised coral atoll that has been a designated nature reserve since 1976.

More than **152,000 Aldabra giant tortoises** live on the atoll

▷ **DOUBLE COCONUT**
The coco-de-mer is endemic to the islands of Praslin and Curieuse. Its double coconut is the heaviest seed on Earth and takes seven years to ripen.

The Great Chagos Bank

The largest coral atoll on the planet, surrounded by some of the cleanest waters in the world

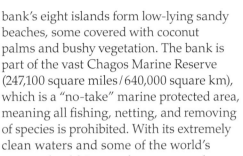

C. Indian Ocean

Located a few hundred miles south of the Maldives (see below), the Great Chagos Bank is Earth's largest coral atoll in terms of area, although most of its 4,880 square miles (12,640 square km) lie underwater. Just 2.2 square miles (5.6 square km) of the bank's eight islands form low-lying sandy beaches, some covered with coconut palms and bushy vegetation. The bank is part of the vast Chagos Marine Reserve (247,100 square miles / 640,000 square km), which is a "no-take" marine protected area, meaning all fishing, netting, and removing of species is prohibited. With its extremely clean waters and some of the world's healthiest reef systems, and comprising seamounts and deep knolls as well as shallow plains, Chagos is one of the world's richest marine environments.

◁ **MARINE REFUGE**
More than 220 species of corals make up the Great Chagos Bank, creating a reef habitat for around 800 fish species, including anemonefish.

▽ **RACE TO THE SEA**
Bird Island is a center of wildlife conservation. Here, endangered hawksbill and green sea turtles lay their eggs, which are protected until they hatch. The hatchlings then head out to sea.

The Maldives

Asia's smallest country, with more than 1,000 low-lying islands, all at the mercy of the sea

C. Indian Ocean

Around 1,190 live coral reefs and sandbar islands make up the double chain of atolls of the tropical Maldives, off the southwestern coast of India. The island count fluctuates, because the Maldives has the lowest sea level of any country. With an average elevation of just 6 ft (1.8 m) above sea level, islands come and go as sea level and climate change. All are projections of the underwater Chagos Laccadive volcanic mountain range. Within each atoll are up to 10 inhabited islands and up to 60 uninhabited ones, most with white sandy beaches that are important nesting grounds for green sea turtles.

△ **GLOWING TIDE**
Bioluminescent plankton light up the surf as waves wash ashore in the Maldives. Marine life in the area ranges from microscopic organisms to whale sharks.

E. Pacific Ocean

The Hawaiian Islands

The longest island chain on Earth, where continual volcanic activity has created a unique tropical habitat

Situated in the Pacific Ocean about 2,360 miles (3,800 km) southwest of the continental U.S. and around 3,850 miles (6,200 km) southeast of Japan, the Hawaiian Islands are some of the most remote in the world. At 1,520 miles (2,450 km) long, they also form the world's longest archipelago.

Expanding landscape

The eight main islands and more than 100 islets are part of the Hawaiian Island–Emperor Seamount chain of undersea volcanoes that began to erupt around 70 million years ago. Each main island has one or more volcanoes, and three are currently active on Hawaii itself (see pp.318–19). The constant outpouring of lava is still adding to Hawaii's landmass: it has gained about ¾ square mile (2 square km) of land since 1983, created by the continual eruption of Kilauea. Hawaiian animals and plants are as unique as their island homes—more than 90 percent are endemic, but this makes them vulnerable, and the islands contain at least one-third of all listed endangered species of the U.S.

▷ **LAVA RIVERS**
Hawaii's Kilauea volcano releases 7,000,000–17,650,000 cubic ft (200,000–500,000 cubic m) of lava each day—enough to resurface 20 miles (32 km) of a two-lane road. Fortunately, most of it flows into the sea.

◁ **STATE FLOWER**
The endangered Hawaiian hibiscus is found on all main Hawaiian islands except Niihau and Kahoolawe, and blooms mainly from spring through early summer.

THE WORLD'S TALLEST VOLCANOES

A mountain's height is usually measured from sea level to summit, but submerged volcanoes are measured from their bases on the sea floor. Hawaii's Mauna Loa is technically the world's tallest volcano. Its base is a depression it created in the seabed, which gives it a total height of 56,332 ft (17,170 m)—more than 27,000 ft (8,230 m) taller than Mount Everest.

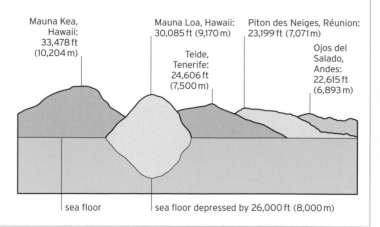

Mauna Kea, Hawaii: 33,478 ft (10,204 m)

Mauna Loa, Hawaii: 30,085 ft (9,170 m)

Piton des Neiges, Réunion: 23,199 ft (7,071 m)

Teide, Tenerife: 24,606 ft (7,500 m)

Ojos del Salado, Andes: 22,615 ft (6,893 m)

sea floor

sea floor depressed by 26,000 ft (8,000 m)

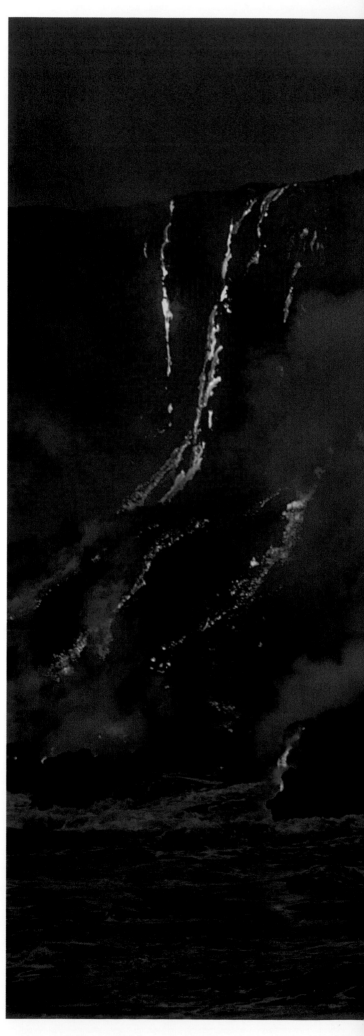

The formation of a
VOLCANIC ISLAND CHAIN

Kilauea has released enough **lava** to **encircle** Earth three times

Like beads in a necklace, volcanic islands often form in arcs, or chains, where two oceanic tectonic plates converge or where an oceanic plate meets a continental plate. The volcanic island chains that encircle the Pacific Ocean, which are known as the Ring of Fire, are formed by both processes, as well as, in the case of the Hawaiian Islands, hotspot volcanism.

Hot marks the spot

Hotspots are areas in Earth's mantle that allow magma to rise through the lithosphere until it erupts, often underwater on the seabed. The hotspot that formed the Hawaiian island chain is located in the middle of the Pacific Plate. While the hotspot is fixed, the plate moves continuously. As the plate migrated over the hotspot, the latter created the shield volcanoes responsible for all 132 islands, atolls, reefs, banks, and seamounts of the Hawaiian archipelago. Currently, the hotspot lies beneath the island of Hawaii, where three volcanoes continue to erupt. It also generates activity at the youngest member of the chain, the submarine volcano Loihi, about 19 miles (30 km) south of Hawaii. Loihi last erupted in 1996, following a series of minor earthquakes.

HOTSPOTS AND FLOOD BASALTS

Hotspots are thought to be stationary. As tectonic plates move over them, chains of progressively older volcanoes form in what is known as a hotspot track. The track follows the direction of plate movement, with the youngest, most active volcanoes and seamounts found nearest the hotspot and the less active and extinct ones farthest away. Hotspot tracks are often marked by large areas of basaltic lava, which flooded the area during peak eruption periods.

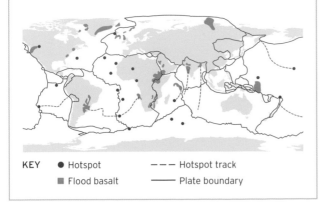

KEY   ● Hotspot   --- Hotspot track
       ■ Flood basalt   — Plate boundary

Niihau
Formed around 4.9 million years ago.

Kauai
Formed around 5.1 million years ago.

Oahu
Formed 3.7–2.6 million years ago.

oceanic crust

asthenosphere

Lithospheric mantle
The crust and uppermost layer of the mantle make up the lithosphere, which is broken up into mobile tectonic plates.

Movement of plate
The Pacific Plate moves over the asthenosphere from southeast to northwest at a rate of 2–4 in (5–10 cm) a year.

◁ **SCULPTED CLIFFS**
The fluted Na Pali cliffs dominate the northwest coast of the island of Kauai, towering up to 4,000 ft (1,200 m) above the Pacific Ocean. The flanks of shield volcanoes often collapse, resulting in jagged, water-permeable surfaces that are further sculpted by rain and wave action.

▲ **THE HAWAIIAN ISLANDS**
This cross-section shows how conditions in Earth's interior have given rise to the Hawaiian Islands. The Pacific Plate moves over a mantle hotspot, setting off volcanic activity at the surface. With increasing distance from the hotspot, the islands cool and activity diminishes.

▷ **GREEN SAND**
One of only a handful of green sand beaches in the world, Papakolea borders Mahana Bay, on the southern tip of Hawaii. The sand gets its color from olivine, a mineral released by eruptions from a now-dormant volcano, whose cinder cone forms three sides of the bay.

▽ **LAND BUILDER**
One of the world's most active volcanoes, Kilauea's eruptions appear violent, but are tame compared to more explosive continental volcanoes. Its basaltic lava flows build layer upon layer of what will one day be fertile soil.

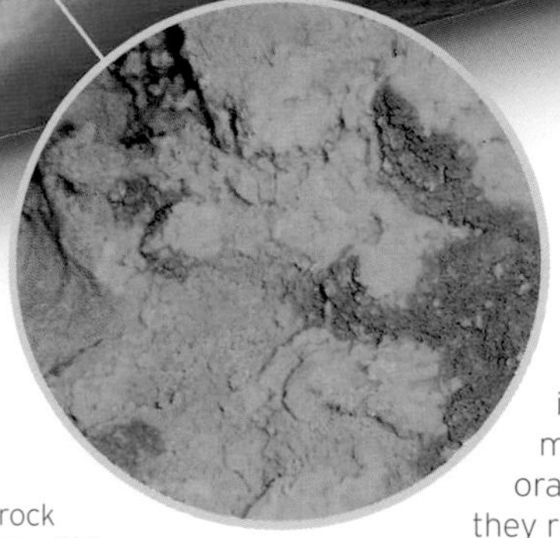

Molokai
Formed 1.9–1.8 million years ago.

Maui
Formed 1.3–0.8 millon years ago.

Hawaii
Also known as Big Island, Hawaii began to form less than 0.5 million years ago and is still forming today due to the activity of three of its five volcanoes.

movement of plate drags head of mantle plume

◁ **IRON SIDES**
Bacteria that feed on iron deposits in volcanic material form yellowish-orange mats, or flocs, as they rust (oxidize) the metal. Many such flocs have been found on the submerged flanks of the Loihi seamount, Hawaii's youngest submarine volcano.

Magma chamber
A magma chamber forms as heat from the mantle plume melts part of the lithosphere. As the plate moves across a magma chamber, the molten rock pushes upward in weaker areas, creating a volcano that punches through the plate.

Magma intrusion
The magma rises, or intrudes, into lower-density rock, creating dikes. Eventually enough dikes are created to allow magma to collect in a chamber.

Mantle plume
Mantle plumes begin as large columns of molten rock (magma), which form deep within Earth's interior and rise through the mantle. Upon reaching the base of the lithosphere, the magma spreads, forming a cap or plume.

E. Pacific Ocean

The Galápagos Islands

A group of Pacific islands, whose unique ecosystem and wildlife provide extraordinary insights into evolution

Clustered around the equator about 620 miles (1,000 km) off the coast of Ecuador, the rocky Galápagos Islands consist of 13 major islands, six smaller ones, and more than 100 small outcrops. All are situated on the Nazca tectonic plate, whose slow but steady movement eastward results in frequent volcanic eruptions. Three major ocean currents converge here, bringing in a variety of marine creatures as well as creating a much cooler climate than expected at the equator.

A window on evolution

Due to the islands' remote location and volcanic conditions, life in the Galápagos evolved in relative isolation, and many species, such as native iguanas and giant tortoises, have not changed radically since prehistoric times. The adaptations of animals on different islands that led to their divergence as separate species inspired the British naturalist Charles Darwin to develop his groundbreaking theory of evolution in the 19th century. Today, the Galápagos is one of the world's most scientifically and biologically important regions, where unique native species make up a very high proportion of the islands' wildlife.

▷ **DAISY TREE**
The shrubby Stewart's scalesia grows only on Santiago and Bartolomé Islands. Its daisy-shaped flower heads release dandelion-like seeds.

Geologically, the Galápagos are **young islands**, the **oldest** having emerged just **4.2 million years ago**

△ SPECIALIZED BILL
The common cactus finch uses its specially adapted bill to feed on the pollen, nectar, fruit, and seeds of *Opuntia* cacti.

GALÁPAGOS VEGETATION ZONES

On the Galápagos, rainfall generally increases with altitude, creating three main vegetation zones. The arid zone harbors cacti and other drought-resistant plants. On larger islands, small trees and shrubs dominate the higher transition zone, before giving way to lush forests in the humid zone.

KEY
Arid · Transition · Humid

◁ GIGANTIC CALDERA
The Sierra Negra volcano is one of six volcanoes that formed Isabela Island. Five are still active, including Sierra Negra, which has an elliptical caldera.

▽ VOLCANIC ISLET
One of the archipelago's youngest islands, Bartolomé Island off the coast of Santiago is home to a breeding colony of Galápagos penguin, the world's second smallest penguin species.

E. Pacific
Ocean

The East Pacific Rise

A volcanic ridge deep on the ocean floor, where life thrives in the harshest of conditions

The East Pacific Rise is part of a volcanic mid-oceanic ridge system that extends along the floor of the southeastern Pacific Ocean. It lies at a depth of about 1½ miles (2.4 km) and rises about 5,900–8,900 ft (1,800–2,700 m) above the sea floor.

Oceanic ridges occur where tectonic plates move apart (see pp.312–13). The process, called rifting, creates fissures along the seabed, releasing lava that forms a ridge as it cools. New basaltic lava is added to either side of its crest as the plates spread apart. Compared to the Mid-Atlantic Ridge, the East Pacific Rise spreads relatively quickly and lacks a pronounced central rift valley. The rifting process produces geyser-like hydrothermal vents that discharge heated, mineral-laden seawater. Despite toxic sulfides, high temperatures, and enormous underwater pressures, the vents support an amazing community of creatures, from single-celled organisms to giant tubeworms.

THE WORLD'S VENT FIELDS

The first hydrothermal vents were discovered in 1977 by an expedition exploring a spreading ridge near the Galápagos Islands. Vent fields have since been recorded in all of the oceans. They are formed in volcanically active areas, which is why so many are found along diverging plate boundaries that coincide with mid-oceanic ridges. Others are created by hotspots (see p.318) or where tectonic plates converge and one pushes beneath (subducts) the other.

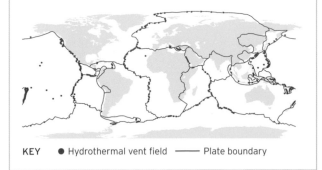

KEY   ● Hydrothermal vent field   —— Plate boundary

◁ **STRANGE FLOWER**
The ocean dandelion is a type of deep-sea siphonophore, a group of animals that includes jellyfish and the Portuguese man o' war. It is not one, but a collection of individuals that share tissues.

Vent clams
These clams anchor themselves into basalt crevices on the seabed in clumps known as clambakes.

Vent mussels
Deep-sea mussels harbor symbiotic bacteria that convert hydrogen released by hydrothermal vents into energy.

Bacterial mats
Bacteria that harvest vent minerals for food form thick mats, which are eaten by other creatures.

▶ **HOW BLACK SMOKERS ARE FORMED**
Hydrothermal vents known as black smokers form when seawater penetrates fissures in new crust created by volcanic activity. Magma superheats the water to 750°F (400°C) or more, dissolving material from surrounding rocks. This superhot vent fluid rises back through openings in the seabed. As they cool, the dissolved minerals precipitate into solid form, creating mineral- and metal-rich chimneylike vents.

Cold seawater descends
Seawater at about 28°F (2°C) seeps down through fissures in the seabed created by volcanic activity, where it is heated by magma.

More than **500 species** have been found living at **hydrothermal vents**

△ BLACK FLUID
Hydrothermal vents emit black or white "smoke"—heated fluids filled with dissolved minerals. Black emissions get their color from iron monosulfide.

Shaft
As minerals accumulate, building the chimney walls, a central shaft allows the vent fluid to escape.

Chimney
As hot vent fluid meets cold seawater, mineral deposits create chimneylike channels up to 50 ft (15 m) high.

Vent crab
This white crab feeds on bacteria, clams, mussels, and tube worms.

△ COLORFUL VENT
Microbial mats also form on the sides of vents, where the action of bacteria on iron compounds released from vent fluid turns the walls a rusty orange.

Tube worms
Giant tube worms, which reach lengths of up to 6 ft (2 m), anchor themselves around a vent. Bacteria inside each worm's body converts vent chemicals into food.

△ TOP PREDATOR
Vent eelpouts are believed to feed on shrimp and crustaceans in vent communities. Due to the water's toxic levels of sulfides, eelpouts have few competitors.

Mound
As dissolved minerals in the vent fluid meet cooler seawater, they fall out of solution, building up a mound of debris as well as the vent chimney.

Heat source
Far below the seabed, upwelling magma superheats the incoming seawater, dissolving minerals such as metals and sulfur from surrounding rocks.

Heated water rises
The superheated vent fluid rises back to the surface of the seabed through fissures in the ocean crust.

W. Pacific

The Mariana Trench

The deepest known point on the seabed of planet Earth, created by the collision of two oceanic plates

On the floor of the western Pacific Ocean, around 200 miles (322 km) southwest of the island of Guam, the Mariana Trench plunges to a staggering 36,201 ft (11,034 m) at a section called the Challenger Deep—the deepest point yet discovered in any of Earth's oceans. If the world's highest mountain, Mount Everest, were placed inside the trench, it would still be covered by 7,162 ft (2,183 m) of water.

When plates collide

The Mariana Trench marks the location of a subduction zone, the point at which one oceanic tectonic plate sinks under another. Here, the huge Pacific Plate has plunged so dramatically beneath the much smaller Philippine Plate and Mariana Microplate that it has created the steep-sided trench, which, at over 1,580 miles (2,540 km) long, is more than five times longer than the Grand Canyon yet averages just 43 miles (69 km) wide.

PARTS OF THE OCEAN FLOOR

Just like landmasses, seabeds contain plains, ridges, mountains, and canyons. From the shoreline, the continental shelf inclines gradually until it drops away sharply into the deep, open waters of the continental slope. Where there is no mid-ocean ridge system, this levels out into the abyssal plain before plunging into deep-sea trenches such as the Challenger Deep.

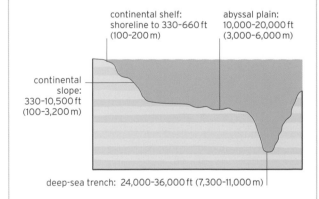

continental shelf: shoreline to 330–660 ft (100–200 m)

abyssal plain: 10,000–20,000 ft (3,000–6,000 m)

continental slope: 330–10,500 ft (100–3,200 m)

deep-sea trench: 24,000–36,000 ft (7,300–11,000 m)

▷ **WHITE SMOKE**
Undersea volcanic activity creates hydrothermal vents along the Mariana volcanic arc, such as this white smoker—the so-called Champagne Vent—which releases bubbles of liquid carbon dioxide.

▶ **SUBDUCTING PLATE**
These cross-sections of the Mariana Trench subduction zone show many of the geological processes involved. As the two oceanic plates converge, one pushes beneath the other, creating the trench itself but also setting in motion the processes that produce chains of seamounts and islands, and new ocean floor.

Remnant arc
The remnant arc once contained active volcanoes, but it has since been pushed away from eruptive areas and is now inactive.

Philippine Plate
Although the Philippine Plate is overriding the Pacific Plate, the upwelling mantle results in spreading, and the plate moves away from the subduction zone.

Active back-arc basin
Back-arc basins are created behind island arcs as a result of rifting and spreading of the sea floor due to the huge forces created by the presence of a deep ocean trench.

Remnant arc
When volcanic activity at an undersea ridge ceases, the ridge is known as a remnant arc.

seamount, an inactive submarine volcano

oceanic crust

lithospheric mantle

asthenosphere

Mariana microplate movement
The Mariana island arc rests on a tectonic microplate called the Mariana Plate. Located between sections of the Philippine and Pacific Plates, it is diverging from the former, while the latter is subducting beneath it at the Pacific Plate's southeastern end.

The Challenger Deep
Despite total darkness, extremely cold water, and crushing pressure, the Challenger Deep is a habitat that supports a host of living creatures.

Subducting plate
As the Pacific Plate slides beneath the overlying plate at a rate of about 1¼ in (3 cm) per year, the plates rub against each other, generating tremors that can be felt across a wide area.

The **pressure** at the bottom of the **Challenger Deep** is **equal** to that underneath **220 Empire State Buildings** stacked on top of each other

Seafloor spreading
Behind the volcanic arc, there is an active spreading ridge where the Philippine and Mariana plates are moving away from each other. Here, intense volcanic activity creates new oceanic crust as magma rises from the underlying mantle.

Mariana Microplate

Active arc
Most volcanoes of the active volcanic arc are seamounts, but some have risen enough to form islands.

Volcanic activity
As the Pacific Plate sinks deep into the mantle, it releases water that triggers melting in the overlying rocks.

Forearc
The region between the volcanic arc and trench is called the forearc. Here, the forces of subduction are evident in the numerous faults that slice up the forearc rock sequences.

Pacific Plate
Faults in the Pacific Plate enable it to bend and plunge almost vertically beneath the Philippine Plate.

△ **ARC ISLAND**
The Mariana island arc parallels the curve of the Mariana Trench. Volcanic island arcs like this are created by the eruption of magma along a line roughly equidistant from the trench, which over time builds up the seafloor to create volcanic structures that rise above the waves. The island of Guam was formed by the union of two volcanoes.

Guyot
A guyot is a seamount that was once above sea level, where wave erosion planed off its top, creating a flat surface. As the seabed subsided, the guyot sank beneath the waves.

Mariana Trench
The sinking Pacific Plate pulls the overlying seafloor down with it, creating a deep trench, or groove, at the bottom of the ocean.

Pacific Plate movement
Both the Pacific and Philippine plates are moving in a northwesterly direction relative to the underlying asthenosphere, but the Pacific Plate is moving faster.

COLORFUL FISH
Small, brightly colored fish such as this groppo are common at depths of 1,150–1,640 ft (350–500 m) around the seamounts closest to the trench.

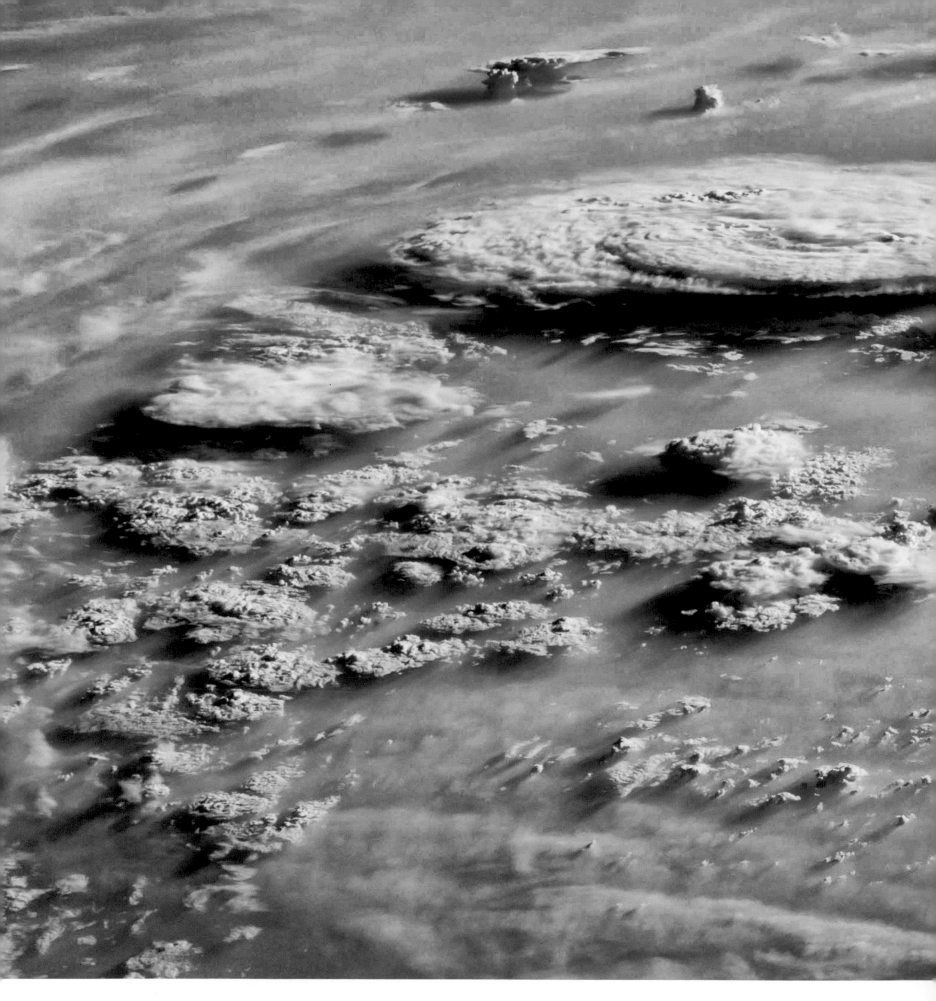

Dust on the move
A storm sends dust into the air across hundreds of miles of
the Sahara. The clouds in the image suggest that the dust is
being driven by a desert wind called a haboob. More than half
of all the dust in the world's oceans blows off the Sahara.

Extreme Weather

A large **tropical cyclone** releases as much **energy** as **half of Earth's electrical generating capacity**

Cyclones

The intensity and scale of cyclones makes them the most destructive phenomena in Earth's atmosphere

A cyclone is a rotating storm system centered around an area of low pressure. It forms as warm, moist air rises, causing a fall in pressure, which draws in more moist air from all directions. An effect known as the Coriolis force causes the air to spiral, counterclockwise in the Northern Hemisphere and clockwise in the Southern Hemisphere.

Tropical cyclones

In temperate regions, cyclones are often called depressions and are only rarely destructive. They are at their most intense in the tropics. They are called hurricanes in the North Atlantic, cyclones in the Indian Ocean, and typhoons in the western Pacific. A tropical cyclone forms over the ocean in the presence of slack winds and surface temperatures of more than 79°F (26°C)—usually in late summer or early fall. It becomes a full cyclone when winds exceed 73 mph (118 kph), although sustained speeds can exceed 200 mph (320 kph). Combined with torrential rainfall, this can cause flooding and mudslides and bring down buildings. During their most vigorous phase, cyclones track west. They begin to die only once they make landfall and their supply of warm ocean water is cut off.

TROPICAL CYCLONE REGIONS

Most tropical cyclones develop in two bands between 10 degrees and 30 degrees north and south of the equator. At higher latitudes, ocean surface temperatures are unlikely to be warm enough to fuel the storm's development. Closer to the equator, the Coriolis force will not be great enough to set circulation in motion.

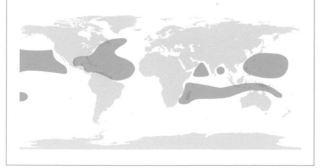

▼ TROPICAL CYCLONE
A section through a tropical cyclone reveals its central eye, where there is an area of sinking air, surrounded by a series of cloudbands.

▷ MATURE STAGE
Seen from overhead, a full-blown tropical cyclone, such as Hurricane Rita, will have a roughly circular cloud pattern, as well as a distinct eye and spiral rainbands.

Flat top
The top of the storm may be as high as 9 miles (15 km) above sea level.

Downpours
The area beneath a cloudband is subjected to heavy rainfall, strong winds, and lightning.

◁ FORCE OF NATURE
Palm trees bend in intense winds as Hurricane Dennis hits the coast of Florida in July 2005, although the storm surge created more damage than the winds.

Cloudbands
Spiralling around the eye are cloudbands that can extend for hundreds of kilometers from the cyclone center, gradually diminishing in height.

▷ **THE EYE OF THE STORM**
Air in the relatively calm center of a tropical cyclone warms as it descends, leaving it clear of cloud. It is possible to see blue sky from the ocean surface as the eye passes.

Cold exhaust
Cooled air at high altitudes spirals away from the eye or descends into it.

Cirrus shield
Cooled air fans out into high-altitude cirrus cloud, spreading 620 miles (1,000 km) or more.

STORM SURGES

Hurricane-force winds blowing onshore bunch ocean water up to abnormal levels in shallow coastal waters. This creates a storm surge—an abnormal rise of water that is higher than the predicted astronomical tides. When combined with normal high tide, this creates a storm tide and can cause extreme flooding in coastal areas.

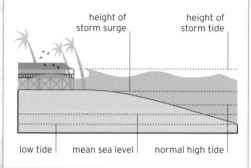

height of storm surge

height of storm tide

low tide

mean sea level

normal high tide

Eye
In the eye, typically 20-40 miles (30-65 km) across, air warms as it descends. The winds are lighter than anywhere else in the storm, but the ocean beneath the eye is rough.

Eyewall
The fastest winds are in the eyewall.

Storm surge
As the storm reaches the coast, a storm surge occurs where winds are blowing onshore.

Moat
Clear rings between concentric cloudbands are where cooled air is slowly sinking.

△ **STORM STRUCK**
Around 250,000 vehicles were destroyed, mainly by flooding caused by the storm surge, when Hurricane Sandy made landfall on the eastern seaboard of the U.S. in 2012.

Thunderstorms

Thunderstorms are rain showers that produce powerful electrical discharges called lightning

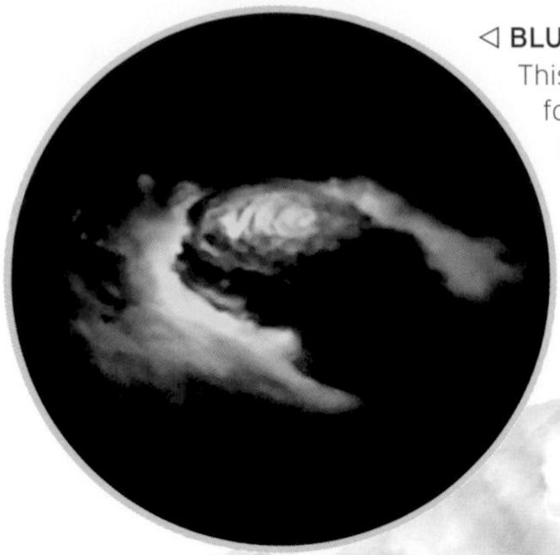

◁ **BLUE JETS**
This type of lightning is found 25–30 miles (40–50 km) above the ground, projecting from the top of a large storm cell to the upper atmosphere.

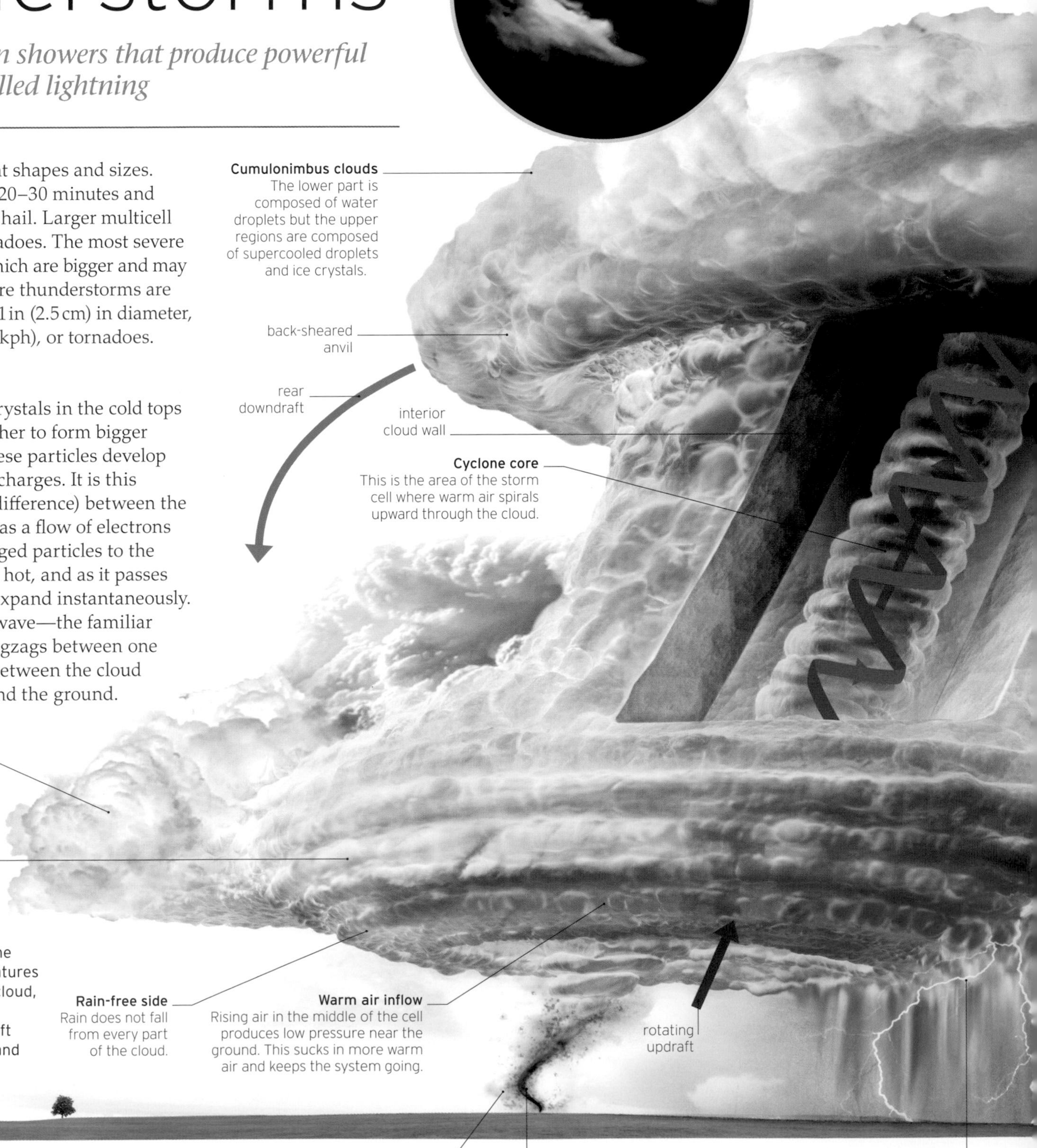

Thunderstorms come in different shapes and sizes. Single-cell storms typically last 20–30 minutes and may produce heavy rainfall and hail. Larger multicell storms may generate weak tornadoes. The most severe thunderstorms are supercells, which are bigger and may produce violent tornadoes. Severe thunderstorms are classed as those with hailstones 1 in (2.5 cm) in diameter, winds of more than 56 mph (90 kph), or tornadoes.

Lightning strikes

As tiny water droplets and ice crystals in the cold tops of the clouds bump into each other to form bigger drops of rain and hailstones, these particles develop positive and negative electrical charges. It is this difference in charge (potential difference) between the particles that creates lightning, as a flow of electrons moves from the negatively charged particles to the positive. Lightning is extremely hot, and as it passes through air it causes the air to expand instantaneously. This expansion creates a shockwave—the familiar rumble of thunder. Lightning zigzags between one part of the cloud and another, between the cloud and air, or between the cloud and the ground.

Cumulonimbus clouds
The lower part is composed of water droplets but the upper regions are composed of supercooled droplets and ice crystals.

back-sheared anvil

rear downdraft

interior cloud wall

Cyclone core
This is the area of the storm cell where warm air spirals upward through the cloud.

flanking line, a mass of developing cumulus clouds extending from the main storm

circulating banks of wall cloud

▶ **SUPERCELL**
No two supercell storms are the same but they do all share most of the features shown here: a huge cumulonimbus cloud, wind blowing around an area of low atmospheric pressure, a large updraft of warm air, heavy rainfall and hail, and downdrafts of cooled air.

Rain-free side
Rain does not fall from every part of the cloud.

Warm air inflow
Rising air in the middle of the cell produces low pressure near the ground. This sucks in more warm air and keeps the system going.

rotating updraft

Tornado
Destructive, rapidly spinning vortices may develop beneath about 30 percent of large supercells.

the rapid vortex of a tornado whips up soil and debris from the ground

Cloud-to-ground lightning
These powerful discharges reach a temperature of 54,000°F (30,000°C).

At **any one time** about **2,000 thunderstorms** are **taking place**, mostly in wet tropical regions

Overshooting top
The very highest part of a large thunder cloud may top 9 miles (15 km).

Anvil
Faster wind at high altitude spreads the top of the cumulonimbus over a very wide area.

◁ **RED SPRITES**
Triggered by lightning within large storm cells, these red discharges occur up in the atmosphere above the clouds at 30-55 miles (50-90 km) above the ground.

outflow

Cloud-to-cloud lightning
Up to 90 percent of lightning discharges are between different parts of the cloud.

Virga
From a distance, falling rain and hail look like lines running from the cloud. These lines are called virga and may obscure parts of the cloud.

▷ **MAMMATUS CLOUDS**
Turbulence often produces these pouch-shaped formations on the undersides of large cumulonimbus clouds. Their formation is not fully understood.

forward flank downdraft

cool outflow

a low, horizontal, wedge-shaped cloud called a shelf cloud usually runs ahead of the main storm

HOW A SUPERCELL FORMS

Several conditions are necessary for a thunderstorm to develop: high levels of moisture in the lower atmosphere, and warm air near the ground overtopped by cooler air so the warm air rises by convection. As it rises, the air cools and its moisture condenses to form cumulonimbus clouds. Rising air in the middle of the cell produces low pressure near the ground. This sucks in more warm, moist air, keeping the system going for hours. In a supercell, the whole system develops a circulation of its own.

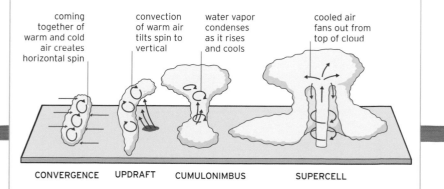

coming together of warm and cold air creates horizontal spin

convection of warm air tilts spin to vertical

water vapor condenses as it rises and cools

cooled air fans out from top of cloud

CONVERGENCE   UPDRAFT   CUMULONIMBUS   SUPERCELL

Tornadoes

Virtually impossible to predict accurately, tornadoes are the most capricious weather phenomena—and some of the most violent

With a tight vortex of air spinning at up to 300 mph (480 kph), a grade F "twister" is the most terrifying sight in nature, capable of destroying almost anything in its path. To meet the definition of a tornado, the vortex must extend from the base of a cloud and be in contact with the ground.

Types of tornadoes

Tornadoes can happen anywhere, but most occur on the Great Plains of the U.S. They occur in different shapes and sizes. Rope tornadoes have a narrow funnel, while wedge tornadoes have a wide diameter at the ground and are even broader at the cloud base. Tornadoes change color depending on the nature of the debris and dust that they pull into their violent vortex and may be near-white, brown, reddish, or nearly black. Their severity is measured on the Fujita scale. An F-0 twister—the most common type—is capable of breaking branches from trees, while an F-5 one is able to carry cars 330 ft (100 m) through the air.

Tornadogenesis

The ideal situation for a tornado to form, a process known as tornadogenesis, is when a cold air mass converges with a warm, moist one, creating instability and towering cumulonimbus clouds. Where there is also wind shear (a change in wind speed or direction with height), a slowly rotating supercell storm may develop (see pp.328–29). If an input of warm, rising air is in contact with the rapidly descending air of the storm's rear flank downdraft, a narrow column of rapidly spinning air develops. The rotating air continues downward, forming a funnel below the cloud. Air drawn into the funnel enters an area of much lower pressure, so it expands and cools, and moisture condenses as a result. The funnel becomes a tornado on touching the ground.

The biggest twister on record is the Tri-State Tornado of March 18, 1925, which was in contact with the ground for at least 219 miles (352 km). It wreaked havoc across Missouri, Illinois, and Indiana, leaving 695 dead. Bangladesh has the highest annual death toll caused by tornadoes, averaging almost 200 each year.

Bangladesh's **Daulatpur-Saturia tornado** of April 26, 1989, **killed 1,300 people**

TORNADO ALLEY

Hundreds of tornadoes are recorded in this region of the central U.S. every year, mostly in spring and early summer. The classic situation for tornadogenesis on the southern Great Plains is a combination of cold, dry air moving southeast from the Rockies and very warm, moist air advancing north from the Gulf of Mexico. This produces atmospheric instability and the growth of large storm cells, some of which produce tornadoes.

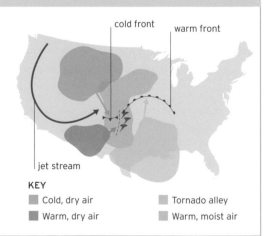

cold front
warm front
jet stream

KEY

▮ Cold, dry air	▮ Tornado alley
▮ Warm, dry air	▮ Warm, moist air

△ FORCE OF NATURE
Professional storm chasers monitor an approaching tornado in Kansas. Tornadoes grow from the base of large storm clouds and pick up dust and debris when they make contact with the ground.

△ ROPE TORNADO
A sinuous rope tornado tears across the Kansas plains and ravages the countryside. Kansas is one of the states most frequently hit by tornadoes.

◁ OVER THE SEA
Waterspouts form when tornadoes pass over water. Most do not suck up water but kick up spray, like this pair over the Mediterranean Sea.

Sandstorms and Duststorms

In arid and semiarid areas, high winds can whip up sand and dust to form powerful storms, blocking out the Sun and stripping away soil

A severe sand- or duststorm can create darkness in the middle of the day and reduce visibility to virtually nothing. These storms are formed when strong winds and dry conditions cause surface materials—dust, soil, or sand—to be lifted in a process called saltation. The wind carries particles a short distance before they fall back to the ground, dislodging other particles and accelerating the process. They remain suspended in the lower atmosphere.

The height and extent of the storm depends on the strength and persistence of the wind and the size of the particles. Wind gusts often exceed 50 mph (80 kph). Dust particles are smaller and are carried higher than sand—exceptionally to 20,000 ft (6,100 m). Some storms are driven by cold fronts (the leading edge of a mass of cold air pushing under warmer air) while others, called haboobs (see panel, below), are created by strong downdrafts from storm cells.

Cold fronts

The most notorious duststorms struck the High Plains of the US on April 14, 1935, turning a warm, sunny afternoon into a choking, zero-visibility blackout in parts of Oklahoma, Texas, Kansas, Nebraska, and Missouri. A combination of an extended drought, poor soil management, and high winds produced by a rapidly moving cold front created several huge duststorms. The great Australian duststorm of 2009, also created by a cold front, carried 2.75 million tons (2.5 million tonnes) of the Australian continent into the ocean.

HOW A HABOOB FORMS

Derived from the Arabic word *habb*, meaning "wind," a haboob is formed when strong winds blow out from a storm cell. The cell forms when warm, moist air rises, but air within that updraft is cooled by evaporating rain. This air may form a powerful downdraft and outflow, led by a gust front. As the front hits dry ground in advance of the storm, it kicks up surface dust or sand. The gust front produces a "wall" of dust, which may spread over a larger area than the storm itself.

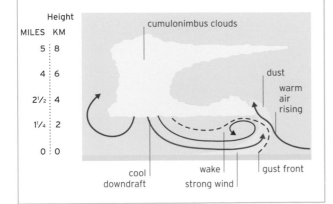

Height
MILES KM
5 : 8
4 : 6
2½ : 4
1¼ : 2
0 : 0

cumulonimbus clouds
dust
warm air rising
cool downdraft
wake
strong wind
gust front

▽ **RED DUST**
Sydney Harbor Bridge was shrouded in red dust during the great duststorm of September 2009. This is recorded to have been the worst to hit eastern Australia in the last 70 years.

▷ **A CITY ENGULFED**
A billowing wall of dust can be seen as it is about to engulf downtown Phoenix, Arizona, in July 2011. The dust cloud was generated by a haboob, one of several to affect the city that summer.

At its greatest extent, the **front** of the **September 2009 Australian duststorm** stretched for **2,140 miles (3,450 km)** from north to south

◁ **PARCHED LAND**
Duststorms increased in frequency in southern China after a severe drought in 2010-2011 dried out farmland and reservoirs, producing huge quantities of surface dust.

△ **ICE SCULPTURE**
Ice storms can produce some beautiful structures. This is St. Joseph's lighthouse, on the shore of Lake Michigan, after it was glazed by an ice storm in January 2015.

▷ **DEEP FREEZE**
This satellite photograph shows the Great Lakes frozen during an ice storm in February 2014. The event left more than a million people without power and grounded thousands of flights.

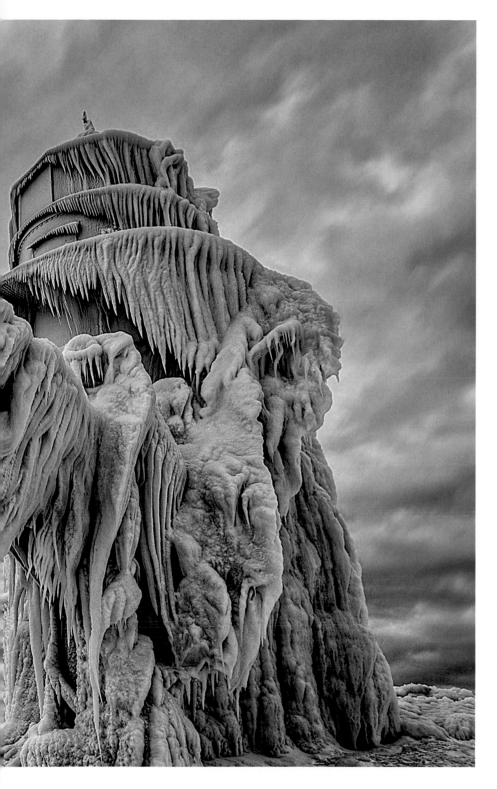

Ice Storms

Unique meteorological phenomena that produce spectacular beauty and destruction in equal measure

Unlike in other storms, conditions are often calm and quiet during an ice storm. But in terms of loss of life, disruption to transportation and power networks, and general mayhem, a major ice storm can produce as much chaos as a tornado or hurricane. The vital elements that cause this phenomenon are a combination of heavy rain falling from a wedge of warm air, an underlying area of supercold air, and freezing ground temperatures. When the rain hits the ground, or any subfreezing structure close to it, it freezes and forms a layer of ice known as glaze ice. As the rain persists, this glaze ice grows thicker.

Collapse under pressure

Glaze ice can be over 2 in (5 cm) thick. In 1961, a layer 8 in (20 cm) thick was measured in parts of Idaho. Since ice is 10 times as heavy as the equivalent thickness of wet snow, the effects are often dramatic. Apart from turning everything white, power lines, trees, and unstable buildings are brought down. Roads become impassable, and aircraft are grounded. In January 1998, an ice storm hit a vast area of New England and southeast Canada—a region particularly susceptible to these events. At least 44 people died, with hundreds more injured. Some 80,000 miles (129,000 km) of power cables collapsed, leaving four million people without power.

◁ **FROZEN FRUIT**
Thousands of apple trees in New York, which were damaged in the January 1998 ice storm, took years to recover.

SLEET, SNOW, AND FREEZING RAIN

If snow passes first through a wedge of warm air then a shallow layer of cold air, it turns into rain, then freezes as it hits the ground. If it passes through a thicker layer of cold air beneath the warm, it partially refreezes as sleet. If it does not pass through the warm air, it lands as snow.

high-altitude cloud

wedge of warm air

warm air moving over cold air | rain freezes on impact | glaze ice on surface | sleet | snow

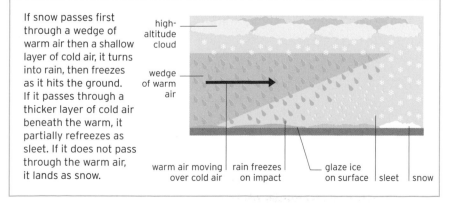

Ice can **increase** the **weight** of **branches** by **30 times**

Aurorae

Spectacular displays of colorful, shimmering lights in the night sky

The aurora borealis, or northern lights (in the Northern Hemisphere), and the aurora australis, or southern lights (in the Southern Hemisphere), appear at latitudes greater than 60° North and South, respectively. They are centered on Earth's magnetic poles and occur when electrically charged particles from the Sun are funneled into Earth's magnetic field above the poles and collide with atoms in the upper atmosphere to trigger reactions that produce streams of light (see panel, right). Some displays are almost static patches of diffuse brightness, but others feature constantly moving lights suggestive of curtains waving in a breeze. Every fold of the curtain consists of many parallel rays, each lined up with the local direction of Earth's magnetic field. The color of an aurora depends on its height and which gas in the atmosphere is being excited. Green is the most common color, but displays of red, blue, violet, or even pink occur occasionally.

Cloudless conditions away from light pollution during the long nights of winter are ideal for watching aurorae. Alaska, Canada, northern Scandinavia, and northern Russia are the best places to see the northern lights. The southern lights can be seen from southern South America, Tasmania, and New Zealand's South Island, although near-inaccessible Antarctica has the best shows.

The **lights** of an aurora are **produced** mostly **56–90 miles (90–150 km) above Earth**, although very high (red) displays may be **up to 620 miles (1,000 km)** away

△ **CURTAIN OF COLOR**
The dazzling display of an aurora can be seen reflected in the waters of the Jökulsárlón glacial lagoon in Iceland.

▷ **A VIEW FROM SPACE**
Astronauts orbiting in the International Space Station have a unique view of the aurora borealis.

△ **AURORA CORONA**
The aurora borealis may sometimes appear overhead as a "corona" of rays. This spectacular display was photographed over Iceland.

WHAT CAUSES AURORAE

Aurorae are caused by charged particles (electrons and protons) that speed out from the Sun. Earth's magnetic field provides a shield against these, but many electrons follow the magnetic field lines toward the poles, where they collide with oxygen and nitrogen atoms and molecules in the upper atmosphere. The electrons excite the atoms, and when the atoms return to their former state, they release energy in the form of photons, which create the dancing patterns of aurorae.

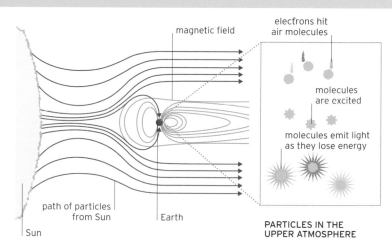

magnetic field

path of particles from Sun

Earth

Sun

electrons hit air molecules

molecules are excited

molecules emit light as they lose energy

PARTICLES IN THE UPPER ATMOSPHERE

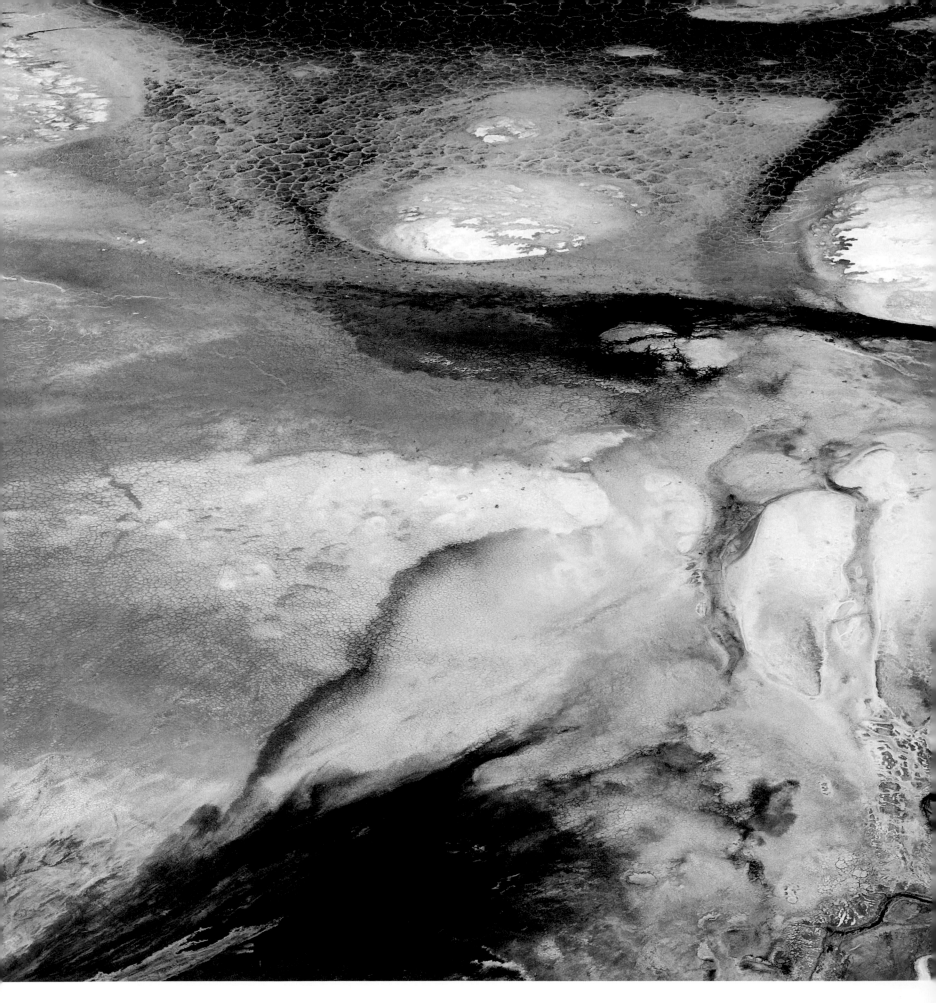

Salty shallows
Water enters Tanzania's Lake Natron through a river delta.
Hot springs provide the lake's other inflow. Because of high
rates of evaporation, the lake is shallow–just 10 ft (3 m)
deep–and extremely salty.

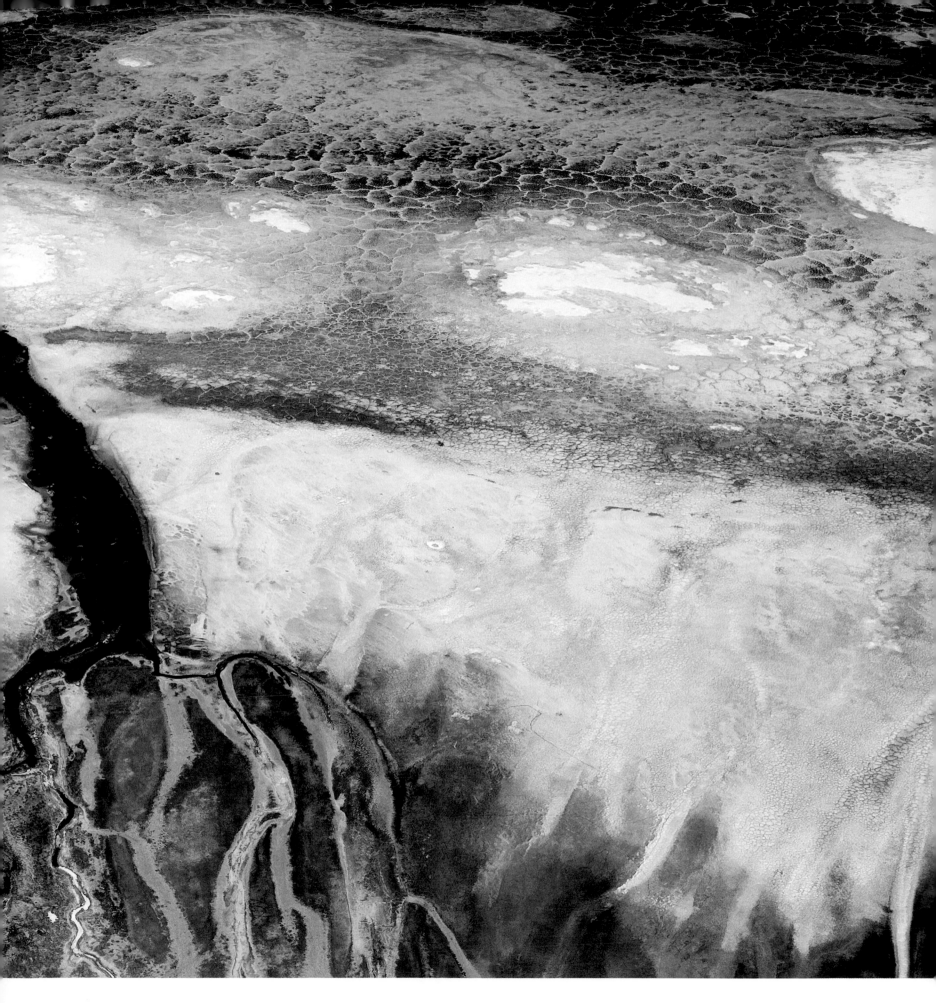

Directory & Glossary

MOUNTAINS AND VOLCANOES

Most mountains are part of chains that formed where two or more plates of Earth's crust collided. A nearly continuous range of mountains (the Circum-Pacific System) runs for 25,000 miles (40,000 km) around most of the Pacific Basin and includes three-quarters of the world's active and dormant volcanoes.

Its mountains include the Andes of South America, the Western Cordillera of North America, and the Japanese Alps. A second, almost unbroken mountain belt (the Alpine–Himalayan System) runs from Morocco through Europe and southern Asia to Southeast Asia, including Earth's greatest mountain range, the Himalayas.

North America

Mount Katmai

Location Alaska, northwest U.S.

This snow-covered volcano is about 6 miles (10 km) in diameter, with a massive, lake-filled caldera at its center. It was once thought to be responsible for the 20th century's biggest eruption, in June 1912, but scientists later showed that the nearby Novarupta lava dome was the actual source. When Novarupta erupted, a magma chamber beneath Mount Katmai emptied, collapsing its summit and creating the caldera. Thousands of steam vents peppered the area for years after the eruption, inspiring the name Land of Ten Thousand Smokes.

The Mackenzie Dikes

Location From the Great Lakes to the Arctic, western Canada

The Mackenzie Dikes are Earth's largest dike swarm—a large group of near-vertical, sheetlike formations of igneous rock. The dikes reach the surface across a vast area that is about 1,900 miles (3,000 km) by 310 miles (500 km). Mostly composed of basalt, they

were intruded into the ancient rocks of the Canadian Shield about 1.3 billion years ago.

The Palisades

Location Along part of the western shore of the Hudson River, eastern U.S.

The Palisades are dramatic, almost sheer cliffs that stretch 20 miles (32 km) along the west side of the Hudson River (see p.363) from near Jersey City to Nyack. The cliffs range in height from 295 to 540 ft (90 to 165 m) and are the eroded edge of the Palisades Sill, a large mass of an igneous rock called diabase. The sill is about 200 million years old.

The Sierra Nevada

Location Mainly in California, western U.S.

The magnificent skyline and dramatic, glacier-eroded valleys of the Sierra Nevada range make it one of the most beautiful regions of the U.S. Fast-flowing rivers, valleys cloaked in pine forest, mountain meadows, and granite peaks are all characteristic of the area. One of the Sierra Nevada's peaks, Mount Whitney, is, at 14,505 ft (4,421 m), the highest in the lower 48 states. The range extends about 400 miles (640 km) from south to north and 70 miles (110 km) west to east.

The Devil's Postpile

Location The Sierra Nevada range, California, western U.S.

The Devil's Postpile National Monument is situated on the western slope of the Sierra

▶ FALL FOLIAGE AT THE PALISADES SILL

Nevada range (see left) at an altitude of 7,500 ft (2,300 m). One of the world's finest examples of columnar basalt, the vertical, hexagonal columns of the Postpile tower 60 ft (18 m) above the valley side; some are more than 3 ft (1 m) in diameter. They were created 100,000–80,000 years ago when a huge outpouring of lava flooded the area to a depth of 410 ft (125 m), later cooling and solidifying. Glacial action subsequently eroded the basalt, exposing the columns in all their glory.

The Coast Ranges

Location Between western British Columbia and southwest California, western North America

Stretching at least 1,680 miles (2,700 km), this series of mountain ranges runs roughly parallel to

the Pacific coast from the Coast Mountains of British Columbia to the Transverse Ranges in southern California. The Olympic Mountains in Washington State are the most spectacular of the ranges. The folded and faulted igneous, metamorphic, and sedimentary mountains are the product of complex tectonic plate movements. Subduction produces volcanism in some regions, while major faults, such as the San Andreas, are active elsewhere.

Parícutin

Location The state of Michoacan, west central Mexico

After several days of small earthquakes and underground rumbling, a crack opened up in a farmer's field in February 1943 and began spewing ash, lava, and pyroclastic bombs. Within 24 hours, its cinder cone was 165 ft (50 m) high, and by the time the eruptions stopped in 1952 the top of the volcano was 1,391 ft (424 m) above the surrounding landscape. Two nearby villages were evacuated before being engulfed in lava. Although Parícutin is now extinct, the crater is still hot and emits steam when rain falls.

Volcán de Colima

Location Mostly in the state of Jalisco, west central Mexico

This massive stratovolcano is one of the most active in North America, having erupted more than 40 times since 1576, most recently in 2017. Eruptions produce flows of viscous lava, explosions of pyroclastic material, and enormous ash clouds. With 300,000 people living nearby, it is one of the most dangerous volcanoes on the continent. Volcán de Colima's summit rises 12,631 ft (3,850 m) above sea level.

▶ STEAM RISING FROM VOLCÁN DE COLIMA

El Chichón

Location The state of Chiapas, southern Mexico

Until late March 1982, El Chichón was thought to be an extinct volcano, with no eruption known for over 600 years. Then, in the space of a few days, three huge eruptions pumped millions of tonnes of gas and ash into the atmosphere, destroying nine villages and killing 2,000 people. The magma was very rich in sulfur, and sulfuric acid droplets forming in the stratosphere produced brilliant sunsets around the globe. A new crater was formed, 0.6 mile (1 km) across, which is now filled with an acidic lake. The peak of the volcano, which has been quiet ever since, is 3,772 ft (1,150 m) above sea level.

Other mountains and volcanoes in North America

- The Alaska Range » p.24
- The Appalachians » p.35
- El Capitan and Half Dome » p.26
- The Cascade Range » p.27
- The Cave of Crystals » p.36
- Crater Lake » pp.28-29
- Devil's Tower » p.35
- Popocatépetl » p.37
- The Rocky Mountains » pp.24-25
- The San Andreas Fault » p.30
- Ship Rock » p.34
- The Sierra Madre » p.37
- Yellowstone » pp.30-33

Central and South America

Mount Pelée

Location Martinique in the Lesser Antilles, Caribbean Sea

Pelée is a stratovolcano (composite volcano) in the Antilles island arc. After many years of inactivity, in 1902 it produced the most devastating eruption of the 20th century in terms of casualties. A series of pyroclastic clouds swept down its slopes to destroy the city of Saint-Pierre, killing about 30,000 people. The Peléan type of eruption, characterized by explosive emissions of ash, gas, and pyroclastic flows, is named after this volcano.

Arenal

Location Northwest Costa Rica, Central America

Arenal is Costa Rica's youngest and most active volcano. Situated over an active subduction zone where the Cocos Plate pushes beneath the Caribbean Plate, this stratovolcano was dormant for hundreds of years before July 1968, when an unexpected and violent eruption destroyed three small towns. Rocks weighing several tons were blasted more than 0.6 mile (1 km) away. Seven

▼ A LAVA FOUNTAIN AT ARENAL

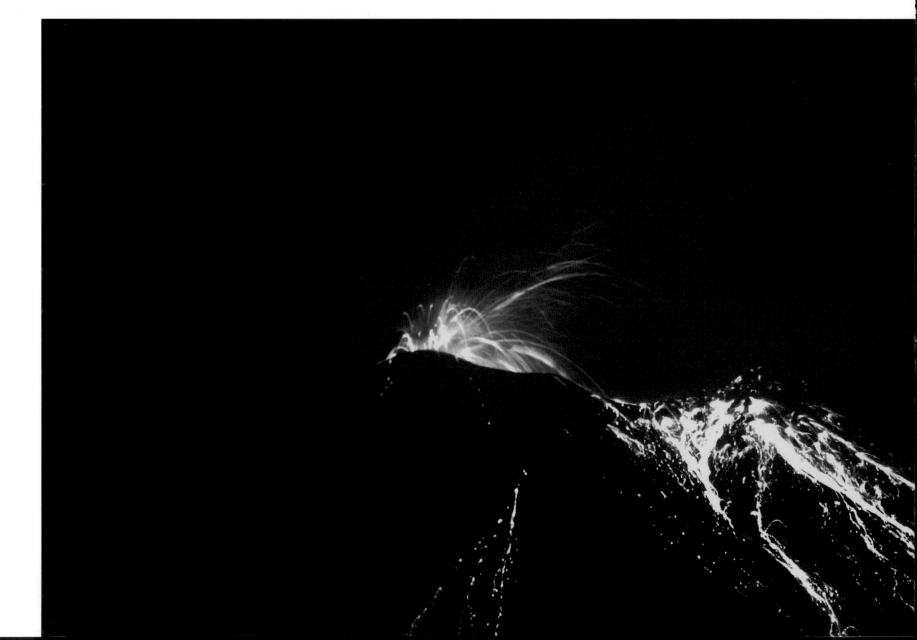

more eruptions occurred over the next 30 years, but the volcano was declared dormant again in 2010.

Galeras

Location Southwest Colombia, northern Andes

Colombia's most active volcano is also potentially its most dangerous, being situated near the city of Pasto's 450,000 inhabitants. It is a large stratovolcano, with a peak that is 14,029 ft (4,276 m) above sea level. The area's volcanism dates back at least 1 million years, with a particularly massive eruption 560,000 years ago. This activity results from the oceanic Nazca Plate plunging beneath the continental crust of the North Andes block.

Ojos del Salado

Location Central Chile-Argentina border, southern Andes

This snow-covered stratovolcano has several claims to fame. At 22,615 ft (6,893 m), it is the highest active volcano in the world and the second highest peak of any description in the Americas. A small caldera not far from the summit is probably the highest lake on Earth. Although there are persistent fumaroles on the volcano, and a gas emission was reported in 1993, the last major eruption took place 1,000–1,500 years ago.

Cerro Azul

Location Central Chile, southern Andes

Cerro Azul is located at the southern end of a volcano chain called the Descabezado Grande–Cerro Azul eruptive system. One of the largest explosive eruptions in South America in recorded history took place in April 1932, when the Quizapa crater of the Cerro Azul stratovolcano released 2.3 cubic miles (9.5 cubic km) of ash and lava in a classic Plinian eruption. At 10,801 ft (3,292 m), Quizapa is one of the world's highest Plinian craters. There has been no eruption since 1932, but small ash clouds are occasionally emitted.

Other mountains and volcanoes in Central and South America

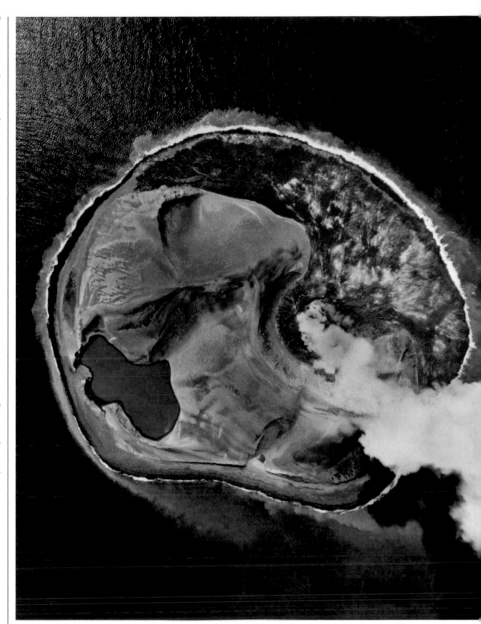

▲ STEAM RISES FROM SURTSEY

Europe

Grímsvötn

Location Southeast highlands, Iceland

Iceland's most active volcano is, unusually, mostly subglacial, covered by the Vatnajökull ice cap. When an eruption starts, large quantities of glacier ice melt and fill Grímsvötn's caldera. Pressure can then build up sufficiently to lift and puncture the ice cap, releasing huge quantities of water in an episode called a jökulhlaup. During a 2011 eruption, Grímsvötn produced an ash cloud 7 miles (12 km) high, and air traffic was disrupted for several days.

Surtsey

Location Off the south coast of Iceland

In November 1963, a series of Earth tremors, the odor of hydrogen sulfide, and finally a column of dark smoke rising from the sea signaled an underwater eruption off the Icelandic coast. Over the course of several weeks, a new volcano built up from the seafloor, and the brand new island of Surtsey was created, largely of a low-density volcanic rock called scoria. Sporadic eruptions continued until the summer of 1967, when the island reached its maximum size. Since then, wave action has gradually eroded it.

The Rhine Rift

Location Southwest Germany and eastern France

The upper Rhine River (see p.393) flows through Europe's most impressive rift, a broad, flat-bottomed valley, for 220 miles (350 km) between Basel and Frankfurt. The rift valley, or graben, is bounded by faults on either side, beyond which are the uplands of the Vosges Mountains to the west and the Black Forest (see p.393) to the east. The rift formed about 30 million years ago, at the start of the Alpine mountain-building period, when the crust was being stretched and faulted on a massive scale.

The Jura

Location Eastern France and western Switzerland

The Jura is a beautiful, part-forested massif of limestone deposits dating from 200–145 million years ago. The range, which runs for about 225 miles (360 km) along the French–Swiss border between the Rhône (see p.372) and Rhine (see p.371), was folded and uplifted during the Alpine orogeny (mountain-building period), which also produced large-scale faulting. The highest peak in the Jura is Cret de la Neige, at 5,636 ft (1,718 m). The elevations of the crests are lower along the outer ridges of the arc.

The Chaîne des Puys

Location The Massif Central, central France

This remarkable concentration of over 70 volcanoes in an area of the Massif Central about 25 miles (40 km) long and 3 miles (5 km) wide includes at least 48 cinder cones, eight lava domes, and 15 maars (broad craters that formed when lava came in contact with groundwater). Volcanic activity in the area spanned a period from about 95,000 to 6,000 years ago and was instigated by thinning of the crust and faulting related to the building of the Alps.

▼ PUY DE PARIOU CONE IN THE CHAÎNE DES PUYS

The Cordillera Cantábrica

Location Northern Spain

The Cantabrian Mountains comprise two ranges extending for 190 miles (300 km) from the western end of the Pyrenees to the Galician Massif. The much smaller coastal range sometimes rises steeply from the sea, and features short, fast-flowing rivers. Inland, the southern range is much higher and more dramatic. It includes the Picos de Europa, a huge massif of mountain limestone, which has been subjected to heavy glaciation, although no glaciers survive. Its highest peak is Torre de Cerredo at 8,688 ft (2,648 m).

▲ HOT SPRINGS AND TERRACES AT TERME DI SATURNIA

The Sierra Nevada

Location Andalucía, southern Spain

This domed massif of metamorphic rocks towers high above the Mediterranean and is dominated by Mulhacén, the highest peak in continental Spain, at 11,421 ft (3,481 m). Mulhacén is just one of 23 tops rising above 10,000 ft (3,000 m) in this 26-mile- (42-km-) long range. Generally, the Sierra's landforms are gentle apart from the heavily glaciated western peaks, which feature steep-sided, ice-sculpted valleys and cirques. No glaciers remain, but winter snow makes the area a popular skiing center. With increased elevation, the vegetation on the massif's slopes changes from subtropical forest to alpine.

The Apennines

Location From the Cadibona Pass to the Egadi Islands, Italy

Among the younger chains created during the Alpine mountain-building period, the Apennines are a series of parallel ridges extending for about 870 miles (1,400 km). They are mostly made of uplifted shales, limestones, and sandstones, with some igneous rocks, and there is plenty of evidence of volcanic activity, including the still-active Vesuvius and Etna. The only glacier here is Calderone, on the slopes of the highest peak, Corno Grande. Earthquake activity is common along the entire range.

Terme di Saturnia

Location Tuscany, Italy

These are the largest and most beautiful of southern Tuscany's hot springs. Geothermally heated groundwater emerges at 99.5°F (37.5°C) and cascades over the waterfalls of Cascate del Mulino and Cascate del Gorello at more than 132 gallons (500 liters) per second into pools formed by the precipitation of travertine (a form of calcium carbonate) from the mineral-rich water. Despite the water's sulfurous odor, people have bathed in the pools since Roman times to take advantage of their curative properties. The groundwater is heated by hot rocks deep underground that are related to the volcanic activity at the nearby lava dome of Mount Amiata.

Solfatara

Location Near Naples, Italy

The mythological home of Vulcan, the Roman god of fire, this shallow volcanic crater is part of the large, multi-cratered Campi Flegrei volcanic area. Solfatara has many fumaroles, which emit jets of steam and sulfurous fumes, as well as hot mud pools, but the volcano itself has been dormant since 1198. Mollusk burrows found 22 ft (7 m) up three Roman columns show that the area has sunk below sea level and subsequently been uplifted again since the columns were erected. This is evidence of the uplifting and deflating of the caldera as a result of the filling and emptying of its magma chamber, a process called bradyseism.

Vulcano

Location Off north coast of Sicily, Italy

As with other volcanoes in the region, activity on this small island has resulted from the collision of the African and Eurasian plates. The volcanic cones here are of three different ages. The oldest are three ancient composite volcanoes, the cones of which have largely collapsed into their caldera. The more recent Fossa cone last erupted in 1888–90. This event was well studied and involved the ejection of blocks of rocks, lava bombs, and ash but no lava flows;

it gave its name to Vulcanian-type eruptions. The most recent cone is Vulcanello, which formed in 183 BCE and was last active in 1550.

The Dinaric Alps

Location From Italy to Kosovo, southeast Europe

Much of this range, which stretches 400 miles (645 km) along the eastern Adriatic coast, is made of limestone and dolomite, and these rocks display some of the most spectacular karst landscape features in the world. These include sinkholes, underground rivers, caves, and gorges where caves have collapsed. One of the biggest underground systems is at Škocjan Caves in Slovenia, through which a 21-mile- (34-km-) long underground section of the Reka River flows.

The Pindus Mountains

Location From the Albanian border to the Peleponnese, Greece

Greece's longest range, the Pindus Mountains, are geologically, an extension of the Dinaric Alps (see above). Some peaks, including the highest—Mount Smólikas— are composed of metamorphic ophiolites, sections of ancient oceanic crust that have been uplifted high above sea level and subsequently eroded. Other

▲ MOUNT SMÓLIKAS IN THE PINDUS MOUNTAINS

▲ HOODOOS AT CAPPADOCIA

areas are made up of limestone and dolomite, and have produced karst features. Vikos Gorge, in the far north of the range, is one of the deepest gorges in the world, plunging 2,950 ft (900 m) in places.

Santorini

Location The Cyclades Islands in the Aegean Sea, Greece

This oval archipelago is all that remains above sea level of a giant volcanic caldera. The islands of Thera, Aspronisi, and Therasia surround a lagoon that is 7 miles (12 km) by 4 miles (7 km). In 1610 BCE, Santorini produced one of the largest eruptions of the past 5,000 years, blasting 24 cubic miles (100 cubic km) of rock particles and ash into the atmosphere and producing a tsunami that devastated the north coast of Crete.

Cappadocia

Location Central Anatolia, Turkey

Ancient volcanic activity on this high plateau has given rise to one of the world's most extraordinary landscapes, the "fairy chimneys" (hoodoos) of Goreme National Park. Volcanoes deposited thick layers of ash, which later consolidated into a rock called tuff, then covered it with basalt. Runoff from heavy rainfall cut gullies through the resistant

basalt before rapidly eroding the less resistant tuff beneath. Over thousands of years, this produced numerous tall, thin spires of tuff topped by protective basalt caps.

The Manpupuner Rock Formations

Location The Komi republic, west of the northern Urals, Russia

These seven huge columns of metamorphic crystalline schist stand up like giants from the western slopes of the northern Urals. Their almost-human shapes have inspired a number of legends about their origin. The formations, which range in height from 100 to 138 ft (30 to 42 m), are flat-topped, almost vertical-sided rock towers, or buttes, created by freeze-thaw and wind erosion over many thousands of years.

Other mountains and volcanoes in Europe

Africa

Pico de Teide

Location Tenerife, the Canary Islands

Measured by its summit elevation above the ocean floor (24,600 ft/7,500 m), this composite volcano on Africa's continental shelf is the highest in the world outside the Hawaiian islands. It rises 12,198 ft (3,718 m) above sea level and is capped by the Las Canadas caldera. Pico de Teide and other vents have erupted several times since the island was settled in 1402, the most recent eruption being in 1909, when lava flows caused some damage.

The Hoggar Mountains

Location Central Sahara Desert, Southern Algeria

This vast lunar landscape in Algeria formed from rocks about 2 billion years old—some of the oldest in Africa. Also known as the Ahaggar Mountains, this range is largely rocky desert, with scant vegetation, at an elevation above 2,950 ft (900 m). Mount Tahat is the highest summit, peaking at 9,541 ft (2,908 m). The bedrock largely comprises ancient metamorphic rocks, although several of the more dramatic peaks result from the erosion of extinct volcanic cones. The droppings of a critically endangered Saharan cheetah were found in the mountains in 2006, and there is a population of Dorcas gazelles in the area.

Kapsiki Peak

Location The Mandara Mountains, northern Cameroon

Close to the village of Rhumski in the volcanic Mandara Mountains, a series of sentinel-like volcanic plugs rise sheer from the grassland. The tallest of them is Kapsiki Peak, which rises to 4,016 ft (1,224 m) above sea level. The plug formed when lava cooled and solidified as resistant rock in a volcano's vent. Subsequent

erosion removed the less-resistant volcanic cone, leaving behind just the hard plug.

Lake Nyos

Location Northwest Cameroon

Situated on the flank of an extinct volcano, Lake Nyos is one of only three "exploding lakes" in the world. Its deep waters are saturated with carbon dioxide (CO_2) from a pocket of magma deep underground. On August 21, 1986, when the lake's waters were disturbed by a landslide, the gas rose into areas of lower pressure, bubbled out of solution, and

escaped. A massive cloud of CO_2—about ¼ cubic mile (1 cubic km)—poured down the mountainside and suffocated 1,746 people in the valley below.

Pico Cão Grande

Location Near the southern end of São Tomé island, São Tomé and Príncipe

Also known as Great Dog Peak, this moss-covered column of rock on the island of São Tomé rises 1,198 ft (365 m) above the surrounding landscape and 2,192 ft (668 m) above sea level. It is a volcanic plug, which formed when magma solidified in the vent of an active volcano. The

volcano's cone has subsequently been eroded, and volcanic activity has long since ceased in the area.

The Ethiopian Highlands

Location The Horn of Africa, mostly central and northern Ethiopia

This range is called the roof of Africa for a good reason: it is the largest area above 4,900 ft (1,500 m) on the continent, and its highest peak, Ras Dashen, rises to 14,928 ft (4,550 m). The source of the Blue Nile, Lake Tana, is in the highlands. Bisected by the East African Rift Valley, they are also home to

▲ CLIFFS AND ESCARPMENTS IN THE ETHIOPIAN HIGHLANDS

several endemic species, including the critically endangered Ethiopian wolf and mountain nyala.

Lake Bogoria

Location Western Kenya, in the East African Rift Valley

This shallow, very saline lake lies in a section of the down-faulted East African Rift Valley between the Maji Moto block to the west and the Bogoria Escarpment to the east. The water depth ranges from 36 to 46 ft (11 to 14 m), with the deepest area being the south basin, a relict volcanic crater. Many hot springs and geysers in and around the lake testify to continued geothermal activity. At times, 1.5 million flamingos feed in the alkaline waters.

The Great Dike

Location Running south-north from east of Bulawayo to just south of the Mozambique border, Zimbabwe

Despite its name, this feature is not a dike but a loppolith, a mass of intruded igneous rocks that is more or less saucer-shaped in cross-section, fed by a vertical dike beneath. Its surface manifestation is a series of narrow ridges, running 340 miles (550 km) from south to north. It was injected into the surrounding rocks about 2.5 billion years ago and is rich in minerals including gold, silver, chromium, platinum, and nickel.

Piton de la Fournaise

Location Réunion Island, in the Indian Ocean, off the east coast of Madagascar

In English, Piton de la Fournaise translates as "Furnace Peak." This is one of the biggest and most active volcanoes in the world, with more than 150 eruptions recorded since the 17th century, the most recent being in 2017. More than 530,000 years old, this shield volcano grew over the Réunion hotspot, which is thought to have been active for 66 million years. For most of its history, its lava flows have intermingled with those of Piton des Neiges. There are many craters within the caldera, which is 5 miles (8 km) across. The volcano's activity is closely monitored at an observatory on its slopes.

Other mountains and volcanoes in Africa

- The Afar Depression » p.183
- The Atlas Mountains » p.182
- The Brandberg Massif » p.188
- The Drakensberg » p.189
- The Great Rift Valley » pp. 184-87
- Table Mountain » p.188

▲ STEAM RISING FROM THE VALLEY OF GEYSERS

Asia

The Siberian Traps

Location Northwest Siberia, Russia

The traps (steplike hills typical of a basalt landscape) are the legacy of the largest known volcanic eruption in Earth's history. About 250 million years ago, numerous fissures and vents in western Siberia released an estimated 720,000 cubic miles (3 million cubic km) of flood basalts over a vast area. A mantle plume may have been responsible for the eruptions, but this is still debated by scientists. The greatest mass extinction in history followed, triggered by dramatic changes in the composition of the atmosphere.

The Valley of Geysers

Location The Kamchatka Peninsula in eastern Siberia, Russia

The world's second-biggest concentration of geysers is in the valley of the fast-flowing Geysernaya River, which carries geothermal waters from a nearby stratovolcano. The valley contains at least 20 large geysers, the most spectacular of which, Velikan, can send out jets of water 130 ft (40 m) high. Some gush every few minutes, others only once every three hours or more. A massive mudslide in 2007 inundated the valley and shut off many of the geysers, but new ones have become active since.

Azerbaijan Mud Volcanoes

Location The eastern Caucasus and Caspian coast, Azerbaijan

More than half the world's known mud volcanoes, possibly more than 400, are located in Azerbaijan. Mud volcanoes, also known as sedimentary volcanoes, occur where groundwater at up to 212°F (100°C) mixes with mineral deposits and is forced upward. Reaching the surface through fissures, the mud-water mixture forms a small cone, usually no more than 12 ft (4 m) high. Some mud volcanoes also emit methane, as shown when the Lok-Batan mud cone spectacularly erupted in flames in 1977 and 2001.

The Arabian Highlands

Location Yemen and southwest Saudi Arabia, Southwest Asia

At 12,028 ft (3,666 m), the rocky peak of Jebal Nabi Shu'ayb in the southern sector of this range is the highest point on the Arabian Peninsula. To the west, running parallel to the Red Sea as far north as Makkah (Mecca), an escarpment rises from the coastal plain of Tihamah. To the east, a high plateau slopes toward the desert interior of Saudi Arabia. In contrast, there are areas of moist deciduous forest, high-elevation agricultural terraces, and sparkling mountain streams within the highlands.

The Deccan Traps

Location Mostly Maharashtra, Gujarat, and Madhya Pradesh provinces, India

The Deccan Traps form one of the largest volcanic features on Earth, a vast area of multiple steplike layers of solidified basalt released about 65 million years ago. In places, the basalt is more than 6,600 ft (2,000 m) thick and covers 190,000 square miles (500,000 square km). One theory proposes that the eruptions were caused by the Réunion hotspot, a mantle plume that once lay under South Asia.

Sigiriya

Location Central Province, Sri Lanka

The near-vertical sides of Sigiriya, or Lion Rock, rise abruptly 590 ft (180 m) from the forest below, with the ruins of an ancient royal palace on its flattened summit. The rock is the solidified magma plug of a volcano whose cone has long since been eroded. As the magma in its vent cooled, the minerals crystallized at different rates, so the composition of Sigiriya changes with height.

Seongsan Ilchulbong

Location Off the east coast of Jeju, South Korea

This World Heritage Site is a tuff cone—made of compacted volcanic rock—that rises from the sea close to the South Korean coast. The sides of the former volcano tower 590 ft (180 m) above the water. Seongsan Ilchulbong was created by a volcanic eruption on the seafloor about 5,000 years ago. The cliffs rise to a circular summit rim, which encloses a crater—an almost perfect bowl 295 ft (90 m) deep and 1,480 ft (450 m) across.

▶ SIGIRIYA'S SUMMIT

Beppu Hot Springs

Location Kyushu island, Japan

Beppu has almost 3,000 geothermal springs, in eight major groups, the second-highest concentration in the world. The springs supply water at temperatures up to 210°F (99°C) for bathing pools (*onsen*). The pool water is a range of different colors, depending on the mineral content. For example, Umi-Jigoku *onsen* is cobalt-blue, while Chinoike-Jigoku is blood-red. The heat source for the geothermal activity is the nearby Tsurumi lava dome.

Mount Unzen

Location Kyushu island, Japan

This volcano complex was responsible for Japan's worst volcanic disaster when, in 1792, the collapse of one of its lava domes triggered a huge tsunami that killed about 15,000 people. Its most recent period of activity was 1990–96, when one eruption generated a pyroclastic flow (a rapid surge of hot gas and lava) that killed 43 people, including three volcanologists. The highest of the Unzen peaks is Heisei-shinzan, at 4,875 ft (1,486 m).

Sakurajima

Location Kyushu island, Japan

Prior to 1914, this composite volcano was an island, but enormous lava flows in an eruption of that year filled the main strait between the island and the mainland. Sakurajima is one of Japan's most active volcanoes,

▲ SANDSTONE COLUMNS IN THE TIANZI MOUNTAINS

erupting almost daily, and regularly sending ash 16,000 ft (5,000 m) into the sky. Its summit has three peaks, but since 2006 activity has centered on the Showa crater, east of the Minami-dake summit.

The Tianzi Mountains

Location Hunan province, China

The Tianzi Mountain Nature Reserve is part of the spectacular Wulingyuan Scenic Area, whose main claim to fame is more

than 3,000 quartzite sandstone columns, many of which are more than 660 ft (200 m) tall. There are also ravines, caves, fast-flowing streams, and waterfalls. Lush forest covers the area, which is often shrouded in mist and fog. The columns are the product of water eroding vertical joints in the 380-million-year-old sandstone.

The Tibetan Plateau

Location Mostly in China (Tibet) and India, Central and East Asia

This is the world's largest and highest plateau, a vast area of flat or undulating steppe about five times the area of France and on average 14,800 ft (4,500 m) above sea level. Bordered to the south by the Himalayas, to the north by the Kunlun Mountains, and to the northeast by the Qilian Range, the plateau supports tens of thousands of glaciers and is the source of the several major rivers, including the Indus and Brahmaputra.

▲ VIOLENT ERUPTION ON THE SHOWA CRATER OF SAKURAJIMA

Baishuitai Hot Springs and Terraces

Location Yunnan province, China

Also known as the White Water Terraces, Baishuitai is located high in the uplands of Yunnan, in the foothills of the Haba Snow Mountains. It has some of the world's most beautiful terraces of travertine (a form of calcium carbonate). Water from geothermal springs pours over hundreds of white semicircular steps and terraces, which cover an area 460 ft (140 m) long and 520 ft (160 m) wide. The high concentrations of calcium carbonate in the water precipitate out at lower temperatures to leave behind the travertine. The terraces are one of the sacred sites of the Dongba culture.

The Huangshan Mountains

Location Anhui province, China

Celebrated for their beauty in art and literature for hundreds of years (for example, in the Shanshui, or "mountain and water," style of painting), the Huangshan Mountains feature oddly shaped pines, towering granite peaks and cliffs,

waterfalls, lakes, and forests, all sometimes draped in mist. There are 77 peaks exceeding 3,300 ft (1,000 m), with Lian Hua Feng (Lotus Flower Peak) reaching 6,115 ft (1,864 m). Many pines grow from fissures in the rock faces, and 19 plant species are endemic to the mountains. Also called The Yellow Mountains, the Huangshan Mountains were listed as a World Heritage Site by UNESCO in 1990.

Mayon Volcano

Location In the south of the island of Luzon, the Philippines

This is the most active volcano in the Philippines, having erupted at least 49 times in the last 400 years. Rising 8,077 ft (2,462 m) above the Abay Gulf, this stratovolcano is considered to be one of the most perfectly symmetrical in the world. Mayon's most violent eruption, in 1814, killed over 1,200 people and ended up devastating several towns. Following another eruption in 2006, heavy typhoon rainfall produced lahars (mudflows) from the recently deposited ash, which killed at least 1,000 people.

▲ TAAL, WITH TAAL LAKE IN THE CALDERA

Tambora

Location On the island of Sumbawa, Indonesia

Tambora's eruption in 1815 was the biggest in recorded history—it killed about 12,000 people. Huge clouds of ash and pumice were emitted into the atmosphere and dimmed sunlight, lowering global temperatures and creating the "year without a summer" in 1816. Tambora lost 4,900 ft (1,500 m) of its height in the eruption, and the present giant caldera formed, 4 miles (6 km) across and 3,640 ft (1,110 m) deep. At sea level, Tambora is 40 miles (60 km) in diameter.

Kelimutu

Location On the island of Flores, Indonesia

This complex volcano has three crater lakes, whose waters are different colors despite being at the summit of the same volcano. Different oxides and salts give the lakes their varying colors. Tiwu Ata Mbupu is usually blue, while Tiwu Ata Polo and Tiwu Nua Muri Kooh Tai, which share the same crater wall, are typically red and green, respectively. The lakes, which are fed by underwater fumaroles, periodically change color.

Taal Volcano

Location In the west of the island of Luzon, the Philippines

The Philippines' second most active volcano after Mayon (see p.355), Taal is known to have erupted 33 times since 1572, with particularly large events in 1754 and 1911. Recent eruptions have been concentrated on Volcano Island, which is in the middle of Taal Lake. The lake fills much of a massive, ancient caldera. Taal is one of a chain of volcanoes that were formed by the subduction of the Eurasian Plate under the Philippines.

Lake Toba

Location In the north of Sumatra, Indonesia

The largest volcanic lake in the world, this was the site of the biggest known eruption of the last 25 million years. The blast, which occurred about 75,000 years ago, covered parts of Sumatra beneath 2,000 ft (600 m) of ash. The collapse of the volcano after the eruption formed a huge caldera, which subsequently filled with water. The lake is now 60 miles (100 km) by 20 miles (30 km) and plunges to 1,657 ft (505 m) deep.

Mount Merapi

Location On the island of Java, Indonesia

This stratovolcano is the most energetic of Indonesia's 129 active volcanoes, and forms part of the Pacific Ring of Fire. It has erupted regularly in the last 500 years, causing many fatalities. Merapi rises 9,738 ft (2,968 m), somewhat lower than before its 2010 eruption, which killed more than 300 people living on its slopes. The volcano is situated over a subduction zone where the Indo-Australian Plate is pushing under the Sunda Plate.

Other mountains and volcanoes in Asia

- **The Altai Mountains** » p.224
- **The Caucasus Mountains** » p.221
- **The Himalayas** » pp.223-23
- **The Karakoram Range** » p.225
- **Kliuchevskoi** » p.229
- **Mount Everest** » pp.226-27
- **Mount Fuji** » p.229
- **Mount Pinatubo** » p.231
- **Pamukkale Hot Springs** » p.220
- **The Tengger Volcanic Complex** » p.230
- **The Tian Shan Mountains** » p.224
- **The Zagros Mountains** » p.220
- **Zhangye Danxia** » p.228

Australia and New Zealand

Rabaul

Location At the eastern end of New Britain island, Papua New Guinea

The large caldera at Rabaul was created about 1,400 years ago. Subsequently, a large breach on the eastern side flooded most of the caldera, which became Blanche Bay, a sheltered inlet of the Bismarck Sea. Three small volcanoes lie just outside the rim of the caldera, and in 1994 (after a long period of inactivity) two of these—Tavurvur and Vulcan—erupted violently, destroying much of Rabaul town, which had developed as a port within the caldera. Tavurvur remains the most active vent, continuously emitting ash.

The Bungle Bungles

Location Northeast Western Australia, Australia

The horizontally striped, beehive-shaped towers of this range form the most impressive sandstone karst landscape in the world. Deposited 360 million years ago, before being uplifted, the sandstone has been worn away by the combined effects of heavy rainfall, chemical weathering, and erosion by the wind. These have cut gullies and gorges, such as Cathedral Gorge. The orange layers of sandstone contain oxidized iron compounds and do not retain water. The gray layers contain more clay and so retain more moisture, supporting cyanobacteria that protect the surface of the sandstone.

▼ STRIATED SANDSTONES AT THE BUNGLE BUNGLES

▲ CRATERS AND CRATER LAKES OF RUAPEHU

Kata Tjuta

Location In the south of Northern Territory, Australia

This spectacular group of 36 domed rock formations towers high above the surrounding countryside and is visible from a great distance. The structures are composed of conglomerate, a mixture of rounded boulders in a sandstone matrix. Their orange–red hue is due to a thin cover of iron oxide. The biggest of the domes is Mount Olga, which stands 1,791 ft (546 m) above the plain.

Waimangu

Location North Island, New Zealand

In June 1886, New Zealand's largest eruption in 700 years created the Waimangu Volcanic Rift Valley, which has been a hotspot of hydrothermal activity ever since and has a number of dramatic features. Acidic Frying Pan Lake is the largest hot spring on Earth. Nearby Inferno Crater Lake contains the largest geyserlike feature in the world— although it can't be seen because it jets into the bottom of the deep lake.

Ruapehu

Location North Island, New Zealand

New Zealand's largest active volcano, and North Island's tallest mountain, Ruapehu has three peaks surrounding a crater, which becomes a crater lake between eruptions. One feature of Ruapehu eruptions has been the breaching of the lake dam and the creation of destructive lahars (mudflows) as water mixes with volcanic sediments and speeds downslope. A lahar in 1953 killed 151 people. The last major eruption was in 1995, although there have been smaller events since.

Other mountains and volcanoes in Australia and New Zealand

- **The Great Dividing Range** » p.274
- **Hyden Rock** » p.278
- **Rotorua** » pp.276-77
- **The Southern Alps** » p.279
- **Tongariro Volcanic Park** » p.275

GLACIERS AND ICE SHEETS

Most of Earth's glacial ice is concentrated in Antarctica and Greenland, but glaciers are found on every major landmass except Australia. Most exist either in the polar regions or in mountainous zones where the climate is cold enough, for at least a part of the year, for snow to accumulate and ultimately become consolidated as ice. The amount of snowfall a glacier receives is crucial for its survival; regions that have suitably cold winter temperatures but little snowfall, such as Siberia, do not have the ice accumulation necessary for glacier formation.

Europe

The Kongsvegen Glacier

Location Svalbard archipelago, on the western side of Spitsbergen

One of more than 1,500 glaciers on Spitsbergen, Kongsvegen shares its lower end (terminus) in the seawater of Isfjorden with another glacier, Kronebreen. Both calve ice blocks into the waters of the fjord. Although Kongsvegen's terminus is now slowly retreating, it is a surge glacier, meaning that periodically it advances rapidly. It last surged in 1948, pushing several miles along Isfjorden. It is a polythermal glacier, being warmer near the base than near the surface.

The Rhône Glacier

Location The Bernese Alps, Switzerland

At the head of an impressive U-shaped valley, which runs roughly northeast to southwest below the giant peak of Finsteraarhorn, the glacier is the source of the Rhône River (see p.372) and one of the main providers of water for Lake Geneva. It is about 5 miles (8 km) long, although its snout has retreated 0.6 mile (1 km) since 1880. Every summer, sections of the ice are covered with white blankets in an attempt to reduce ice melt.

The Pasterze Glacier

Location The eastern Alps, Austria

At just over 5 miles (8 km) from head to terminus, Pasterze is the longest glacier in the eastern Alps. It is also located in one of the most dramatic settings, with Austria's highest mountain, Grossglockner, towering above it. The glacier's head is Johannisberg, at 11,329 ft (3,453 m), and at its snout it gives rise to the Moll River. Pasterze is currently retreating at a rate of 33 ft (10 m) a year, and its volume has halved since the first measurements were taken in the mid-19th century.

> **Other glaciers and ice sheets in Europe**
> - **The Aletsch Glacier** » p.151
> - **The Jostedalsbreen Icefield** » p.149
> - **Mer de Glace** » p.150
> - **The Monaco Glacier** » p.148
> - **Vatnajökull** » pp.146–47

▲ ICE CAVE AND MELTWATER UNDERNEATH THE PASTERZE GLACIER

▲ GLACIATED PEAKS NEAR THE SIACHEN GLACIER

Africa

The Kilimanjaro Ice Cap

Location At the southern end of the Great Rift Valley's eastern fork, northeast Tanzania

Despite lying virtually on the equator, low temperatures at the top of Kilimanjaro (19,340ft/5,895m) ensure that snow accumulates, and, in the recent past, glacier ice has formed. Global climate change reduced the area of permanent ice by 85 percent between 1912 and 2011. The northern icefield at the summit is the biggest area, but only 0.4 square mile (1 square km) of this remains, having shrunk by 29 percent since 2000.

Asia

The Inylchek Glacier

Location The Tian Shan Mountains of Kyrgyzstan, Kazakhstan, and China, Central Asia

The world's sixth longest nonpolar glacier has two branches—North and South. The glacier's source is in the western slopes of the Khan Tengri massif, part of the Tian Shan mountain range. At 24,406ft (7,439m), the range's highest peak, Jengish Chokusu, towers over the southern branch of the Inylchek Glacier, which is 40 miles (60 km) long, covers an area of 7 square miles (17 square km), and is up to 660ft (200m) thick. Inylchek is the longest and fastest-moving glacier in the Tian Shan, and it feeds the seasonal Lake Merzbacher—named after Austrian explorer Gottfried Merzbacher, who discovered the lower branch in 1903. The lake drains annually into the Inylchek River below when its ice dam melts.

The Siachen Glacier

Location The Karakoram Mountains of Jammu and Kashmir state, India

This glacier, situated between the main Karakoram and Saltoro ranges, falls from 18,875ft (5,753m) to 11,877ft (3,620m) over the course of its 47-mile (76-km) length. It is the second longest nonpolar glacier and, including tributary glaciers, covers an area of 270 square miles (700 square km), but it is retreating rapidly. The Siachen Glacier is supplied by annual snowfall of

snow-covered mountains of the Zanskar Range. It is an immense mountain glacier 14 miles (23 km) long, with an average thickness of 490 ft (150 m), which feeds the Doda (or Stod) River, and so is important for irrigating the fertile farmland downslope.

The Rongbuk Glacier

Location Southern Tibet, China

Two tributary glaciers (the East and West Rongbuk) converge to form one flow of ice that is 14 miles (22 km) long—a fine example of a valley glacier moving north from Mount Everest onto the Tibetan Plateau. Climbers have used East Rongbuk's impressive medial moraine as a way to access Everest's North Col, describing it as the "magic highway." On either side of the moraine, the ice has been churned into fields of seracs (towering, jagged ice pinnacles)— some up to 100 ft (30 m) high— with ice bridges and pillars adding to the chaotic impression.

> **Other glaciers and ice sheets in Asia**
>
> - **The Baltoro Glacier** » p.233
> - **The Biafo Glacier** » p.234
> - **The Fedchenko Glacier** » p.232
> - **The Khumbu Glacier** » p.235
> - **The Yulong Glacier** » p.232

Australia and New Zealand

The Fox Glacier

Location The Southern Alps of South Island, New Zealand

Fed by four alpine glaciers, this heavily crevassed maritime glacier plunges down 8,500 ft (2,600 m) over the course of its 8-mile (13-km) flow from Mount Tasman to the coast of New Zealand's South Island. Its outflow, 980 ft (300 m) above sea level, is the Fox River, and it is one of the few glaciers to terminate in lush temperate rain forest. Its recent history of advance and retreat is fascinating. During the last ice age, it extended beyond the present coastline, since which time it has retreated a long way. Between 1985 and 2009, it advanced again, sometimes rapidly, but since then the retreat has continued.

> **Other glaciers and ice sheets in Australia and New Zealand**
>
> - **The Franz Josef Glacier** » p.280
> - **The Tasman Glacier** » p.281

▼ CREVASSES IN THE FOX GLACIER

up to 394 in (10,000 mm) and avalanches in an environment where winter temperatures may plunge to -58°F (-50°C).

The Kolahoi Glacier

Location Jammu and Kashmir state, India

Below the pyramidal rock mass of Kolahoi Peak, this glacier flows from an icefield that also feeds three other glaciers. Kolahoi, which is the source of the Lidder River, has an average elevation of 15,400 ft (4,700 m). Between

1963 and 2005, its area contracted from 5 square miles (14 square km) to 4 square miles (11 square km), and it is still retreating at the rate of 10 ft (3 m) a year. The interior of the glacier is believed to be hollowing out.

Durung Drung

Location Jammu and Kashmir state, India

One of the most beautiful Himalayan glaciers, Durung Drung is a gently meandering stream of ice flowing between the towering,

RIVERS AND LAKES

Every continent—even Antarctica, where most water is locked up in ice—has rivers and freshwater lakes. Rivers are fed by groundwater springs, lakes, glaciers, and surface runoff. Measured by the amount of water flowing through its channels, the Amazon–Orinoco basin in South America is the largest, while the Nile, in Africa, closely followed by the Amazon, is the longest. The largest lake by volume is Lake Baikal in Russia, which formed as a result of Earth's crust being pulled apart. Lakes also form in basins with no outlet, in volcanic craters, and behind natural and artificial dams.

North America

Great Bear Lake

Location Northwest Territories, western Canada

The eighth largest lake in the world by surface area, the Great Bear Lake is also deep, plunging to 1,355 ft (413 m). It drains to the west, via the Great Bear River, to the mighty Mackenzie River. Its waters are cold and, being both far from oceanic influences and close to the Arctic Circle, it is frozen from November to July. It is the largest inland body of water to exist at such a high latitude. During the last glaciation, about 10,000 years ago, it was part of the even larger Lake McConnell, another glacial lake.

Spotted Lake

Location British Columbia, western Canada

Also known as Lake Okanagan, this salty alkaline basin contains some of the most mineral-rich waters on Earth—and the most colorful—with high concentrations of magnesium sulfate, calcium, and sodium sulfate, as well as smaller amounts of silver and titanium. Each summer, as the lake's waters evaporate, about 300 small, shallow pools are all that remain, each colored

▲ COLORFUL MINERAL DEPOSITS ON SPOTTED LAKE

according to the minerals concentrated within—a mix of blue, green, yellow, and orange. The lake is a traditional healing site for the First Nations of Okanagan, who use its mud and water to treat aches, pains, and other medical problems.

The Columbia River

Location From British Columbia to Oregon, western North America

This river has the largest volume of any river flowing into the Pacific Ocean from the Americas. It rises in the Rocky Mountains of British Columbia and flows 1,200 miles (2,000 km) through a drainage basin the size of France before entering the Pacific between Oregon and Washington. The river's steep

gradient and huge flow make it ideal for hydroelectric generation, and almost half of the US's hydroelectric power is generated in its basin.

Manicouagan Lake

Location Quebec province, eastern Canada

This ring-shaped crater lake has been called the eye of Canada after its rough-edged, circular shape on a map. It was formed by the impact of a massive meteor, which measured 3 miles (5 km) across, about 214 million years ago. The original crater was about 60 miles (100 km) across, although the lake now has a diameter of 45 miles (72 km). Mount Babel, on the island in the middle of Manicouagan Lake, marks the crater's central peak, which formed after Earth's crust rebounded from the original impact. The lake drains into Manicouagan River to the south.

▼ TUFA TOWERS ON MONO LAKE

Lake Seneca

Location New York, eastern U.S.

This is the largest of New York's 11 freshwater finger lakes and has a reputation as a superb trout fishery. The lakes are in north-south valleys that were originally eroded by rivers but were later greatly deepened by south-flowing glaciers about 2 million years ago. The deepest point in Lake Seneca is 617 ft (188 m), well below sea level. It is fed by large underground springs, whose currents keep the lake mostly free of ice.

The Hudson River

Location From the Adirondack Mountains to New York, eastern U.S.

Although not especially long, at 315 miles (507 km), this is a river of great contrasts, rising in the sparsely populated Adirondack Mountains and flowing south to reach one of the world's busiest ports and greatest cities, New York City. A rise in sea level since the last ice age has "drowned" the lower reaches, meaning that almost half of the Hudson's length is tidal. When sea levels were lower, its estuary was 200 miles (320 km) south of New York City, at the edge of the continental shelf.

Mono Lake

Location California, western U.S.

Mono Lake is a shallow, salty water body with no outlet, lying in a basin just east of the Sierra Nevada. The basin is probably 3–4 million years old, and the lake is known to have existed for about 750,000 years. Although no more than 165 ft (50 m) deep now, it has been much deeper in the past. Along parts of the shoreline close to underwater springs, towers of tufa (a kind of limestone) have formed. The lake's saline water supports a huge population of brine shrimps, which provide good feeding for up to 2 million water birds.

The Rio Grande

Location Southwest U.S. and U.S.-Mexico border

This iconic river forms much of the U.S.–Mexico border. Near its source in the San Juan Mountains of Colorado, it flows as a clear, snow-fed stream before passing through a series of deep canyons near Big Bend, Texas, then sluggishly meandering across a broad floodplain to reach the Gulf of Mexico. At 1,900 miles (3,100 km) long, it is the 20th longest river on Earth, and its drainage basin is vast, but heavy water extraction for agriculture, industry, and domestic use means only 20 percent of its natural discharge reaches the ocean. Nowhere is the river deep, and only small boats can navigate it.

The Arkansas River

Location From Colorado to Arkansas, southern U.S.

The second longest tributary of the Mississippi–Missouri system, the Arkansas River rises in the Rocky Mountains, and its headwaters descend 4,600 ft (1,400 m) in 120 miles (193 km) through many systems of rapids. As it flows out of the mountains and is joined by numerous tributaries, the river widens significantly before joining the Mississippi near Napolean, Arkansas, 1,469 miles (2,364 km) from its source. Water extraction has greatly reduced the flow of the river, but it is still an important waterway for barge transportation in eastern Oklahoma and Arkansas.

The Red River of the South

Location From Colorado and Texas to Louisiana, southern U.S.

Apart from being one of North America's longest rivers, this river's main distinction is that the Prairie Dog Town Fork, one of the two major forks that form its headwaters, flows through the Palo Duro Canyon in northern Texas—the US's second largest canyon after the Grand Canyon. Half of the river's total length of 1,360 miles (2,190 km) forms the border between Texas and Oklahoma. The river once flowed into the Mississippi River but now joins the Atchafalaya River. Most of the river's traffic is concentrated in Louisiana.

Mammoth Cave

Location The Appalachians of central Kentucky, eastern U.S.

The world's longest known cave network, Mammoth Cave consists of 405 miles (652 km) of tunnels, shafts, and underground caverns in limestone that is more than 320 million years old. A resistant cap of sandstone covers the limestone. But where water has eroded through this, sinkholes have formed and water has flowed underground, dissolving the limestone and creating stalactites, stalagmites, and a spectacular accumulation of calcium carbonate that has been named the Frozen Niagara because it looks like a waterfall.

The Great Dismal Swamp

Location Virginia and North Carolina, eastern U.S.

Despite its discouraging name, the Great Dismal Swamp is a beautiful complex of wetland habitats. This poorly drained area of forest, with Lake Drummond at its center, is one of the US's largest swamps, with an area of more than 170 square miles (450 square km). Groundwater moves into the swamp from higher ground to the west but is held close to the surface by impermeable clay beneath. In winter and spring, the water table rises and the swamp floods, then in summer evaporation dries much of it out. Biodiversity is rich, with bald and Atlantic white cypress, maple, tupelo, and many other tree species, a diverse bird and reptile fauna, and populations of black bears and bobcats.

▼ THE HEADWATERS OF THE
ARKANSAS RIVER

Ox Bel Ha

Location Quintana Roo state, southern Mexico

Meaning "three paths of water" in Mayan, this is the longest known undersea cave network in the world, comprising 170 miles (270 km) of tunnels close to and beneath the east coast of the Yucatan Peninsula. Created as water dissolved limestone, the caves later flooded as the sea level rose. They were inundated both by saltwater from the Caribbean Sea and fresh groundwater. The lighter freshwater flows on top of the heavier, near-static saltwater layer. The caves are accessed through several sinkholes.

Sistema Huautla

Location Oaxaca state, southern Mexico

The ancient limestone highlands of the Sierra Mazateca mountains contain one of the world's finest collections of caves. First discovered in 1965, Sistema Huautla has subsequently been

found to be the deepest in the Americas, extending to 5,069 ft (1,545 m) beneath the surface. Explorers have to dive through long submerged sections to reach the deepest point of the 40-mile- (64-km-) long system. When dye was put in the system's water, it was found to discharge at a spring in the Santa Domingo Canyon, at a much lower level, indicating that more passages are yet to be discovered. As the canyon has cut deeper, it has triggered changes throughout the cave system. Fresh stream passages have been formed under now-abandoned tunnels.

Sac Actun

Location Quintana Roo state, southern Mexico

Translated from the Mayan language, Sac Actun means White Cave. With a total surveyed length of 205 miles (331 km), it is the longest cave system in Mexico and the second longest on Earth. While the upper parts of this coastal system are above water, most of the system is flooded with a mixture of freshwater from inland and seawater from the Caribbean. The cave can be reached through about 170 sinkholes. Some deep flooded shafts are yet to be explored, suggesting that there are more tunnels yet to be found.

Other rivers and lakes in North America

▶STALACTITES AND STALAGMITES IN THE DRAPERY ROOM AT MAMMOTH CAVE

Central and South America

Windsor Caves

Location Jamaica, Caribbean Sea

These are some of the biggest underground caverns of the forested limestone hills of Jamaica's Cockpit Country. "Cockpits" refer to the steep-sided depressions up to 330 ft (100 m) deep that dominate the landscape of this area. Thick layers of the White Limestone Formation have been weathered and eroded by the action of rainwater and groundwater to form underground caverns and other karst features. Windsor Great Cave is 2 miles (3 km) long and has an underground river, stalactites, stalagmites, and some beautiful roof scalloping. It also has a high-level remnant of an ancient drainage system, now largely choked with calcite deposits, and houses the island's largest bat roost.

▼ THE LONG DROP OF KAIETEUR FALLS

Lake Nicaragua

Location The Pacific Coast of Nicaragua

Along with its neighbor, Lake Managua, this giant freshwater lake formed in a part of the bottom of a graben—a faulted section of crust that slipped down between the land on either side. Lake Nicaragua is the largest lake in Central America, 110 miles (177 km) long and 36 miles (58 km) wide. Although only a thin strip of land separates it from the Pacific Ocean, the San Juan River drains the lake into the far more distant Caribbean Sea.

The Orinoco

Location Venezuela and Colombia, from the Guiana Highlands to the Atlantic Ocean

Along with the Amazon, the Paraná, and the Tocantins, the Orinoco is one of the four great rivers of South America. It rises in the forested Guiana Highlands and flows through them in a great arc before crossing the seasonally flooded plains of the Llanos. Finally, it splits into many distributaries in a vast delta, its waters reaching the Atlantic Ocean 1,700 miles (2,740 km) from its source. The Orinoco's basin incorporates 80 percent of Venezuela and 25 percent of Colombia. The river's rate of flow varies dramatically—at Ciudad Bolívar, for example, its depth varies from 50 to 165 ft (15 to 50 m) according to the season.

Kaieteur Falls

Location Mazaruni-Potaro Shield, Guyana

A combination of height, width, and water volume makes Kaieteur one of the world's largest single-drop waterfalls. In the pristine rain forest of the Guiana Highlands, the Potaro

▲ THE TURQUOISE WATERS OF LAGUNA 69

River plunges vertically 741 ft (226 m) over a resistant conglomerate lip atop a sandstone cliff. The water splashing at the bottom is eroding the less resistant sandstone, creating an overhang. The river then thunders down a further series of cascades into a 20-mile- (32-km-) long gorge. Although many waterfalls are higher, what makes Kaieteur remarkable is its volume of flow, averaging 23,400 cubic feet (663 cubic meters) per second.

Laguna 69

Location Huascaran National Park, Ancash region, central Peru

Probably the most beautiful of Huascaran National Park's 400-plus lakes, what Laguna 69 (Laguna Sesentineuve) lacks in size it more than makes up for in its setting. The lake's turquoise color is produced by fine sediment that has been eroded and transported by glaciers. It lies at an elevation of 15,100 ft (4,600 m) and is surrounded by snow-clad mountains. In summer, the lake is replenished by a waterfall plunging down the slopes of Chakraraju. In winter, it is frozen. Most of the national park's lakes did not have traditional names, hence the numeric designation.

Tres Hermanas Falls

Location Otishi National Park, Junin, central Peru

After Angel Falls in Venezuela and Tugela Falls in South Africa, this is the third highest waterfall in the world and one of only three to attain a total drop of 3,000 ft (915 m). Tres Hermanas, or "Three Sisters," carries an unnamed Andean stream over a three-stepped drop in an area of undisturbed rain forest. Water plunges down the first two tiers into large, natural catch basins, before the final drop deposits water into the Cutivireni River.

The Cotahuasi River

Location Arequipa region between the Altiplano and the Ocona River, southwest Peru

One of the best whitewater kayaking rivers on Earth, the Cotahuasi's modest source is high on the Altiplano. It descends quickly over rapids and waterfalls through the Cotahuasi Canyon —one of the world's deepest canyons, reaching a maximum depth of 11,486 ft (3,501 m)—before merging with the Maran River to form the Ocona, Peru's largest river, flowing into the Pacific. In 150 miles (240 km), the Cotahuasi descends 11,500 ft (3,500 m).

The Colca River

Location Arequipa region between the Andean Altiplano and Pacific Ocean, southwest Peru

On the Colca river's 280-mile (450-km) journey from the high mountains of the western Andes to the Pacific Coast, it descends 14,800 ft (4,500 m) and has three different names—Colca, Majes, and Camana. The Colca's main claim to fame, though, is flowing through the Colca Canyon, one of the world's deepest. At its deepest point, the river is at least 11,500 ft (3,500 m) below the summits on either side. The canyon is 70 miles (120 km) long and is famed for its whitewater rafting.

▲ THE ALGAE-RICH WATERS OF OF LAGUNA COLORADA

Laguna Colorada

Location The Andean Altiplano, southwest Bolivia

Situated at an elevation of 14,035 ft (4,278 m) on the arid Altiplano of Bolivia, this hypersaline lake is also called the Red Lake because of the high concentrations of algae that color it. White islands of the mineral borax dot its vast expanse. This endorheic lake covers 23 square miles (60 square km) but only averages 12 in (30 cm) in depth. It sits on ignimbrite, a volcanic rock. Large numbers of endangered James's flamingos feed in the lake's shallow water, along with smaller numbers of Chilean and Andean flamingos.

The Paraguay River

Location From Brazil through Bolivia, Paraguay, and Argentina, southern South America

South America's fifth largest river has its source in the Mato Grosso uplands of Brazil and flows through a variety of landscapes for 1,629 miles (2,621 km) before its confluence with the Paraná River near Corrientes. For much of its length, seasonal flooding is characteristic. North of Corumba, its maximum discharge is in February, but to the south the greatest flow is in July. After leaving the Mato Grosso, the river runs south through the vast Pantanal wetland before

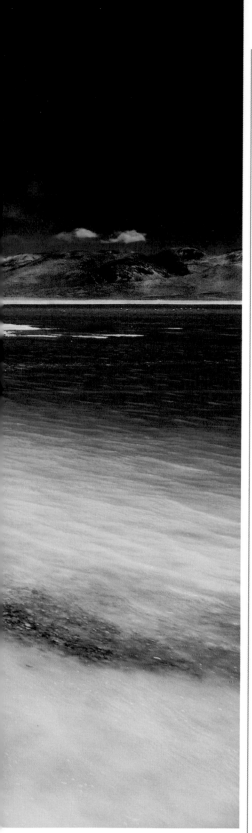

Gruta do Janelão

Location Minas Gerais state, eastern Brazil

This impressive limestone cave system in the valley of the Peruaçu River has many interesting features but is most famous for its rock art, which is more than 4,000 years old. The Peruaçu runs through part of the 3-mile- (4.7-km-) long system. Spectacular features include the tallest free-standing stalagmite in the world, which is 92 ft (28 m) from top to bottom, and one chamber that is 330 ft (100 m) from floor to ceiling. In places, sinkholes admit enough light through the cave roof for vegetation to grow within.

Other rivers and lakes in Central and South America

- The Amazon » pp.106–109
- Caño Cristales » p.105
- General Carrera Lake » p.113
- Iguaçu Falls » pp.110–11
- The Llanos » p.104
- The Pantanal » p.112
- Lake Titicaca » p.105

Europe

Dettifoss

Location Vatnajökull National Park, northeast Iceland

Fed by meltwater from the Vatnajökull Glacier, Dettifoss is often described as the most powerful waterfall in Europe, based on its drop of 144 ft (44 m) and width of 330 ft (100 m). It is fed by the sediment-rich waters of the Jökulsá River, which flows from Vatnajökull across columnar basalt before thundering over the sheer cliff of Dettifoss. The water ranges from grayish white to brownish white, depending on the sediment load. The average flow is 6,180 cubic feet (175 cubic meters) per second.

Seljalandsfoss

Location South Region, Iceland

This waterfall carries the waters of the Seljalands River 200 ft (60 m) over an ancient sea cliff into a pool and then to the coast. Unusually, it is possible to walk behind the fall as it pours over an overhang where resistant solidified lava lies on top of less resistant tillite (a glacial deposit). The river is fed by meltwater from the Eyjafjallajökull ice cap, lying on the caldera of the active volcano of the same name (which last erupted in 2010).

Skógafoss

Location South Region, Iceland

This is one of Iceland's most famous waterfalls, where the Skógá River thunders down a drop of 200 ft (60 m), producing clouds of spray—and rainbows on sunny days. Skógafoss is the most impressive of many falls that plunge down the basalt cliffs marking the position of the 3,000-year-old former coastline. This relic shoreline, now 3 miles (5 km) inland, dates from the last ice age, when Iceland was covered in ice; the coastline has subsequently fallen as the land has rebounded upward following the melting of the ice.

▼ WATER CASCADING FROM BASALT CLIFFS AT SKÓGAFOSS

dividing Paraguay into two, with alluvial savanna (the Gran Chaco) to the west and wetter, forested regions to the east. The river is navigable for a greater length than any other river on the continent apart from the Amazon. Fishes such as the salmonlike dorado, the piranha, and the basslike pacu inhabit the Paraguay River.

The Jagala Waterfall

Location Jõelähtme municipality, Estonia

Estonia's largest waterfall, on the Jagala River not far from where it enters the Gulf of Finland, is 25 ft (8 m) high and more than 165 ft (50 m) wide. Downstream, the river runs through a small gorge 980 ft (300 m) long, charting the slow movement of the waterfall upstream as a result of erosion of the tough limestone bedrock. In winter, the waterfall freezes, creating a tunnel between it and the rock behind.

Loch Ness

Location The Great Glen of Scotland, northern UK

Loch Ness is a deep, narrow freshwater lake that runs 23 miles (37 km) along the bottom of the Great Glen, a straight valley created by the Great Glen Fault and later scoured by glaciation. The greatest depth of Loch Ness is 750 ft (230 m). The Great Glen Fault is an ancient strike-slip fault, with crust to the north moving northeast relative to the south. It is now mostly inactive.

Lough Derg

Location Clare, Galway, and Tipperary counties, Republic of Ireland

This long, thin lake is the second biggest in the Republic of Ireland. It is both fed and emptied by the Shannon River. The lough is 24 miles (39 km) long and covers an area of 50 square miles (130 square km). While its average depth is only 25 ft (8 m), the deepest point is 118 ft (36 m)—the result of the deepening of the Shannon valley by glaciers during the last ice age.

The Severn

Location From the Cambrian Mountains to the Bristol Channel, UK

This is the longest river in the United Kingdom, at 220 miles (354 km). It also has the largest discharge, averaging 2,100 cubic feet (60 cubic meters) per second but occasionally many times more than this. Its source is the upland of Plynlimon, in Wales, and it discharges into a tidal estuary and the Bristol Channel. The latter has a large tidal range, and during the highest tides, rising water is funneled up the lower reaches of the river, against the river current; this phenomenon is known as the Severn Bore.

The Thames

Location From the Cotswold Hills to the North Sea, UK

The Thames is the second longest river in the United Kingdom, generally accepted as 215 miles (346 km) long. Its source is in the Cotswold Hills, and it flows roughly east, through London, to empty into a broad estuary and the North Sea. Huge quantities of water are removed for domestic and industrial use, so its discharge is significantly less than that of the Severn (see left). However, it poses a considerable threat. Heavy rains combined with exceptionally high tides have flooded the city on many occasions. The lowest 55-mile- (89-km-) long stretch of the river, to Teddington Lock, is tidal.

▲ FROZEN WATER AT THE JAGALA WATERFALL

The Rhine

Location From the Alps to the North Sea, central Europe

Second only to the Danube as the greatest river of west and central Europe, the Rhine rises from two headstreams (the Vorderrhein and the Hinterrein) high in the Swiss Alps and flows for 764 miles (1,230km). Along the way, it passes through Lake Constance (see p.373), the Upper Rhine graben (a block of crust that has moved downward between two faults), the Rhine Gorge, and across the North German Plain before reaching its delta near Rotterdam. It is one of the most important arteries of industrial transport in the world.

The Seine

Location From Burgundy to the English Channel, northern France

France's second longest river has its source in the limestone hills

▲ THE RHINE WINDING THROUGH THE RUINAULTA CANYON

of Burgundy, from where it flows 483 miles (777km) northwest through the agricultural heartland of the Île de France region, then through Paris to the English Channel between Le Havre and Honfleur. The Seine carries most of France's inland waterway traffic, and canals link it to the Rhine (see left), Loire (see right), and Rhône (see p.372) navigations. The estuary experiences a bore, called the *mascaret*, when the highest tides send a surge of water sweeping upriver.

The Loire

Location From the Massif Central to the Bay of Biscay, France

The longest river in France rises from three springs near the base of Mont Gerbier de Jonc, a lava dome in the Massif Central. The Loire flows north to the Orleans region, where it swings to the west, eventually discharging via a tidal estuary into the Bay of Biscay. It is 629 miles (1,012km) long, and drains more than 20 percent of France.

▲ LIMESTONE STALACTITES AT VERCORS CAVES

flows for 505 miles (813 km) to the Mediterranean west of Marseilles. Its course follows large glaciated valleys to Lake Geneva, then southwest to Lyon, where it converges with its biggest tributary, the Saône. The Rhône changes direction to the south and flows between the Alps and the Massif Central before splitting into several branches of its delta at Arles.

Vercors Caves

Location Provence-Alpes region, southwest Alps, southeast France

This large limestone massif comprises a series of high plateaus divided by steep-sided valleys, including the Bourne and Furon gorges. Beneath the surface are many cave systems that quickly flood when the winter snow melts, or after heavy rain, and become treacherous. Highlights include the Grotte de Bournillon, which has an entrance 260 ft (80 m) high and 100 ft (30 m) wide—said to be the biggest in Europe. The Hall of the Thirteen, the main passage in the Gouffre Berger system, has massive stalagmites and delicate stalactites. Gouffre Berger was the first cave on Earth to be explored to a depth of more than 3,300 ft (1,000 m).

Grotte Casteret

Location The Pyrenees of Aragon, Spain

At an elevation of more than 8,500 ft (2,600 m), the limestone cave system at Grotte Casteret contains a huge chamber called Grande Salle. The subterranean hall, which is 230 ft (70 m) long and 200 ft (60 m) wide, has an extraordinary frozen floor of clear ice covering 22,000 square feet (2,000 square meters); the deeper layers of ice are believed to be very old. There are also spectacular ice columns and a 65-ft- (20-m-) high ice wall. The cavern was discovered in 1926.

Lascaux

Location Dordogne, southwest France

The walls of dozens of caves in the valley of the Vézère River, which flows through a limestone karst landscape, have ancient illustrations painted or etched on them. Those in the cave at Lascaux, a UNESCO World Heritage Site, include a huge depiction of fighting aurochs (an extinct species of wild cattle) and are believed to be about 17,000 years old. The main cavern at Lascaux is 65 ft (20 m) wide and 16 ft (5 m) high.

The Rhône

Location From the Alps to the Mediterranean, Switzerland and southeast France

Meltwater from the Rhône Glacier (see p.359), high in the Swiss Alps, is the source of this river, which

The Lakes of Covadonga

Location The Picos de Europa, Asturias, Spain

The permeable limestone landscape of the high Picos de Europa massif in the Cantabrian Mountains is renowned for its amazing karst features, including gorges, caves, sinkholes—and a virtual absence of standing surface water. Two exceptions are Lakes Enol and Ercina, at an elevation of more than 3,600 ft (1,100 m), which retain water because of the presence of impermeable glacial deposits left behind after the last ice age.

The Tagus

Location The Sierra de Albarracín to Lisbon, Spain and Portugal

The longest river of the Iberian Peninsula (626 miles/1,007 km long) drains the second most expansive drainage basin (after that of the Ebro River). Rising in the Sierra de Albarracín, in eastern Spain, the Tagus flows through limestone ravines and gorges before reaching the vast semiarid plains of Castille. There it has been dammed in several places to provide water and hydroelectric power before it reaches the lowlands of Portugal, discharging into the Atlantic Ocean near Lisbon.

Doñana

Location Andalucía, Spain

This vast area at or close to sea level was given national park status for its rich tapestry of marshland, reedbeds, shallow streams, pools, and sand dunes— and the fauna that these habitats support, especially breeding and migrating birds. In the past, Doñana has been covered by both freshwater and saltwater, but it is now slowly drying out, mainly as a result of the extraction of water from its aquifer for agriculture. It is protected from inundation by a wide barrier of sand dunes.

Lake Constance

Location Western Austria, eastern Switzerland, and southern Germany

This freshwater body lies in a *zungenbecken*, or tongue basin, a long hollow scoured out and left behind by the Rhine Glacier as it retreated. The hollow filled with water and is now 39 miles (63 km) long and up to 827 ft (252 m) deep. The Rhine (see p.371) and other rivers are constantly depositing sediments in the lake, extending its coastline and reducing its depth.

▼ SHALLOW WATERS OF LAKE CONSTANCE

The Po

Location From the Alps to the Adriatic Sea, northern Italy

The Po, Italy's longest river at 405 miles (652 km), takes water from the country's largest drainage basin, including its most fertile plain. It descends steeply from its source in the Cottian Alps, but from Turin follows a more gentle descent as it passes across a broad, flat plain. It is connected to Milan by a series of channels. Then it flows more or less due east to its delta, the most complex of any European river, with at least 14 mouths entering the Adriatic Sea.

The Plitvice Lakes

Location The Velebit Mountains, Croatia

This stepped series of 16 lakes and interconnecting waterfalls cascades down the Korana River in the forested limestone landscape of the Velebit Mountains. Over the course of 5 miles (8 km), the river falls in elevation by 440 ft (133 m). The waters of the river contain high concentrations of calcium carbonate, some of which is deposited to form travertine barriers, which dam the lakes. As water flows over the top of the dams, it forms waterfalls, the highest of which—Veliki Slap—drops 256 ft (78 m). The river runs through a gorge, and its valley includes other karst features, such as caves and underground streams.

▼ WATERFALLS AND LIMESTONE HILLS AT PLITVICE LAKES NATIONAL PARK

Bigar Waterfall

Location The Anina Mountains, southwest Romania

Located in the Caraș-Severin County, Romania's most famous waterfall is neither spectacularly high nor thunderously powerful, but what it lacks in size it more than makes up for in beauty. The Bigar Waterfall is a stream flowing into the Miniș River gorge over a 25-ft- (8-m-) high domed rock overhanging the river below.

As the water flows over the moss-covered rock, it divides into myriad tiny rivulets to form a delicate fringe dropping into the underlying pool.

Optimisticheskaya

Location Ternopil Oblast, western Ukraine

With 140 miles (230 km) of passageways, this is the longest gypsum cave system in the

world—and the longest cave of any description in Eurasia. Despite this, it exists in layers of gypsum just 65 ft (20 m) thick. Gypsum is a soft sulfate mineral, which is more soluble than limestone and readily dissolves when in contact with slightly acidic rainwater to form underground features similar to those in limestone caves. Optimisticheskaya has an extraordinarily dense network on several levels, giving rise to its name of "the maze cave." It is accessible only through a single entrance, dug through the floor of a doline in 1966 by local cavers.

The Black Sea

Location Bulgaria, Romania, Ukraine, Russia, Georgia, and Turkey

This saline water body is described as a trapped ocean basin. It is almost completely surrounded by land but underlain by oceanic crust that was trapped by tectonic plate collisions. To the southwest, it connects to the Mediterranean Sea through the Bosporus Strait, Sea of Marmora, and Dardanelles Strait. To the north, it connects to the Sea of Azov. Excluding the latter, the Black Sea has an area of 168,300 square miles (436,000 square km) and a maximum depth of 7,218 ft (2,200 m). Many rivers discharge into it, including the Danube and Dnieper.

Other rivers and lakes in Europe

▲ ACCUMULATION OF CYANOBACTERIA ON LAKE NATRON

Africa

Lake Chad

Location The Chad Basin, Nigeria, Niger, Chad, and Cameroon

This freshwater lake is situated in the Chad Basin, the second largest endorheic (closed) drainage basin on Earth. Between 1963 and 1998, it shrank from 10,000 square miles (26,000 square km) to 520 square miles (1,350 square km), although a small increase followed. Mostly fringed by marshes, it averages just 5 ft (1.5 m) in depth, although this varies from year to year. Lake Chad is fed by rivers such as the Ngadda and Komadugu Yobe and has no outlet. The lake is a vital source of water and fish for local communities.

Lake Nakuru

Location The eastern arm of the Great Rift Valley, western Kenya

One of the ultra-alkaline soda lakes in the bottom of the down-faulted East African Rift Valley, this water body is famous for supporting hundreds of thousands of flamingos that come to feed on the lake's algae. Nakuru varies in size from 2 to 17 square miles (5 to 45 square km), but its maximum depth is just 10 ft (3 m). About 10,000 years ago, Nakuru and its neighboring lakes (Elmenteita and Naivasha) formed a single deep freshwater lake, which shrank dramatically as the climate became drier, leaving behind three separate remnant lakes.

Lake Natron

Location Arusha region, northern Tanzania

Fed by mineral-rich hot springs and the Ewaso Ng'iro River, Lake Natron is unusual for its deposits of the evaporite minerals natron and trona, its brightly colored waters, and the huge flocks of flamingos that feed and breed there. With water temperatures as high as 140°F (60°C), this shallow soda lake at the bottom of the Gregory Rift (the eastern branch of the East African Rift Valley) is inhospitable for most organisms, but abundant cyanobacteria color its waters deep red or orange.

The Zambezi

Location Northwest Zambia to the coast of Mozambique

This is the fourth longest river in Africa, after the Nile, Congo, and Niger, flowing 1,599 miles (2,574 km) through six countries from its source to the ocean. It rises in a marshy miombo woodland close to the watershed with the Congo Basin and discharges into the Indian Ocean at a large delta on the Mozambique coast. After plunging down the Victoria Falls, the river is dammed to form lakes Kariba and Cahora Bassa, in Zambia and Mozambique, respectively.

Blyde River Canyon

Location Drakensberg Mountains of Mpumalanga, South Africa

The Blyde River Canyon is also called the Motlatse Canyon. Where the Blyde River meanders through the Drakensberg Mountains, it has carved one of the largest canyons on Earth. Some 15 miles (25 km) long and averaging 2,460 ft (750 m) deep, it features some remarkable rock formations, including three buttes capped by ultra-resistant Black Reef quartzite, known as the Three Rondavels (rondavels are traditional huts whose shape they mimic). The canyon is also the site of a quartzite column called the Pinnacle and, at the bottom of the canyon, the Kadishi waterfall, the second highest waterfall flowing over tufa.

Cango Caves

Location Swartberg Mountains of Western Cape, South Africa

This limestone cave system in a ridge of the Swartberg Mountains is probably the best known in

◀ BOURKE'S LUCK POTHOLES
AT BLYDE RIVER CANYON

▲ THE FRESH AND SALINE WATERS OF LAKE BALKHASH

Africa. Although relatively small—about 2 miles (4 km) of passageways and caverns are known—it is famous for cave paintings dating back to the Early Stone Age and for impressive screens of stalactites, large stalagmites, and other dripstone features. Some of the last bear a resemblance to various figures or structures and have been given individual names, including Madonna and Child, Cleopatra's Needle, Dried Tobacco Leaves, and Leaning Tower of Pisa.

Other rivers and lakes in Africa

Asia

The Amur

Location From northeast China to the Sea of Okhotsk, eastern Asia

One of Asia's longest rivers, the Amur proper begins at the confluence of the Shilka and Argun (Ergun) Rivers, then flows 1,755 miles (2,824 km) to the Tartar Strait. If the headwaters are included, it is 2,761 miles (4,444 km) long. Its remarkable drainage basin encompasses desert, steppe, taiga, and tundra, and covers an area of 700,000 square miles (1.8 million square km). Fed largely by monsoon rains in summer and fall, the river's waters often rise 46 ft (14 m) above normal at this time, overflowing its banks in the marshy country downstream of Khabarovsk. For much of its course, the Amur forms the border between China and Russia.

Lake Balkhash

Location Eastern Kazakhstan

The freshwater western section of this vast, shallow lake is separated from the saline eastern section by the Saryesik Peninsula, which almost divides Balkhash in two. The lake narrows and deepens to the east, and there is little mixing between the two parts. Most of the lake's input comes from the Ili River, one of several that drain the mostly arid Balkhash-Alakol Depression (a closed, or endorheic, basin) and discharge into the lake. Balkhash has a surface area of 6,330 square miles (16,400 square km).

The Aral Sea

Location Northern Uzbekistan and southern Kazakhstan

In the 1960s, this endorheic lake was the fourth largest on Earth, with an area of 25,500 square miles (66,000 square km), but it shrank rapidly after its feeder rivers were diverted for irrigation projects. By 2007, its waters had become so shallow that it had split into four lakes, and its surface area was reduced to just one-tenth of what it had once been. Much of what was once water is now desert. Efforts to replenish the waters of the northern section (the North Aral Sea) have succeeded in increasing its surface area and depth, and fish populations are increasing.

Krubera Caves

Location The Arabika Massif of the western Caucasus Mountains, Georgia

There are several major cave systems in the thick beds of folded limestone of the Arabika Massif, including Krubera—the world's deepest known cave. It descends through a series of vertical pits and steep meandering passageways to a level 7,208 ft (2,197 m) below its entrance in a glacial valley. For most of this depth, it is relatively dry, with no indication of major flooding, but the very deepest part, which was explored in 2012, is submerged.

Jeita Grotto

Location Nahr al-Kalb valley, Lebanon

This is the longest cave system in the Middle East, stretching for 6 miles (9 km). It has two separate, but interconnected, sections, one 200 ft (60 m) below the other. The upper galleries contain the White Chamber and the Red Chamber. The former houses the world's longest stalactite, measuring 25 ft (8 m), while the limestone walls of the latter have been stained red by iron oxide. An underwater river flows through the lower galleries, and there is a subterranean lake. The chamber known as Grand Chaos Hall, is 1,640 ft (500 m) long.

Sambhar Salt Lake

Location Rajasthan, northwest India

This is India's largest saline lake, with a surface area of 75–90 square miles (190–230 square km) and a depth varying from 10 ft (3 m) after the monsoon to 2 ft (0.6 m) at the end of the dry season. It is set in an endorheic basin and receives water mainly from the Mendha and Rupangarh Rivers. When salinity levels reach a sufficient concentration, water is released through a dam into evaporation pools for commercial salt production. Many flamingos spend the winter at the lake.

Lonar Lake

Location Maharashtra, western India

India's largest impact-crater lake is almost circular in shape, with an average diameter of ¾ mile (1.2 km) and a depth of about 20 ft (6 m). The lake's waters are saline and are fed by one stream. The lake is surrounded by a crater rim, created when a meteorite impacted the Deccan Traps plateau. The date of impact is debated—estimates range from 570,000 to 52,000 years ago.

Borra Caves

Location The Eastern Ghats of Andhra Pradesh, eastern India

These are some of the deepest caves in India, containing formations such as stalactites, stalagmites, columns, and shapes reminiscent of mushrooms, a human brain, and a mother and child. The main chamber is 660 ft (200 m) long and 39 ft (12 m) high. Sulfurous springs discharge into the cave system, which is the source of the Gosthani River.

◄ THE LOWER CAVERN OF JEITA GROTTO

Dudhsagar Falls

Location The Western Ghats of
Karnataka-Goa, southwest India

One of India's highest waterfalls,
this four-tiered cascade is produced
by the Mandovi River as it plunges
1,020 ft (310 m) from the Deccan Traps
(see p.353) in the Western Ghats.
The falls are ordinary in the dry
season, but their character changes
completely during monsoons,
between June and September.
Dudhsagar means "sea of milk"
in the local Konkani language,
a reference to the creamy-white
appearance of the water as it pours
down the near-vertical rock face.

Hogenakkal Falls

Location Karnataka-Tamil Nadu border,
southern India

Hogenakkal means "smoking rocks,"
a reference to the clouds of spray
that hang in the air above the falls.
While the 65 ft (20 m) that the
Kaveri River drops does not sound
great, the width of the waterfall is.
As the river approaches "India's
Niagara," it forks into multiple
streams that gush into a narrow
gorge from both sides. Hogenakkal
is at its most spectacular during
monsoons (July–August).

Yarlung Tsangpo Grand Canyon

Location The Eastern Himalayas,
China (Tibet) and India

Arguably the world's deepest
canyon, the difference in
elevation between the Yarlung
Tsangpo River and the mountain
peaks on either side of it exceeds
16,000 ft (5,000 m) in places.
Much of the river's course is
in open valleys on the Tibetan
Plateau, but in southeast Tibet
it enters a deeply incised canyon
through the eastern Himalayas,
switching its course from
northeast to southwest as it
flows around Namcha Barwa.
After 315 miles (505 km), it
emerges into the broad
Brahmaputra valley.

▼ MULTIPLE STREAMS OF
HOGENAKKAL FALLS

▲ THE SACRED WATERS OF GURUDONGMAR LAKE

The Irrawaddy

Location From the eastern Himalayas to the Andaman Sea, Myanmar

This major river arises from the confluence of the N'mai and Mali rivers, both of which have their sources in Himalayan glaciers. As it bisects Myanmar (formerly Burma), the Irrawaddy flows 1,350 miles (2,170 km) through rain forest, dry forest, extensive areas of rice cultivation, and—in the delta zone—freshwater swamp forest and mangroves before entering the Andaman Sea. The river's massive sediment load extends the delta by 165 ft (50 m) each year. At peak times in the monsoon season, the

river's discharge is more than 40,000 cubic meters (1.4 million cubic feet) per second.

Gurudongmar Lake

Location The Himalayas of northern Sikkim, northeast India

Situated close to the summit of Mount Khangchen Gyao, this lake is one of the highest on Earth, at an elevation of 17,815 ft (5,430 m). Surrounded by snow-covered mountains, Gurudongmar is fed by glacial meltwater and drains into a small stream that later becomes the Teesta River. The lake is frozen in the winter months. It is considered sacred by both Buddhists and Sikhs.

Loktak Lake

Location The Manipur Valley of Manipur, northeast India

The largest freshwater lake in northeast India is best known for its *phumdis*, islands of vegetation in various stages of decomposition, some of which are inhabited. The lake has a surface area of 111 square miles (287 square km), and the largest of the *phumdis* covers 15 square miles (40 square km). Water enters and leaves the lake through the Manipur River. Discharge from the lake is restricted by a 25 ft (8 m) rocky barrier called Sugnu Hump, which reduces the river's flow.

Akiyoshi-do

Location Southwest Honshu, Japan

Akiyoshi-do is Japan's longest cave, a product of the action of water on the thick Akiyoshi Group reef limestone over hundreds of thousands of years. This cave is 6 miles (9 km) long, has a ceiling up to 260 ft (80 m) above the floor, and has a river flowing through it. There is also a beautiful complex of 500 terraced pools. Akiyoshi-do is one of more than 400 caves in the area, which is Japan's most important karst region.

Red Beach

Location Liaoning, northeast China

Large parts of the saltmarsh at the Liaohe River delta near Panjin have been colonized by a succulent halophyte (a salt-tolerant plant) called seepweed, which thrives in salty alkaline soils. During the summer months, this plant is an unremarkable green color, but it turns deep crimson from August to October, coloring a vast expanse of beach. Seepweed flats provide an important nesting location for the endangered red-crowned crane and Saunders' gull.

Qinghai Hu

Location The Qilian Mountains of Qinghai, western China

This vast saline lake is the largest in China and the most expansive mountain lake without an outlet in Central Asia. It covers

1,667 square miles (4,317 square km) in a large depression in the Qilian Mountains and lies at 10,515 ft (3,205 m) above sea level. Most of the rivers that once discharged into Qinghai Hu now run dry before they reach it, mostly because water is diverted from them for irrigation. Consequently, the lake is shrinking.

Jiuzhaigou Valley National Park

Location The Minshan Mountains of Sichuan, central China

This extraordinarily beautiful valley has been afforded World Heritage Site status for its superb landscape features. Notably, these include numerous blue, green, and turquoise lakes with clear, mineral-rich water, spectacular waterfalls, and limestone terraces set below high peaks. The valley also boasts classic limestone features such as caves and a variety of old-growth forest habitats, which are home to a population of giant pandas.

Xiaozhai Tiankeng

Location Chongqing municipality, central China

Xiaozhai Tiankeng is the world's largest and deepest sinkhole. On the surface, it has a diameter of more than 1,640 ft (500 m), and apart from one sloping ledge, its walls descend vertically 2,172 ft (662 m) through the bedrock to Di Feng Cave, into which the limestone of the sinkhole dissolved and was washed away by the subterranean Migong River. Although local people have long known about Xiaozhai, naming it the Heavenly Pit, it was only discovered by western explorers in 1994.

▼ FALL LANDSCAPE AT JIUZHAIGOU VALLEY NATIONAL PARK

▲ LIMESTONE PINNACLES AT GUNUNG MULU NATIONAL PARK

The Pearl River

Location Guangdong and Guangxi provinces, southern China

This name usually encompasses the three great rivers of Guangdong—the Xi, Bei, and Dong—that share a common delta. Measured from the source of the longest of these (the Xi) to the South China Sea, the Pearl River is the third longest in China at 1,365 miles (2,197 km). Its basin covers an area of 158,300 square miles (410,000 square km). The river is named after the pearl-colored shells found on its bed in Guangzhou.

Puerto Princesa Subterranean River National Park

Location The western coast of Palawan, Philippines

The most extraordinary feature of this World Heritage Site is its underground river, which is 5 miles (8 km) long, discharges directly into the South China Sea, and is tidal in its lower reaches. Underground chambers, waterfalls, stalactites, and stalagmites are other features of the cave system. Measured by volume, the 1,200-ft (360-m) Italian's Chamber is one of the largest in the world.

Gunung Mulu

Location Sarawak, Borneo

Located beneath the rain forests of the Gunung Mulu National Park, everything about this cave system is on a massive scale. It contains the Sarawak Chamber, which at 2,000 ft (600 m) long, 1,360 ft (415 m) wide, and 260 ft (80 m) high is the largest on Earth. Gunung Mulu has a total of 183 miles (295 km) of explored caves, including Clearwater Cave, the longest in Asia. Dating of sediments in Clearwater Cave shows that the caves have evolved over a period of 2 million years. The surrounding landscape has many other features of a limestone karst environment, including deep canyons, underground rivers, and limestone pinnacles. Gunung Mulu is home to millions of bats and cave swiftlets.

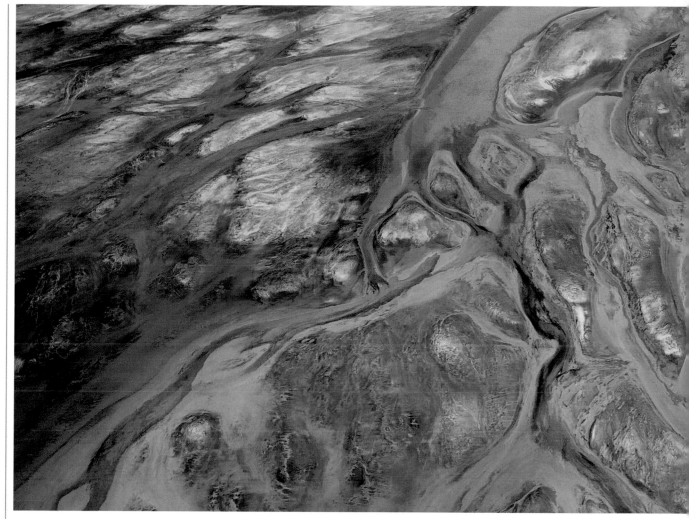

▲ LAKE EYRE, COLORED RED BY BACTERIA

Australia and New Zealand

Koonalda Caves

Location The Nullarbor Plain, South Australia

The main significance of this cave in the Nullarbor Plain, a vast area of limestone, is its Aboriginal art. A sinkhole 100 ft (30 m) deep leads through relatively soft limestone to a steep passage that connects to a large chamber with a domed roof 150 ft (45 m) above the floor. Two more passages lead off from the chamber, one connecting to three underground lakes. The wall art, which is more than 20,000 years old, consists of concentric circles, parallel lines,

and herringbone patterns. In the first chamber, archaeologists have found plenty of evidence that flint mining was practiced here 24,000–14,000 years ago. Hearths and charcoal were presumably used to provide firelight by which to work.

Lake Eyre

Location The Great Artesian Basin, South Australia

When it is full, this endorheic lake covers at least 3,670 square miles (9,500 square km) of the Great Artesian Basin. Most of its water comes from monsoon rains falling on Queensland, in the northeast part of the basin, and entering the lake via the Diamantina River and Coopers Creek. On average, a small flood occurs once every three years, but Lake Eyre fills completely much more rarely.

Occasionally, such as in 1984 and 1989, local rainfall topped up the lake. Now officially known as Kati Thanda-Lake Eyre, it also has the lowest point in Australia— 50 ft (15 m) below sea level.

Waituna Lagoon

Location The South coast of South Island, New Zealand

Important as a feeding area for a range of breeding and migratory waterbirds, especially shorebirds, Waituna Lagoon is a freshwater lake pooled behind a coastal gravel barrier. The lake covers an area of 14 square miles (36 square km) and is fed by three streams. It lies over gravel deposited by glacial meltwater during the last ice age and is bordered on its landward side by saltmarsh and bogs. The wide coastal barrier was created by longshore drift.

Huka Falls

Location Waikato River, North Island, New Zealand

This series of waterfalls is on New Zealand's longest river, the Waikato, which drains Lake Taupo. Just upstream of the falls, the river narrows from 330 ft (100 m) wide to just 65 ft (20 m) as it passes through a canyon that has been incised into hard volcanic rock. The effect of this narrowing, and a rapid descent over a series of small falls, is to accelerate the water as it approaches the final 36-ft (11-m) drop. Consequently, the water thunders over the falls with immense power.

COASTS, ISLANDS, AND REEFS

The world's continents and islands are all surrounded by coastlines. North America has the longest coast of any continent with 193,000 miles (310,000 km), and Canada is the individual country with the longest at 130,500 miles (210,000 km). Coastal features include gentle shores with a broad intertidal zone, sand dunes, vertical cliffs, rock stacks and arches, mangrove swamps, salt marshes, and wide estuaries. Some are deeply indented, especially where sea-level rise has drowned coastal valleys. Coasts advance where deposition exceeds erosion, but recede where marine erosion is greater than deposition.

North America

Ilulissat Icefjord

Location Near Ilulissat, west coast of Greenland

This tidal fjord opening into the Davis Strait is the outlet for the Ilulissat Glacier (also known as Sermeq Kujalleq), the fastest-moving glacier outside Antarctica, pushing into the fjord at 65–115 ft (20–35 m) per day. As the glacier has retreated, the fjord's open water has extended farther inland, and it is now about 30 miles (50 km) in length. It cuts through gneiss, granite, and mica schist 2.5–1.6 billion years old. The ice-carved fjord is up to 3,300 ft (1,000 m) deep, but much shallower at the iceberg bank, where it opens into the ocean; many of the larger icebergs run aground there.

Puget Sound

Location Washington, western U.S.

This huge complex of tidal basins and channels is the US's second largest estuary, stretching 100 miles (160 km) from north to south. Its waters connect with those of the Strait of Juan de Fuca and the Pacific Ocean at Admiralty Inlet. Puget Sound's deep basins, which plunge to 920 ft (280 m) below sea level, were carved and scoured by glaciers. Shallower parts of the estuary mark where there are glacial deposits on the seafloor.

▲ ICEBERGS AT THE ILULISSAT ICEFJORD

Dungeness Spit

Location North coast of Washington state, western U.S.

Stretching 6 miles (9 km) into the Strait of Juan de Fuca, this is the longest natural spit in the U.S. Prevailing westerly winds drive longshore drift, which has extended the sandy spit by an average of 15 ft (4.5 m) per year over the past 120 years. It is vulnerable to the destructive action of the ocean and was temporarily broken into three sections by a storm in 2001.

Mount Desert Island

Location Maine, eastern U.S.

This is the second largest island on the U.S.'s eastern seaboard. Much of it is granite, the remnants of an ancient igneous intrusion. More recently, it has been shaped by the erosive action of glaciers, which carved out Somes Sound, a deep inlet that almost cuts the island in two. This is the only fjordlike inlet on the east coast of the U.S. Most of Mount Desert Island is in Acadia National Park.

Long Island

Location New York, eastern U.S.

The largest island on the east coast of the U.S., Long Island stretches 120 miles (190 km) from east to west and is separated from the mainland by the East River and Long Island Sound. Long Island is also incredibly diverse, ranging from the New York City boroughs of Brooklyn and Queens to the wilderness of the Pine Barrens temperate coniferous forest. Much of the surface geology is made up of moraines and other glacial deposits.

Lighthouse Reef

Location Belize coast, Caribbean Sea

Oblong in shape, 22 miles (35 km) from north to south, and generally less than 20 ft (6 m) deep, Lighthouse Reef is a large atoll, part of the Belize Barrier Reef. Divers consider it one of the best dive sites on Earth. Near the center is the Great Blue Hole—a large sinkhole that plunges more than 330 ft (100 m) below sea level. It was formed during the last ice age when the sea level was much lower and the surrounding limestone was dry land. The atoll has an amazing variety of corals and fish, including sharks, and sea turtles, which breed here.

Other coasts, islands, and reefs in North America

- Acadia Coastline » p.65
- The Bay of Fundy » p.64
- Big Sur » p.63
- The Hawaiian Islands » pp.316-19
- Monterey Bay » p.62
- Thor's Well » p.62

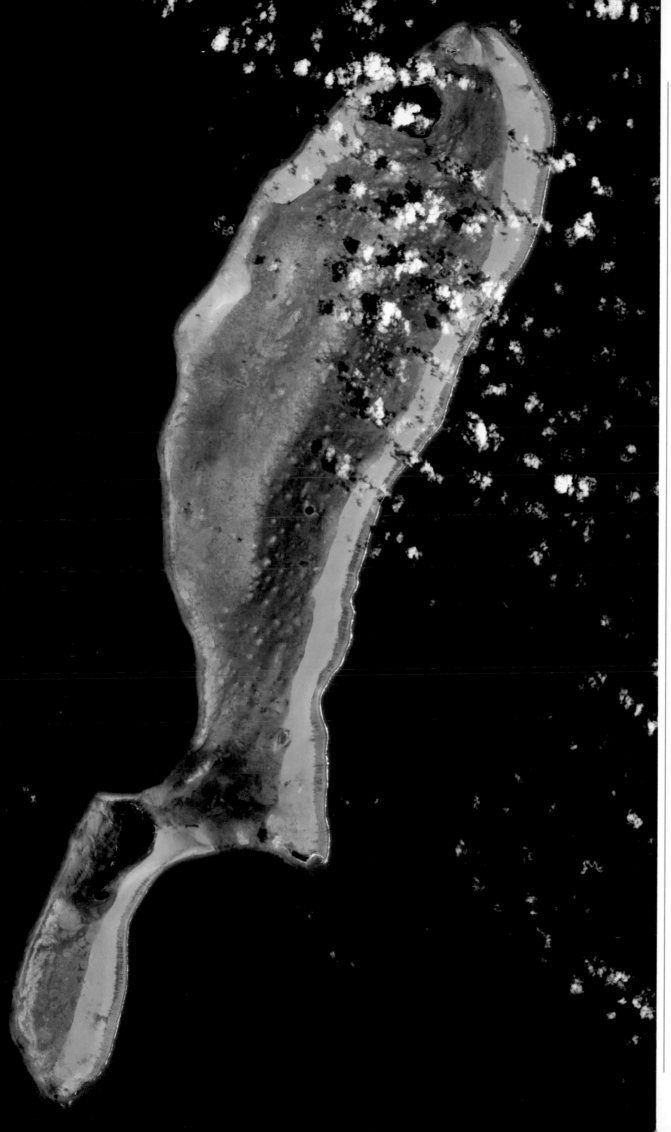

Central and South America

Les Pitons

Location Soufrière Bay, St. Lucia, Lesser Antilles

Translated as "the Peaks," Les Pitons are two adjacent forest-clad volcanic lava domes that rise from the Atlantic Ocean and tower over St. Lucia. Linked by Piton Mitan Ridge, steep-sided Gros Piton (2,526 ft high/770 m) and Petit Piton (at 2,438 ft/743 m) are cloaked in tropical and (at higher elevations) subtropical moist forest. Sulfurous fumaroles, hot springs, and deposits of ash, pumice, and lava flows are all indications of the volcanic origins of the peaks, which are the remains of a collapsed stratovolcano. The region was declared a World Heritage Site in 2004.

The Paracas Peninsula

Location Pacific coast of Peru

Sometimes referred to as the Galápagos Islands of Peru, the Paracas Peninsula is part of Peru's only marine reserve. This arid peninsula extends into the Pacific Ocean toward the cold, nutrient-rich Humboldt Current, which is extremely rich in fish and other oceanic life forms and attracts huge numbers of seabirds, including Humboldt penguins, as well as whales, dolphins, and sea turtles. A massive prehistoric design (or geoglyph), the Paracas Candelabra, has been carved into the north side of the peninsula; pottery found nearby dates from 200 BCE and could be of the same age.

◄ THE OBLONG-SHAPED ATOLL OF LIGHTHOUSE REEF

▲ THE LOFOTEN ISLANDS RISING FROM THE GREENISH BLUE WATERS OF THE NAPPSTRAUMEN STRAIT

The Bahia Coastline

Location Bahia state coast, eastern Brazil

Stretching for about 620 miles (1,000 km), this is the longest coastline in Brazil, featuring sandy beaches, river estuaries, small deltas, large bays, and coral reefs. Much of the Bahia coast is backed by expanses of Atlantic forest. There are fringing reefs at Pintaunas, Praia do Forte, and Itacimirim. Just south of the idyllic sandy bay at Trancoso is Monte Pascoal National Park, an old-growth Atlantic forest with a population of the endangered red-browed Amazon parrot, along with scarce mammals such as jaguar, cougar, and giant armadillo.

Other coasts, islands, and reefs in Central and South America

Europe

The Lofoten Islands

Location Nordland region, northern Norway

This archipelago of five large islands and many small ones stretches for 100 miles (160 km), and is characterized by steep-sided mountains of ancient metamorphic gneiss and quartzite, rising up from the sea. The landscape of deep valleys and fjords was crafted by the action of glaciers. The Rost Reef, discovered in 2002, is the world's largest deepwater coral reef. Built by Lophelia corals, it is 22 miles (35 km) long. One of the world's most powerful tidal eddies occurs close to the island of Mosken.

The North Jutland Dunes

Location Between Frederikshavn and Skagen, northern Denmark

For centuries, windblown sand caused problems for farmers in the Skagen area. Persistent southwesterly winds shifted dunes onto farmland and even buried buildings. Most of the dunes have been stabilized by the planting of vegetation, but one—the Rabjerg Mile—has been left unrestrained. Up to 130 ft (40 m) high and containing an estimated 140 million cubic feet (4 million cubic meters) of sand, this huge dune migrates up to 60 ft (18 m) northeast every year. It began its movement from the coast of Skagerrak, a strait separating Denmark from southwest Sweden, 300 years ago.

The Devon Ria Coast

Location Plymouth and Exmouth, Devon, UK

The coast of south Devon has several drowned valleys, or rias, partially submerged by a rise in sea level of up to 80 ft (25 m)

layers of white and pale gray chalk are 86–96 million years old. They form vertical cliffs up to 335 ft (102 m) high. Marine erosion has produced three natural arches and a spikelike structure called L'Aiguille, which rises 230 ft (70 m) from the sea.

Cap Ferret

Location Gironde, southwest France

This gravel-and-sand barrier spit almost completely separates the Bay of Biscay from the Arcachon Lagoon, into which the Leyre River discharges. The spit formed as a result of the longshore drift of sediments from the north. Cap Ferret is known to have built seaward and southward since the early 18th century, pushing the entrance to the lagoon south as it did so. However, it has undergone erosion at its southern tip since 1970.

Cathedral Beach

Location Cantabrian coast, northern Spain

Described as one of the most beautiful beaches in the world, Cathedral Beach has a series of rock arches that are reminiscent of a cathedral's nave. The dramatic caves, arches, and stacks are best appreciated at low tide. The incessant wave action has exploited faults in the metamorphic quartzite and slate strata to open up caves, which later form arches and then stacks when their roofs collapse.

since the last ice age. The most important rias are the valleys of the Tamar–Lynher, Kingsbridge, Dart, Teign, and Exe. The drowned section of the Kingsbridge valley is a broad estuary that, despite being tidal 5 miles (8 km) inland, is fed only by small streams. Longshore drift has built bars across parts of the mouths of the Teign and Exe estuaries, and sedimentation occurs behind the bars.

The Étretat Cliffs

Location Normandy coast, northern France

These are part of an 80-mile (130-km) stretch of chalk cliffs on the English Channel coast of Normandy. Étretat's horizontal

The Galician Coast

Location Northwest coast of Spain

This dramatic coastline, facing the Atlantic Ocean to the west and the Bay of Biscay to the north, is about 900 miles (1,500 km)

long and very irregular, with dozens of inlets and headlands, and at least 316 islands and islets. The cliffs at Vixia Herbeira rise 2,037 ft (621 m) above the sea and are among the highest in Europe. Drowned estuaries, or rias, such as those of the Arousa, Vigo, and Pontevedra Rivers, are the result of rising sea level since the last glaciation.

Cape de Creus

Location Catalonian coast, northeast Spain

Many of the classic features of crustal folding, faulting, shearing, and deformation are exhibited in the cliffs of this headland as well as anywhere on Earth. Located in the northeastern part of the Iberian Peninsula, Cape de Creus marks the point where the Pyrenees reach the editerranean. The rock structures displayed here demonstrate the massive pressures and strains applied to the crust during the formation of these mountains. Folded layers of schist and other metamorphic rocks have been intruded into the rocks by dikes of pegmatite, an igneous rock with large crystals. The Cape de Creus peninsula and its marine environments were designated a National Park by the Catalan government in 1998.

▼ AN ARCH IN THE ÉTRETAT CLIFFS

▲ DAZZLING WHITE TRUBI MARL OF THE TURKISH STEPS

The Amalfi Coast

Location Sorrento Peninsula, west coast of Italy

The southern coastline of the Sorrento Peninsula is one of the most beautiful in the world. Mountains of limestone and volcanic deposits plunge into the sea in a stunning scene of precipitous crags, with narrow valleys descending steeply through them to the shore. The seismic activity associated with Vesuvius, an active volcano not far to the north, produces frequent earthquakes, and the rocks of the peninsula are extensively faulted. The Amalfi Coast experiences more flash floods and landslides than any other region of Italy.

The Turkish Steps

Location Southern Sicily, Italy

These cliffs of slightly inclined strata of white Trubi marl have been eroded into a series of long, parallel steps. Trubi marl is a relatively soft calcareous sedimentary rock, which was deposited on the ocean floor

about 5 million years ago. It contains the tracks of animals, probably worms that lived within it while it was being deposited.

Other coasts, islands, and reefs in Europe

- The Algarve Coast » p.170
- The Cliffs of Dover » p.167
- The Cliffs of Moher » p.166
- The Dalmatian Coast » p.171
- Dune du Pilat » p.167
- Fingal's Cave » p.166
- The Giant's Causeway » pp.164–65
- The Great Chagos Bank » p.315
- The Jurassic Coast » pp.168–69
- Norwegian Fjords » pp.162–63

Africa

Legzira

Location Atlantic coast of southern Morocco

Until 2016, there were two massive arches of red sandstone on the beach at Legzira. The layers of sandstone are 100–150 million

years old and were deposited on an unconformity over much older granite. They derive their coloration from iron oxide in the natural cement binding the grains in the rock. The arches were eroded by the action of waves, with one of them collapsing onto the beach in September 2016.

Banc d'Arguin

Location Golfe d'Arguin, west coast of Mauritania

A national park covers 4,630 square miles (12,000 square km) of this coastline—which includes sand dunes, seagrass flats, intertidal mud, sand, and shell fragments—as well as the shallow Golfe d'Arguin stretching up to 40 miles (60 km) from the shore. An offshore upwelling zone generates huge quantities of invertebrate food for fish and birds, with up to 50,000 pairs of colonial birds breeding and more than two million wintering in the park.

Socotra

Location Arabian Sea

One of Earth's most isolated islands, which originated as part of a continent rather than as new oceanic crust, Socotra is 150 miles (240 km) from the African mainland. It is also a major center of endemic plants and animals. Socotra has an area of 1,415 square miles (3,665 square km), divided between coastal plains, a limestone plateau with caves and other karst features, and a mountainous core. The climate is arid or semiarid, apart from high elevations, which experience more rainfall. Once part of the supercontinent of Gondwanaland, it became separated when the Gulf of Aden opened up. More than one-third of its plants are found nowhere else, including the bizarre-looking dragon's blood tree, and there are also endemic birds, reptiles, and invertebrates.

Coffee Bay

Location Eastern Cape province, South Africa

Near this coastal settlement is the spectacular Hole in the Rock formation, where wave action has punched a hole through an offshore cliff of shale and sandstone. Erosion has created a rock arch, rather than dividing the cliff into two stacks, because a thick intrusion of hard volcanic dolerite lies above the less resistant shale and sandstone, which is 260 million years old. The arch is large enough for a yacht to sail through at high tide.

▶ DRAGON'S BLOOD TREES AT SOCOTRA

Jeffreys Bay

Location Eastern Cape province, South Africa

One of the best surfing locations in the world, Jeffreys Bay has ideal wave conditions, which are attributed to a combination of swell size, direction, and frequency; wind direction; tides; and the shape of the seabed. Just offshore, the warm Agulhas Current passes west before turning back on itself at the eastern edge of the Agulhas Bank, where it meets cold water from the west. Here, upwelling has produced a huge marine ecosystem.

Cape Point

Location Western Cape province, South Africa

This headland lies just east of the Cape of Good Hope and forms the western entrance to False Bay. Hard quartz sandstones form cliffs that rise more than 660 ft (200 m) from the sea. There are two lighthouses, although only one is operational. The headland is part of the Cape Floral Region, an area with an extraordinary variety of plants that grow nowhere else in the world. Offshore are the cold, nutrient-rich waters of the Benguela Current, which fuel high rates of phytoplankton growth and sustain a very productive marine ecosystem.

Other coasts, islands, and reefs in Africa

- **Los Gigantes Cliffs** » p.198
- **The Red Sea Coast** » p.199
- **The Seychelles** » p.314

Asia

The Yaeyama Islands

Location East China Sea, southwest of Okinawa, Japan

This archipelago of 32 islands and numerous islets has volcanic origins. More than 90 percent of Iriomote, the largest island, is covered by tropical rain forest, with mangroves around the coast. It is home to the critically endangered and endemic Iriomote cat. Ishigaki is the second biggest island. Its offshore coral reefs provide prime habitat for dolphins, sea turtles, manta rays, and whale sharks, although dugongs are now virtually extinct here.

▲ BRILLIANT WHITE SILICA SANDS OF WHITEHAVEN BEACH

El Nido

Location Northern Palawan Island, Philippines

Spanish for "the nest," El Nido is the gateway to the famous Bacuit Archipelago. El Nido comprises the northern tip of Palawan Island as well as 45 islands and islets. Erosion of thick layers of limestone has produced a spectacular landscape of coastal karst, with cliffs towering vertically from the sea, combined with caves, white sand beaches, and coral reefs. The Secret Lagoon on Miniloc Island is a sea-filled cavern that can only be accessed via a narrow entrance. Experienced divers access a submarine tunnel beneath Dilumacad Island through an entrance that is 39 ft (12 m) below the water surface. The fauna of the offshore reefs include pipefish, pufferfish, scorpionfish, lionfish, and many other beautiful creatures.

Other coasts, islands, and reefs in Asia

- Andaman Sea Reefs » p.253
- Dragon Hole » p.251
- Ha Long Bay » p.250-51
- The Krabi Coastline » p.252-53
- The Maldives » p.315
- Nusa Tenggara » p.253
- Shiraho Reef » p.251

Australia and New Zealand

Horizontal Falls

Location Talbot Bay, Buccaneer Archipelago, western Australia

These tidal waterfalls are produced as ocean water pours through narrow gorges cutting through the two parallel rock barriers of the McLarty Ranges. On a rising tide, water builds up on one side of the barriers more quickly than it can push through the gorges. The resultant difference in water level reaches 16 ft (5 m) and produces the waterfall.

Whitehaven Beach

Location Whitsunday Island, Queensland, Australia

The origin of Whitehaven's 4 miles (7 km) of fine, bright white sand has long been debated. Whitsunday Island is part of an ancient volcanic caldera, but its sand—which is 98 percent quartz—could not have come from the island's igneous rhyolite and dacite rocks. In fact, it is derived from eroded granite on the Queensland mainland and was deposited by ancient longshore drift. This conveyor belt of coastal movement stopped at least 6,500 years ago, but the sand remains.

Ninety Mile Beach

Location West coast of Aupouri Peninsula, North Island, New Zealand

Despite its name, this strip of sand, which faces the Tasman Sea, actually stretches 55 miles (88 km) between the rocky headlands of Scott Point and Ahipara. Even more remarkable are the sand dunes—probably the largest dunes in the Southern Hemisphere and a renowned sandboarding site—that lie behind the beach. The dunes increase in height from south to north, culminating in summits more than 460 ft (140 m) high.

Other coasts, islands, and reefs in Australia and New Zealand

- The Great Barrier Reef » pp.284-87
- Moeraki Beach » p.290
- New Zealand Fjords » p.291
- Shark Bay » pp.288-89
- The Twelve Apostles » p.290

FORESTS

About 30 percent of Earth's land surface is covered by forest, which provides some of the richest of all habitats. At high latitudes, between 53° and 67°N, are the huge boreal forests of North America and Eurasia, which are dominated by coniferous trees. Between 10°S and 10°N are the world's great tropical rain forests, including the Amazon and Congo, regions of great biodiversity. In between are a wide variety of subtropical and temperate forests, some of the latter being rain forests. Russia has the greatest area of forest and Suriname, the greatest proportion of cover.

North America

The Tongass National Forest

Location Alaska, northwest U.S.

Covering most of southeast Alaska, this is the largest national forest in the U.S., at 26,640 square miles (69,000 square km). It is part of the Pacific Northwest Rain Forest, Earth's largest temperate rain forest. The dominant trees are red cedar, sitka spruce, and western hemlock. Logging of old-growth forest is strictly controlled. Brown and black bears live in the forest. The region also supports large populations of breeding birds.

The Redwood National Forest

Location California, western U.S.

Half of the world's remaining old-growth coast redwood trees — the tallest on Earth—are protected within this forest reserve. One, Hyperion, is 380 ft (116 m) tall. Coast redwoods are native to coastal northern California and southern Oregon. Black bears, cougars, beavers, flying squirrels, and more than 400 species of birds have been recorded in the forest, which covers 61 square miles (158 square km).

▶ SNOW-COVERED TREES AT THE TONGASS NATIONAL FOREST

Californian Chaparral and Woodland

Location California and Baja California, western U.S. and northern Mexico

Many coastal areas of central and southern California, the Central Valley, and Baja California are characterized by this scrub-dominated environment. Scrub oaks, sages, and *Ceanothus* are typical plants, and the flora is well adapted to recover from frequent wildfires. The climate is a Mediterranean type, with warm, wet winters and hot, dry summers.

The Sierra Madre Pine-Oak Forest

Location Between Jalisco state, northwest Mexico, and New Mexico, southwest U.S.

Large areas of these subtropical mixed forests survive at high altitudes in the dramatic mountains and steep-sided valleys of the Sierra Madre Occidental, although much has also been logged. Typical trees in the region include several species of pine, Douglas fir, Emory oak, and Mexican blue oak. The forests are home to black bears, jaguars, and more than 300 breeding birds, including some that are found nowhere else on Earth— the thick-billed parrot, eared quetzal, military macaw, Sinaloa martin, golden eagle, and tufted jay.

▲ CARACOL FALLS SURROUNDED BY RESTINGA FOREST

The Jalisco Dry Forest

Location Nayarit, Jalisco, and Colima states, west to central Mexico

Although relatively small, this ecoregion along the Pacific coast of central Mexico is one of the most biodiverse on Earth. Of 1,200 plant species, 16 percent are found nowhere else, and of 733 vertebrates, 29 percent are endemic, including Merriam's desert shrew, the Mexican parrotlet, and the San Blas jay. The Jalisco Dry Forest also serves as an important stopover for migrating birds from North America on their route toward South America. A wet season lasts from June to September, but the region is dry for most of the year.

Other forests in North America

- Cherokee National Forest » p.70
- The Giant Forest » pp.68-69
- Green Mountain National Forest » p.71
- North American Boreal Forest » p.66
- Pando Aspen Grove » p.70
- Pacific Northwest Rain Forest » p.67

Central and South America

Chocoan Rain Forest

Location Eastern Panama and northern Ecuador, northwest South America

This biome of ultrawet tropical rain forest is one of the most species-rich lowland areas on Earth. More than 11,000 types of plants have been recorded, about a quarter of them endemic, and there are 650 species of birds, many of them found nowhere else on Earth. This rain forest covers about 70,000 square miles (180,000 square km) of the zone between the Pacific coast and an elevation of about 3,300 ft (1,000 m) in the western Andes. The annual rainfall here exceeds 512 in (13,000 mm).

Atlantic Coast Restinga Forest

Location Three enclaves in Atlantic Forest, eastern Brazil

This is now restricted to three relatively small fragments of tropical forest in the north and, farther south, subtropical forest on sandy, nutrient-poor soils that have often formed on stabilized sand dunes. The forest typically comprises closed-canopy broadleaf trees of medium height, 16–50 ft (5–15 m) tall, but there is also an open-canopy variant, more like savanna. Several animals that are dependent on restinga are threatened with extinction. Much of the biome has been destroyed for urban development, and this still poses a major threat.

South American Dry Forest

Location Western Paraguay, southern Bolivia and Brazil, and northern Argentina

Occupying a huge lowland area of sandy and silty sediments west of the Paraguay River, south of the Pantanal, and east of the Andes, this forest type generally grades from parklike savanna in the east through palm savanna to thorn scrub in the west. The change is related to rainfall, which generally decreases moving west, and may be less than 2 in (50 mm) in the west. The plants are tolerant of low water levels. This forest is one of the last refuges of the flightless rhea and other animals, including jaguars, ocelots, pumas, tapirs, and spiny anteaters.

Chilean Matorral

Location Between the Pacific Ocean and the Andes, central Chile

This biome of shrubland and forest covers 57,100 square miles (148,000 square km) of the central Chilean coastal strip. It is transitional between the arid Atacama Desert to the north and the wetter Valdivian forest to the south, with a Mediterranean climate of rainy winters and dry summers. It has an enormous variety of plant species, about 95 percent of which are endemic to Chile, including many sclerophytic shrubs (with tough evergreen leaves to reduce water loss), cacti, and fuschias.

Other forests in Central and South America

- The Amazon Rain Forest » pp.120-21
- The Andean Yungas » p.119
- The Monteverde Cloud Forest » pp.118-19
- The Valdivian Temperate Forest » p.119

Europe

The Caledonian Forest

Location The Scottish Highlands, UK

Scots pine forest colonized the UK after the last ice age, 9,000 years ago. Wetter and windier conditions later reduced its extent, and grazing deer and sheep have also contributed to the Caledonian Forest being reduced to its present 70 square miles (180 square km), divided into 35 remnants. In addition to the pines, there is birch, rowan, aspen, juniper, and oak. The UK's only endemic vertebrate species, the Scottish crossbill, feeds on pine cones in the forest.

The Crooked Forest

Location Western Pomerania, Poland

This is a bizarre-looking group of 400 pine trees that are bent at a 90-degree angle close to the ground and have subsequently readjusted to grow vertically again. Even stranger, the trees are surrounded by a forest of normal, upright trees. Several theories have been advanced to explain the phenomenon, but the most likely one is that the farmers who planted the trees in the 1930s were responsible. They may have wanted trees with curved trunks to use for shipbuilding or the manufacture of furniture.

Hainich National Park

Location Thuringia, central Germany

This national park contains part of the largest continuous area of deciduous forest in Germany, Hainich Forest, which is dominated by European beech. The park is being managed with the intention of allowing the forest, which was formerly used for military training, to return to its natural state. In spring, before the thick canopy closes, an understory includes masses of wild garlic. Later, 16 species of orchid grow here, which provide a habitat for wildcats, bats, woodpeckers, and thousands of species of invertebrates.

The Black Forest

Location Baden-Württemberg, southwest Germany

This large upland area is called the Black Forest because the spruce trees here grow so close together that parts of it are very dark. The area is an uplifted fault block (horst) of sandstone, gneiss, and granite, which is bounded by the Rhine River to the south and west. The forest is roughly 100 miles (160 km) by 40 miles (60 km). The original cover of mixed forest was largely cleared in the 19th century to be replaced by spruce monoculture for forestry, but recent years have seen an increase in deciduous tree cover once more. Spruce and silver fir mingle with stands of beech in the hills and valleys of the forest. Germany's two largest nature parks are in the Black Forest, and there is also a national park.

▼ PINE TREES IN THE CROOKED FOREST

▲ FROST-COVERED TREES IN THE CARPATHIAN MOUNTAIN FORESTS

Kalkalpen National Park

Location The Northern Limestone Alps, northern Austria

More than 80 percent of this protected park is forest. Cloaking hills, ridges, and valleys on limestone, this is central Europe's largest continuous tract of forest. Within the forest are several enclaves of virgin spruce, fir, and beech forest, thought not to have been cleared since the end of the last ice age. Larger areas are now being left unmanaged, so dead wood is left to rot—providing habitats for many invertebrates and woodpeckers. Almost 1,000 species of plant have been recorded. Lynx, brown bear, and endangered bats inhabit the region.

Carpathian Mountain Forests

Location The Carpathian Mountains, western Ukraine and eastern Slovakia

Ten areas of primeval beech forest along a 115-mile (185-km) stretch of the Carpathians were designated a UNESCO World Heritage Site in 2007 for the quality of the woodland and their biodiversity. About 70 percent of the 300 square miles (780 square km) of forest, which includes oak and hornbeam in addition to beech, is in Ukraine, the remainder in Slovakia. The site includes old-growth and virgin forest and supports 64 species of mammals, including brown bear, lynx, wolf, and wild boar.

The Ardennes

Location The Ardennes Mountains of Belgium, Luxembourg, Germany, and France

This upland region comprises a landscape of ridges up to 2,277 ft (694 m) above sea level and steep-sided valleys on sandstone, quartzite, slate, and limestone bedrock. At least half of its area is covered by forest, with a rich mix of oak, beech, ash, hazel, maple, aspen, spruce, and pine. Generally scarce bird species – such as hazel grouse, hawfinch, Tengmalm's owl, black and gray-headed woodpeckers, and nutcrackers—breed in the forest.

Mediterranean Evergreen Forest

Location The Mediterranean Basin, including Spain, southern France, Italy, Greece, Morocco, Tunisia, and Algeria

This broadleaf evergreen forest, which is transitional between Mediterranean pine forest and

Africa

Cameroonian Highlands Forest

Location Eastern Nigeria and western Cameroon, West Africa

This moist broadleaf forest grows at elevations above 2,950 ft (900 m) on the fertile soils of a range of extinct volcanoes in an area roughly 390 miles (625 km) by 110 miles (180 km). Because of its altitude, the climate is cooler than most of tropical Africa, and parts of the forest experience heavy rainfall. Afromontane trees, including brittle-wood, yellow-wood, African cherry, and Cape beech, are plentiful. The forest is home to numerous endemic plants, reptiles, birds, and mammals, and there are also several species of endangered primates, including western gorillas and chimpanzees.

East African Coastal Forests

Location Coastal regions from southern Mozambique to southern Somalia, East Africa

The tropical and subtropical moist forest extends over 43,200 square miles (112,000 square km), stretching in a relatively narrow strip just inland from the Indian Ocean. Its mosaic of closed-canopy woodland, more open savanna, grassland, and wetland varies from south to north. Much has been cleared for farmland, but it remains home to more than 600 species of resident and seasonal migrant birds.

scrubland and cool-temperate deciduous forest, grows to an elevation of 4,600 ft (1,400 m). The forest is dominated by small-stature holm oaks, usually 16–39 ft (5–12 m) tall. These trees are slow growing, cast a deep shadow, and have tough, drought-resistant leaves. Although the forest is much reduced in area, some large tracts survive, for example in the lower Pyrenees of Aragon and the Sierra Morena of Andalucía, both in Spain.

> **Other forests in Europe**
> - The Bavarian Forest » p.173
> - The Hallerbos » p.172

> **Other forests in Africa**
> - The Congo Rain Forest » p.200
> - Madagascan Dry Forest » pp.202-203
> - Madagascan Rain Forest » p.201

Asia

Russian Far East Temperate Forest

Location The Sikhote-Alin mountains of Primorsky Krai and Khararovsky Krai, Russian Far East

After escaping glaciation during the last ice age, this region of mixed broadleaf and conifer forest covering the Sikhote-Alin mountain range became a refuge for many animal species. Today, it is an area of extraordinary biodiversity, with a small population of Amur tigers and Amur leopards. This is the only forest on Earth where tigers, leopards, and brown bears coexist. Plant diversity is varied, with at least 2,500 species known. Broadleaf trees are common at lower elevations but conifers become dominant as the altitude increases. Winters are bitterly cold,

summers are mild, and rainfall can be heavy during summer and fall.

The Forest of the Cedars of God

Location Mount Makmel, northern Lebanon

Protected since 1876, this is the most famous surviving remnant of the cedar forest that once covered much of the Lebanese mountains. For thousands of years, a succession of civilizations used cedar wood for shipbuilding and the construction of important buildings, so the forest is now only a fraction of its former size. The 375 trees of the forest grow at an elevation of more than 6,200 ft (1,900 m) on Mount Makmel, whose slopes are blanketed by snow in winter. Four of the cedars are more than 115 ft (35 m) tall.

▼ WOODED SLOPES IN THE FOREST OF THE CEDARS OF GOD

Western Himalayan Temperate Forest

Location Nepal and far northern Pakistan and India, on the southern slopes of the western Himalayas

Comprising broadleaf forest at an altitude of about 2,000–8,500 ft (600–2,600 m) and subalpine conifer forest at higher elevations up to some 12,500 ft (3,800 m), this biome is more fragmented than its eastern Himalayan counterpart and receives less moisture from the Bay of Bengal monsoon. Birch and rhododendron are common broadleaf trees, and conifers such as blue pine, spruce, yew, and fir are plentiful. One area of forest, the Palas Valley, has the richest flora in Pakistan. The forest is home to endangered mammals, including the Himalayan tahr, and birds such as the western tragopan and Himalayan monal.

▼ THE SOUTHWESTERN GHATS MOIST FOREST

The Mudumalai Forest

Location The Western Ghats mountains in Tamil Nadu, southern India

Up to 80 tigers live in the protected section of this forest, which is a 120-square-mile (320-square-km) national park. Other endangered species include the Indian leopard, sloth bear, and gaur. More than 200 bird species live here, including the endemic black-and-orange flycatcher. The forest can be divided into tropical moist deciduous, tropical dry deciduous, and tropical dry thorn forest—primarily reflecting differences in rainfall.

Southwestern Ghats Moist Forest

Location The Western Ghats mountains in southwest India

This ecoregion is one of extremely rich biodiversity, including more than 4,000 types of flowering plants. Many animals are endemic to these tropical and subtropical moist broadleaf forests, including lion-tailed macaque, Nilgiri tahr, 90 species of reptiles, and 85 species of amphibians. This forest grows at elevations of 820–3,300 ft (250–1,000 m), between the Malabar coast moist forest below and montane forest at higher altitudes. It is characterized by high rainfall, most of which falls during the southwest monsoon. Protected areas include the Indira Gandhi and Periyar national parks.

Sri Lankan Moist Forest

Location Southwest Sri Lanka

Covering about 6,000 square miles (15,500 square km), this tropical and subtropical moist broadleaf forest is one of the world's hotspots of endemism, with many animals and plants found nowhere else on Earth. The nature of the forest changes with elevation. At lower levels, more than 300 tree species are known, with some individuals in protected areas such as Sinharaja Forest towering to 165 ft (50 m). The region is wet—with annual rainfall exceeding 197 in (5,000 mm) in places—conducive to the greatest diversity of amphibians on Earth.

The Sagano Bamboo Forest

Location Near Kyoto, the Japanese island of Honshu

Established by a monk in the 14th century, this forest now covers 6 square miles (16 square km). It is harvested regularly, but the bamboo regenerates rapidly, growing up to 80 ft (25 m) tall. A walk through this small area of *Phyllostachys* (moso) bamboo is an amazing experience aurally as well as visually as even the slightest breeze knocks the stems, or culms, together.

► TALL STALKS OF BAMBOO AT SAGANO BAMBOO FOREST

Nansei Shoto Archipelago Forest

Location Ryukyu archipelago, southwest Japan

This archipelago consists of a long chain of larger volcanic and smaller coral islands between southern Japan and Taiwan and China. It has many tracts of subtropical moist broadleaf forest, which benefit from warm winters, hot summers, and high rainfall. The forest has a unique mix of fauna and flora, including many species that are found nowhere else on Earth. The Iriomote Islands contain the world's only habitat for the rare and endangered Iriomote cat. The Amami rabbit is another species endemic to the archipelago. The largest tract of forest is on Okinawa and is home to a woodpecker and a rail, both endemic to the island and bearing its name.

Xishuangbanna Rain Forest

Location Near Jinghong, Yunnan, China

This is the largest area of tropical China still covered by native forest. Despite the forest's high elevation (above 1,640 ft/500 m), it has most of the characteristics of the lowland rain forest found in southeast Asia. This is partly because it is protected from cold northerly winds in winter by the Hengduan Mountains. An extraordinary variety of more than 3,300 species of trees and other plants grow at Xishuangbanna. Some trees grow to a height of more than 260 ft (80 m).

Other forests in Asia

- **Bornean Rain Forest** » pp.258–59
- **Eastern Himalayan Forest** » p.256
- **Taiheiyo Montane Forest** » p.256
- **Siberian Taiga** » pp.254–55
- **Upper Yangtze Forests** » p.257

GRASSLANDS AND TUNDRA

This varied biome includes tropical savannas and temperate prairies and steppes. Steppe is virtually treeless, but savanna can have an open woodland canopy. Despite their variety, grasslands are united in receiving too little rainfall to support forest but not so little that a desert forms. Rainfall may be seasonal or spread through the year. Other factors influencing the distribution and nature of grasslands include: seasonal flooding, which prevents forest growth; geology, which heavily influences soil type and hence vegetation; and heavy grazing that has created a grassland from a former forest.

Central and South America

The Cerrado

Location Between the Pantanal, Amazon Rain Forest, and Atlantic Forest, central Brazil

This vast savanna stretches across more than 20 percent of Brazil's total area. It has a semihumid tropical climate, with a wet season and a dry season, and comprises different types of savanna crossed by closed-canopy gallery forest along watercourses. More than 10,000 species of plants, including 800 trees, have been noted here. Mammals include the giant anteater, giant armadillo, jaguar, maned wolf, and marsh deer.

▼ MORICHE PALM TREES SPREAD ACROSS THE BRAZILIAN CERRADO

> **Other grasslands and tundra in Central and South America**
>
> ● **The Pampas** » pp.122–23

Africa

The Central and Eastern Miombo Woodlands

Location From southern Zambia to northern Tanzania, and from eastern Angola to northern Mozambique

Broadly divided into central Zambezian miombo woodland and eastern miombo woodland, these tropical and subtropical grasslands, savannas, and shrublands cover an area of about 745,000 square miles (1,930,000 square km). Although there are regional variations, shared features include nutrient-poor soils, the dominance of many types of miombo tree, and a long, hot dry season. Kafue National Park, Zambia's largest protected area, and the Niassa Reserve in Mozambique are examples of these ecoregions, which have a high diversity of large mammals.

> **Other grasslands and tundra in Africa**
>
> ● **The Cape Floral Region** » p.205
> ● **Ethiopian Montane Grassland** » p.204
> ● **The Serengeti** » pp.206–07
> ● **The Sudanian Savannas** » p.204

Asia

The Mongolian-Manchurian Steppe

Location Extending in a crescent north of the Gobi Desert, Mongolia to northeast China

Extensive temperate grasslands spread inland from northeastern China's coastal hills toward the boreal forests of southern Siberia, covering almost 350,000 square miles (900,000 square km). The rolling grasslands experience warm summers and bitterly cold, windswept winters, with most rain falling in the weak summer monsoon. Feathergrasses are dominant in many areas, with drought-resistant species being more common closer to the Gobi Desert. There are also small, spiny shrubs. Native wild mammals include Mongolian gazelles and Asiatic wild asses, and Przewalski's horses have been reintroduced. The brown eared-pheasant is the only endemic bird of this ecoregion; it uses the grassland and shrub habitat as winter refuge.

> **Other grasslands and tundra in Asia**
>
> ● **The Eastern Steppe** » p.260
> ● **Siberian Tundra** » p.261
> ● **The Terai-Duar Savannas** » p.260

DESERTS

About one-third of Earth's land surface is desert, receiving under 10 in (25 cm) of precipitation annually. Deserts can be hot, cold, or polar. The Sahara is the largest hot desert, lying under a vast area of sinking, warming air, meaning rainfall is rare. The Gobi is the largest cold desert; hot in summer, cold in winter, it receives little precipitation because it lies in the rain shadow of the Himalayas. The world's biggest desert is polar—the Antarctic, where the air is too cold to hold much water vapor for rain or snow.

North America

The Colorado Plateau

Location Arizona, Utah, New Mexico, and Colorado, southwest U.S.

This high, arid plateau made up of sedimentary rock layers boasts some of the most spectacular scenery in the world. Ranging in elevation from 2,000 ft (600 m) at the base of the Grand Canyon, which has been cut by the Colorado River, to 12,700 ft (3,870 m) in the La Sal Mountains, it owes its character to remarkable geological stability. There are many deep canyons, flat-topped mesas, towering buttes, and natural rock arches within the plateau's 130,000-square-mile (337,000-square-km) expanse. The walls of the formations reveal rocks ranging in age from billions to just a few hundred years old.

Other deserts in North America

- Antelope Canyon » p.80
- Bryce Canyon » p.78
- The Chihuahuan Desert » p.81
- The Great Basin Desert » p.74
- Mesa Arch » p.79
- Meteor Crater » p.79
- The Mojave Desert » p.75
- Monument Valley » pp.76–77
- The Sonoran Desert » p.76

Central and South America

La Guajira Desert

Location Guajira Peninsula, northeast Colombia

This arid zone of xerophytic (drought-resistant) plants lies in the rain shadow of the Serrania de Macuira, a range of low mountains at the eastern end of the Guajira Peninsula, which intercepts the northeast trade winds. La Guajira is a land of thorny bushes, cacti, and other succulents. Large numbers of American flamingos breed at the Parque Natural los Flamencos, and limited-range bird species include the vermilion cardinal, white-whiskered spinetail, and chestnut piculet.

The Sechura Desert

Location Piura and Lambayeque provinces, northern Peru

Covering 73,350 square miles (190,000 square km) of low-lying coastal Peru, the Sechura Desert extends up to 60 miles (100 km) inland to the secondary ridges of the Andes mountains. To the north, it transitions into tropical dry forest while to the south it becomes the Peruvian Coastal Desert. Rain falls for only a few days each year. It owes its aridity to the atmospheric subsidence caused by the upwelling of cold water offshore. However, during the 1998 El Niño event, much higher than usual rainfall inland swelled rivers flowing through the desert and flooded a large part of the region, creating, for a time, Peru's second largest lake in the Bayovar Depression.

▼ THE GLEN CANYON REGION OF THE COLORADO PLATEAU

The Caatinga

Location Northeast Brazil

This vast ecoregion covers almost 10 percent of Brazil's territory, some 328,000 square miles (850,000 square km) of drought-resistant shrubland and thorn forest. The Caatinga experiences two seasons: a very hot dry season, when soil temperatures may reach 140°F (60°C) and plants drop their leaves to reduce transpiration; and a shorter, hot rainy season. After the first rain falls, the cerrado transforms in just a few days from a dead-looking grayish landscape to one that is bright green and full of life.

▲ HEXAGONAL PATTERNS AT SALINAS GRANDES SALT DESERT

Salar de Arizaro

Location Puna de Atacama, northwest Argentina

The world's sixth largest salt flat covers 600 square miles (1,600 square km) and lies at an elevation of about 11,350 ft (3,460 m) in the Andes. Formed as a result of the evaporation of surface water and the deposition of its constituent salts, Salar de Arizaro also features yardangs (ridges of rock aligned with the prevailing northwesterly winds), rocks that have been weathered by the repeated growth of salt crystals, relic lake shores, and the small volcanic cone of Cono de Arita, which rises about 390 ft (120 m) above the salt.

Talampaya National Park

Location La Rioja province, northwest Argentina

This World Heritage Site protects 830 square miles (2,150 square km) of arid scrub desert in the rain shadow of the Andes mountains. The Talampaya National Park includes dramatic landscapes shaped by water and wind, including the 469-ft- (143-m-) deep Talampaya Gorge and many red sandstone cliffs. The area's main claim to fame is its fossils—the most complete known range of life forms dating from 250–200 million years ago, including the early dinosaur *Eoraptor* and mammals, fish, amphibians, and plants.

Salinas Grandes

Location Jujuy and Salta provinces, Puna de Atacama, northwest Argentina

This large salt flat lies at the bottom of an endorheic basin (one with no outlet) at an elevation of 10,990 ft (3,350 m) on the arid Andean puna. Its dazzling white salt deposits, which cover more than 120 miles (200 km), were precipitated from evaporated runoff from the surrounding mountains. Rectangular pools of water mark places where salt extraction is taking place, and local artisans carve salt figures to demonstrate their skills.

Other deserts in Central and South America

▼ LARGE DUNES AT THE ERG CHEBBI

Europe

The Tabernas Desert

Location Almeria, Andalucía, southern Spain

Protected from humid winds coming from the Mediterranean by the Sierra de Alhamilla, this area of 110 square miles (280 square km) is semiarid rather than a true desert. Annual rainfall of 6–9 in (150–220 mm) is restricted to a few days. Occasional torrential rain has cut steep-sided gullies, whose river beds are usually dry, and natural underground pipes, through which water permeates below the surface at the top of a slope, passes underground, and

emerges lower down the slope. The drought-resistant flora includes oleanders, tamarisks, and the endangered sea lavender.

The Stone Desert

Location Varna depression, northeast Bulgaria

This area of low rainfall and sparse vegetation is Bulgaria's only desert. It is best known for the Pobiti Kamani (hammered stones)—18 groups of upright limestone columns up to 22 ft (7 m) tall and 10 ft (3 m) in diameter. The most popular theory that explains this phenomenon is that methane-bearing fluids seeping up shafts to an ancient ocean floor were

oxidized by bacteria to form cylinders of calcium carbonate. Subsequent uplift of the region and erosion of the softer, sandy sediments around the cylinders has left them standing prominently.

Africa

The Erg Chebbi Sand Dunes

Location Erfoud, Sahara Desert, eastern Morocco

Erg Chebbi is one of two large ergs—seas of dunes composed of windblown sand with little or no vegetation—in the Moroccan Sahara. Its biggest dunes rise 490 ft

(150 m) above the surrounding desert. Erg Chebbi extends about 17 miles (28 km) from south to north and 4 miles (7 km) from west to east. Daytime temperatures soar to 104°F (40°C) in July, but at night in winter they may fall to 37°F (3°C). The wettest month of the year is November, with just ½ in (10 mm) of rain. Despite the aridity, reptiles and nocturnal mammals such as jerboas and fennec foxes live here.

The Nyiri Desert

Location South of Nairobi, southern Kenya

Also known as the Nyika Desert and the Tarudesert, the Nyiri

Desert lies north of Mount Kilimanjaro and encompasses Amboseli National Park, including the northern half of Lake Amboseli. The annual rainfall averages 14 in (350 mm), but outside of the brief rainy season in April and May, water is scarce in this region, being restricted to a few springs and riverbeds. Vegetation comprises a mixture of grassland and thorn scrub (including some poisonous plants), with scattered baobab trees—some more than 2,000 years old, their circular trunks often as much as 10 ft (3 m) in diameter. Large mammals include giraffes, lions, leopards, and the lesser kudu and impala. More than 400 species of birds are residents or migrants, including 47 different types of raptors.

The Makgadikgadi Pan

Location Kalahari Desert, northeast Botswana

This is actually a group of enormous salt pans, together making up one of the largest in the world, with an area of 6,200 square miles (16,000 square km). The Sua, Nwetwe, Nxai, and other pans are relics of the formerly huge Lake Makgadikgadi, which dried up several thousand years ago. Almost lifeless dry, salty crusts for most of the year, they are fed fresh water seasonally by the Nata, Boteti, and other rivers. This encourages the growth of vegetation around their margins and attracts migrant ducks, flamingos, and other birds. The Makgadikgadi pan is home to one of only two breeding populations of greater flamingos in southern Africa.

> **Other deserts in Africa**
> - The Adrar Plateau » p.210
> - The Kalahari Desert » p.211
> - The Karoo » p.211
> - The Namib Desert » pp.212-13
> - The Sahara Desert » pp.208-209
> - The White Desert » p.210

Asia

The Kyzylkum Desert

Location Between the Amu Darya and Syr Darya rivers in Uzbekistan, Kazakhstan, and Turkmenistan

Sloping gently northwest toward the Aral Sea between the Amu Darya and Syr Darya rivers, this arid region of 116,000 square miles (300,000 square km) of sand dunes is occasionally punctuated by highlands, such as the Bukantau, Tamdytau, and

◀ THE MAKGADIKGADI PAN

Auminzatau mountains. Another feature of the desert is the presence of takirs: shallow hollows that are submerged after seasonal rains and dry out to form cracked crusts when the water evaporates. Annual rainfall averages just 6 in (150 mm), while temperatures fluctuate greatly from 86°F (30°C) or more in summer to 16°F (-9°C) in winter.

The Syrian Desert

Location Southern Syria, western Iraq, northeast Jordan, and northern Saudi Arabia

Covering a vast area of true desert and steppe, the Syrian Desert comprises an elevated plateau in southern Syria and eastern Jordan, and a plain that gently slopes northeast toward the Euphrates River. The whole area is dissected by wadis (dry valleys). Annual rainfall is as low as 4 in (100 mm) in some areas. Daytime temperatures reach 113°F (45°C) in summer, and a hot, dry wind called the khamsin frequently sweeps across the desert from the south or southeast, blowing up duststorms. The southern part of the desert is inhabited by several nomadic tribes and breeders of Arabian horses.

The Dasht-e-Kavir Desert

Location Southeast of Tehran, in northern Iran

Extending 240 miles (390 km) from the Alborz mountains southeast to the Dasht-e-Lut desert, this extremely arid region is surrounded on all sides by mountains. Although it receives virtually no rainfall, runoff from the mountains flows into seasonal lakes and the salt marshes of Kavir Buzorg, where evaporation leads to the formation of a salty surface crust. Rig Jenn is a vast area of sand dunes where summer temperatures can reach 122°F (50°C). The Kavir

National Park has a small population of the critically endangered Asiatic cheetah.

The Indus Valley Desert

Location Between the Chenub and Indus rivers, northwest Punjab province, Pakistan

This inhospitable plain, which covers 7,700 square miles (20,000 square km), bakes in temperatures up to 113°F (45°C) during the day in summer, but can be freezing in winter. Annual rainfall ranges from 24 in to 31 in (600 mm to 800 mm), and a variety of vegetation reflects this. Drought-resistant *Prosopis* shrubs and trees are plentiful. Five large mammal species are native to the desert: the Indian wolf, striped hyena, caracal, Indian leopard, and urial (a form of wild sheep).

> **Other deserts in Asia**
> - The Arabian Desert » pp.262-63
> - The Dasht-e-Lut » p.263
> - The Gobi Desert » pp.266-67
> - The Karakum Desert » p.264
> - The Taklamakan Desert » p.264
> - The Thar Desert » p.265

Australia and New Zealand

The Little Sandy Desert

Location Between Newman and Wiluna in Western Australia

Covering an area of 42,500 square miles (110,000 square km), this undulating country of red dune fields is vegetated by acacia and drought-resistant spinifex grass. The landscape is sometimes broken by escarpments of ancient sandstone, such as the Calvert Range. In the north of the desert is Lake Disappointment, an endorheic salt lake, which holds water only in very wet periods.

> **Other deserts in Australia and New Zealand**
> - The Gibson Desert » p.298
> - The Great Sandy Desert » p.295
> - The Great Victoria Desert » p.299
> - The Pinnacles » p.298
> - The Simpson Desert » pp.296-97

▲ SALT MARSHES OF THE DASHT-E-KAVIR DESERT

GLOSSARY

A

A'A LAVA Basaltic lava with a rough surface of broken lava blocks or clinker. See also *block lava, pahoehoe lava.*

ABLATION ▼ The loss of ice from a glacier due to melting, evaporation, sublimation, calving, or erosion by the wind. The ablation area is the region of a glacier where there is a net loss of ice due to these processes. See also *accumulation area.*

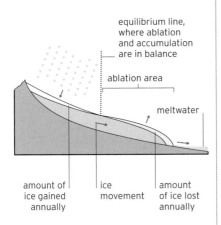

equilibrium line, where ablation and accumulation are in balance

ablation area

meltwater

amount of ice gained annually | ice movement | amount of ice lost annually

ABYSSAL Referring to the deep ocean floor and its environment, beyond the continental slopes. The abyssal zone is the part of the water column between a depth of 6,600 ft (2,000 m) and the abyssal plain. It is deeper than the bathyal zone but not as deep as the deep-sea trenches. See also *bathyal, deep-sea trench.*

ABYSSAL PLAIN The relatively flat, sediment-covered plain that forms the bed of most of the oceans, beyond the continental slopes. It lies at a depth of 13,000–20,000 ft (4,000–6,000 m). See also *continental slope.*

ACCUMULATION AREA ▼ The part of a glacier where snowfall exceeds losses by melting, evaporation, sublimation, calving, and wind erosion. See also *ablation.*

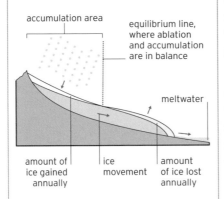

accumulation area

equilibrium line, where ablation and accumulation are in balance

meltwater

amount of ice gained annually | ice movement | amount of ice lost annually

ACID RAIN Precipitation (rain or snow) containing dissolved acids. Since it contains carbon dioxide, most rain is slightly acidic, but atmospheric pollution or the gases released by volcanic eruptions can make rain more acidic.

ADVECTION FOG A kind of fog that forms as moist air passes over cooler surfaces. It can form over land or the ocean. See also *radiation fog.*

AEROSOL A tiny particle suspended in the air. It may be dust or liquid, and is usually no larger than one millionth of a millimeter in diameter.

AIR MASS A body of air with relatively uniform characteristics derived from the surface region where it forms and distinct from surrounding air masses. Examples include polar maritime and tropical maritime air masses.

ALBEDO The proportion of the Sun's incoming radiation reflected by Earth's surface. For example,

fresh snow may reflect up to 90 percent, and its albedo will then be expressed as 0.9.

ALLUVIAL DEPOSIT River deposits that include sand, silt, mud, gravel, and organic matter. They may also be rich in minerals. See also *alluvium.*

ALLUVIAL FAN A cone-shaped deposit of sediment built up by rivers or streams. Fans are typically found where steep-flowing watercourses, rich in deposits, emerge from mountains onto a plain. See also *alluvium.*

ALLUVIUM Any sedimentary material deposited by rivers, including sand, silt, mud, gravel, and organic matter. Any accumulation of alluvium is called an alluvial deposit. See also *alluvial deposit, alluvial fan.*

ANTICLINE A fold of stratified rocks in which the oldest strata (layers) are at the core. Anticlines result from massive horizontal compression. See also *fold (panel), syncline.*

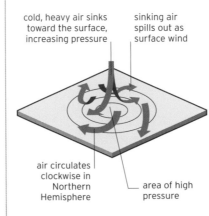

cold, heavy air sinks toward the surface, increasing pressure

sinking air spills out as surface wind

air circulates clockwise in Northern Hemisphere

area of high pressure

ANTICYCLONE ▲ A weather system in which atmospheric pressure is highest in the center. Wind circulates around the system in a clockwise

direction in the Northern Hemisphere, counterclockwise in the Southern Hemisphere. See also *cyclone, pressure system.*

AQUIFER ▼ An underground layer of permeable rock from which groundwater can be extracted. See also *groundwater.*

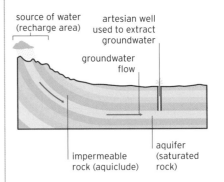

source of water (recharge area)

artesian well used to extract groundwater

groundwater flow

impermeable rock (aquiclude)

aquifer (saturated rock)

ARCHIPELAGO A chain or cluster of islands. See also *island arc.*

ARÊTE ▼ A sharp-crested ridge separating two adjacent cirques. See also *cirque.*

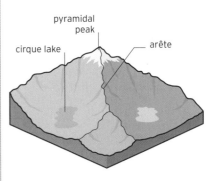

pyramidal peak

cirque lake

arête

ASTHENOSPHERE The viscous layer of the upper mantle immediately below the rigid lithosphere. Rock in this layer moves slowly in a solid state and is key to the movement of tectonic plates. See also *lithosphere (panel), mantle, tectonic plate.*

ASYMMETRICAL FOLD A fold whose axial plane is inclined, dipping in the same direction as the more gently dipping flank of the fold. See also *anticline, fold (panel), syncline*.

ATMOSPHERE The gaseous layers that surround Earth. The atmosphere is made up of four distinct layers—the troposphere, stratosphere, mesosphere, and ionosphere. See also *stratosphere, troposphere*.

ATOLL ▼ A ring-shaped coral island, or a ring of small coral islands, surrounding a lagoon. Atolls are created as coral reefs build up in shallow water around subsiding volcanic islands. See also *coral reef, fringing reef*.

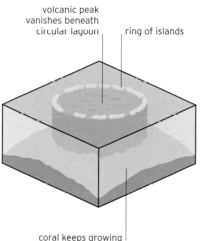

volcanic peak vanishes beneath circular lagoon
ring of islands
coral keeps growing as bedrock sinks

AUREOLE An area of rock around an igneous intrusion whose composition, structure, or texture has been thermally altered or metamorphosed. See also *country rock, intrusive igneous rock, metamorphism*.

AURORA Colored streamers of light visible in the night sky at certain latitudes and at certain times of the year. They are caused by the interaction between high-energy particles from the Sun and Earth's magnetic field. In the Northern Hemisphere, displays are called the aurora borealis; the equivalent in the Southern Hemisphere is the aurora australis.

B

BACK-ARC BASIN An undersea basin formed behind an island arc near where two tectonic plates meet. See also *island arc*.

BACKSHORE An area of the shore that is above the average high-water mark. This region is covered by water only during storms and the highest tides. See also *foreshore*.

BACKWASH The flow of water down a beach after a wave has broken on the shore. See also *swash*.

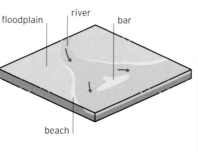

floodplain
river
bar
beach

BAR ▲ A linear deposit of gravel or sand deposited by a river or the sea. Bars may be exposed at low tide or permanently submerged. See also *barrier bar, longshore bar, point bar*.

BARCHAN A crescent-shaped dune formed by wind action in a sandy desert. The steeper slope, or slip face, is on the concave side. It is also called a barkhan dune or crescentic dune. See also *dune* (panel).

BARRIER BAR A longshore bar of gravel or sand roughly parallel to the coastline, whose surface is below mean sea level. See also *barrier island, longshore bar*.

BARRIER ISLAND A long, thin island of sediment roughly parallel to a coastline, whose surface is usually exposed. See also *barrier bar, longshore bar*.

BASALT A common, fine-grained or glassy volcanic rock that usually originates from solidified lava. The oceanic crust is composed of basalt, and basaltic lava can also erupt on continents. See also *extrusive igneous rock, flood basalt*.

BASIN A low-lying area or depression on land or in an ocean where sediment often accumulates. See also *river basin*.

BATHOLITH A very large igneous intrusion, 60 miles (100 km) or more across, which originates deep underground. Although they form deep underground, batholiths may be exposed by erosion. See also *igneous intrusion* (panel).

BATHYAL Referring to the zone of the ocean between about 660 and 6,600 ft (200 and 2,000 m) deep. The bathyal zone is not as deep as the abyssal zone. See also *abyssal*.

BAYMOUTH BAR A spit or bar that extends completely across the mouth of a bay. See also *barrier bar, spit*.

BEACH See panel, below.

BEACH FACE The sloping section of a beach, which is exposed to the wash of the waves. Also called the foreshore. See also *beach* (panel), *swash*.

BEDDING The sequence in which layers of sedimentary rock were originally laid down, one on top of another. See also *bedding plane, sedimentary rock*.

BEACH

A beach is an accumulation of sand, gravel, shell debris, or other deposits along the shoreline of an ocean, lake, or river. Ocean beaches are constantly changing as tides and wave action deposit and carry away sediment. Every beach has a distinct profile above and below the water line. The intertidal zone lies between the low- and high-water marks.

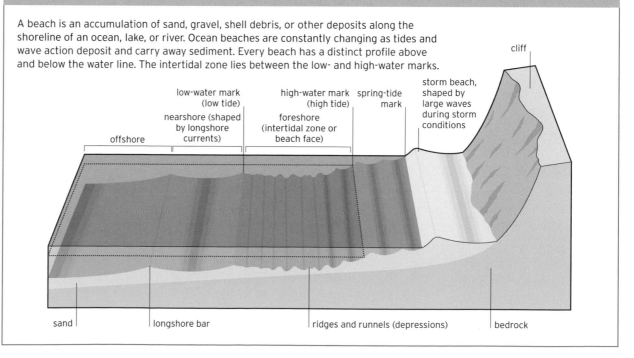

low-water mark (low tide)
high-water mark (high tide)
spring-tide mark
cliff
nearshore (shaped by longshore currents)
foreshore (intertidal zone or beach face)
storm beach, shaped by large waves during storm conditions
offshore
sand
longshore bar
ridges and runnels (depressions)
bedrock

BEDDING PLANE The surface separating a layer of a sedimentary rock from the layer above or below. See also *bedding, sedimentary rock*.

BERGSCHRUND A deep crack or crevasse that forms where the moving ice of a glacier pulls away from its headwall. See also *crevasse* (panel), *glacier*.

BERM A ridge of shingle on the upper part of a beach, above the beach face, usually marking the point of the highest high tides. See also *beach face*.

BIODIVERSITY The variety of living plants and animals in a given area or habitat.

BIOLUMINESCENCE The production of light by living life forms. Bioluminescent organisms include some bacteria, fungi, cephalopods, jellyfish, and fish.

BIOME A large-scale biological community defined primarily by its vegetation. Examples include tropical rain forest, temperate grassland, taiga, and tundra.

BLACK SMOKER A hydrothermal vent in which the emerging, very hot water is colored black with dark minerals—mostly iron sulfide. See also *hydrothermal vent, white smoker*.

BLOCK LAVA Volcanic lava with a surface of smooth-sided angular fragments. This lava usually solidifies to form the rock andesite.

BLOCKFIELD An area of broken rocks, usually in a mountainous region. Also called felsenmeer.

BOG A peat-accumulating wetland. Bogs are acidic, with low levels of nutrients, and their soils are composed almost entirely of dead plant matter. See also *fen, marsh*.

BORE See *tidal bore*.

BOREAL A term used for parts of the Northern Hemisphere between the Arctic and temperate zones—for example, the boreal forest of Eurasia.

BOSS An igneous intrusion that is roughly circular in horizontal cross section. See also *igneous intrusion* (panel).

BOUNDARY See panel, below.

BRACKISH Water that has a salt content less than that of ocean water but greater than that of freshwater.

BRAIDED STREAM A watercourse where water flows in several shallow channels that constantly separate and rejoin. This system is common downstream of many glaciers. See also *glacier, meltwater*.

BREAKER ZONE The part of a beach where waves break. See also *beach* (panel), *beach face, swash*.

BRECCIA A sedimentary rock made up of angular fragments of minerals and other rocks.

BROADLEAF A term describing trees that have broad leaves, as distinct from needlelike leaves. Examples of broadleaved trees include chestnuts, maples, and oaks. See also *coniferous, deciduous, evergreen*.

C

CAATINGA A kind of thorny forest found in semiarid areas of northeast Brazil.

CALCITE A common transparent or opaque mineral form of calcium carbonate.

CALDERA A large, bowl-shaped volcanic depression, typically more than 0.6 mile (1 km) in diameter. It forms when a volcano collapses into its magma chamber after it has been emptied by a volcanic explosion. See also *magma chamber, volcano* (panel).

CALVING The process of ice blocks shedding from a glacier or ice sheet and falling into a lake or the ocean. See also *glacier, ice sheet*.

CANOPY The crowns of the tallest forest trees, which receive full sunlight. The canopy may be closed, with crowns touching each other to form a continuous layer, or open. See also *rain forest*.

CANYON A deep, steep-sided, relatively narrow valley. See also *gorge*.

CATCHMENT AREA The total area drained by one river and its tributaries. See also *drainage basin, watershed*.

CERRADO A type of savanna dotted with small trees and shrubs, found mostly in central Brazil. See also *savanna*.

BOUNDARY

The margin between two or more tectonic plates is called a boundary. At a divergent (constructive) boundary, two plates move apart, creating a rift. Magma rises from the mantle beneath the rift, pouring out through long fissures. A convergent (destructive) boundary forms where two plates collide, often where oceanic crust is pushing against continental crust and is being forced beneath it, a process known as subduction. The continental crust often buckles up into a mountain range, while the subducting oceanic crust forms a deep seafloor trench. Volcanoes form above the area of subduction. A transform plate boundary involves two plates pushing past each other. Although no magma reaches the surface, earthquakes are regular features of transform fault locations.

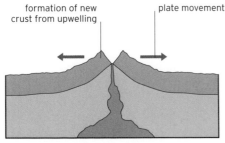

DIVERGENT (CONSTRUCTIVE) BOUNDARY

CONVERGENT (DESTRUCTIVE) BOUNDARY

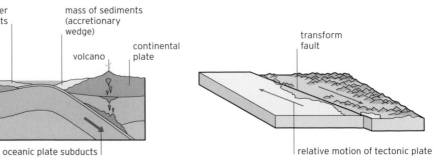

TRANSFORM (CONSERVATIVE) BOUNDARY

CINDER CONE A steep-sided volcano made up of ash, cinders, or clinker (rough, irregular lava fragments). Also called a scoria cone. See also *volcano* (panel).

CIRQUE ▼ A steep-sided hollow at the head of a valley or on a mountainside carved out by a glacier. Many glaciers originate in mountain cirques, from which they flow to lower ground. See also *glacier*.

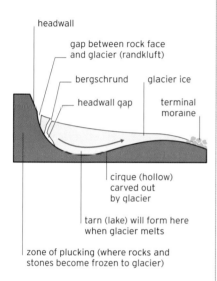

headwall

gap between rock face and glacier (randkluft)

bergschrund

glacier ice

headwall gap

terminal moraine

cirque (hollow) carved out by glacier

tarn (lake) will form here when glacier melts

zone of plucking (where rocks and stones become frozen to glacier)

CIRQUE GLACIER A glacier that originates in, and is mostly restricted to, a cirque. If a cirque glacier advances far enough, it may become a valley glacier. See also *cirque, glacier, valley glacier*.

CLAST A fragment of gravel, sand, or other sediment incorporated into another, newer rock. For example, sandstone is made up of many clasts of sand. See also *grain*.

CLAY An earthy material with mineral particles smaller than about 0.00008 in (0.002 mm) in diameter. Clay is porous, but water flows through it only slowly.

CLOUDFOREST A damp forest that is often shrouded in cloud. Cloudforest is generally found high in mountains.

COLD DESERT A desert that becomes very cold for at least part of the year, due either to high elevation or high latitude. Examples of cold deserts are Antarctica and the Gobi Desert.

CONDENSATION The conversion of a substance from a vapor to a liquid. Clouds form when water vapor condenses to form tiny droplets of water.

CONFLUENCE The place where two streams, rivers, or glaciers join.

CONIFEROUS Cone-bearing. Firs and pines are examples of coniferous trees.

CONSERVATIVE BOUNDARY A margin along which two tectonic plates slide past each other in opposite directions or at different speeds. See also *boundary* (panel).

CONSTRUCTIVE BOUNDARY See *divergent boundary*.

CONTACT METAMORPHISM The process by which very hot intrusive igneous rocks alter the country rocks around them. See also *country rock, igneous intrusion* (panel), *metamorphic rock*.

CONTINENTAL CRUST The layer of mainly sedimentary and metamorphic rocks that forms the continents and the relatively shallow oceans close to their shores. See also *continental shelf, crust, oceanic crust*.

CONTINENTAL DIVIDE A continent's drainage divide. Water on either side of the divide drains into different oceans or seas. See also *watershed*.

CONTINENTAL MARGIN The zone of the ocean floor that separates thin oceanic crust from thick continental crust. The continental shelf, continental slope, and continental rise together make up the continental margin. See also *continental shelf, continental slope*.

CONTINENTAL RISE The lowest part of the continental slope, adjacent to the abyssal plain. See also *abyssal plain*.

CONTINENTAL SHELF The submerged, gently sloping portion of continental crust between the coast and the continental slope. See also *crust*.

CONTINENTAL SLOPE The sloping ocean floor between the continental rise and the continental shelf. See also *continental rise, continental shelf*.

CONVECTION The movement of gases, liquids, and molten rocks in response to differences in temperature. One example is when air is warmed as it passes over hot ground and rises high in the atmosphere.

CONVERGENT BOUNDARY The line along which two or more tectonic plates move toward each other and collide. The result is either a subduction zone or a continental collision. See also *boundary* (panel), *subduction, tectonic plate*.

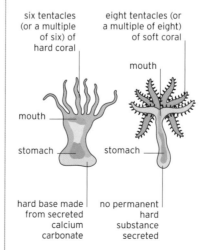

six tentacles (or a multiple of six) of hard coral

eight tentacles (or a multiple of eight) of soft coral

mouth

mouth

stomach

stomach

hard base made from secreted calcium carbonate

no permanent hard substance secreted

CORAL ▲ An ocean-dwelling invertebrate. Corals often live in huge colonies and can secrete calcium carbonate skeletons to support themselves. The accumulation of these corals eventually produces a coral reef. See also *coral reef*.

CORAL BLEACHING ▼ The process whereby corals expel the color-giving algae (zooxanthellae) from their tissues and consequently turn white. One of the main causes is an increase in the temperature of seawater. Bleached corals are not necessarily dead, but bleaching can lead to death.

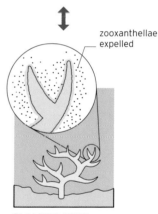

polyps   zooxanthellae

HEALTHY CORAL

zooxanthellae expelled

BLEACHED CORAL

CORAL REEF ▼ A structure built up over many years from the skeletons of corals. Examples include fringing reefs, barrier reefs, and atolls.

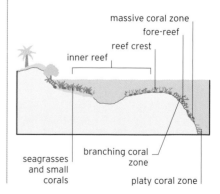

massive coral zone

fore-reef

reef crest

inner reef

seagrasses and small corals

branching coral zone

platy coral zone

CORE Earth's innermost layer. It consists of a solid inner core and a liquid outer core, both made of nickel-iron.

CORIOLIS EFFECT The tendency for the direction of winds and currents to be deflected. Due to Earth's rotation, winds are deflected to the right in the Northern Hemisphere and left in the Southern Hemisphere. Without the Coriolis effect, air would simply flow directly from areas of high pressure to low pressure. See also *anticyclone, cyclone*.

CORROSION A form of chemical erosion in which water dissolves minerals in rocks and washes them away.

COUNTRY ROCK Existing rock into which an igneous intrusion invades. See also *igneous intrusion* (panel).

CRATER ▼ A bowl-shaped depression through which a volcano erupts gases, lava, ash, or volcanic bombs. Another form of crater is caused by a large meteorite impact.

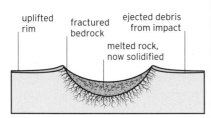

uplifted rim | fractured bedrock | ejected debris from impact

melted rock, now solidified

CRATON See *shield*.

CRESCENTIC DUNE See *barchan*.

CREVASSE See panel (below).

CRUST Earth's rocky outermost layer. The continents and their margins are made of thicker but less dense continental crust, while thinner, denser oceanic crust underlies the floor of the deep oceans. See also *continental crust, oceanic crust*.

CRYOTURBATION The churning up of different layers of soil by repeated freezing and thawing. Also called frost-churning.

CRYSTAL Any solid in which the individual molecules are arranged in a regular geometrical pattern. Calcite and quartz, for example, form crystals.

CURRENT The flow of water in an ocean, lake, or river.

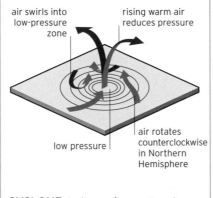

air swirls into low-pressure zone

rising warm air reduces pressure

low pressure

air rotates counterclockwise in Northern Hemisphere

CYCLONE ▲ A weather system in which winds circulate around an area of low pressure. A tropical cyclone is

another name for a hurricane or typhoon. See also *anticyclone, depression, hurricane, pressure system*.

D

DECIDUOUS Describing trees and shrubs with leaves that die and fall at a particular time of year. See also *coniferous, evergreen*.

DEEP-SEA FAN A fan-shaped deposit of sediment on the ocean floor, usually at the bottom of the continental slope, deposited by turbidity currents. It is also known as an abyssal fan. See also *turbidity current*.

DEEP-SEA TRENCH A canyonlike depression in the ocean floor. Such trenches occur where one tectonic plate is pushing under another at a subduction zone. They are the deepest regions of the oceans. See also *boundary* (panel), *subduction*.

DEFORMATION The change in shape of rocks caused by the pressure of geological movements.

DELTA The area of gently sloping silt, sand, and other sediments built up by a river as it enters the ocean or a lake.

DENDRITIC DRAINAGE The pattern of streams and rivers in a drainage basin resembling the twigs and branches of a tree. See also *drainage basin*.

DEPOSITION The laying down of silt, sand, gravel, and other sediments by rivers, ocean currents, moving ice, or the wind.

DEPOSITIONAL COAST A coast where there is net deposition of sediments. This may result from longshore drift or the deposition of sediments in estuaries and deltas. See also *longshore drift*.

DEPRESSION A weather system where winds circulate around an area of low pressure. Also called a cyclone. See also *cyclone, pressure system*.

DESERT PAVEMENT ▼ A surface layer of closely packed rock fragments in a desert.

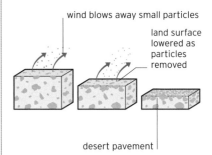

wind blows away small particles

land surface lowered as particles removed

desert pavement

DESERT VARNISH A glossy orange, brown, or black coating sometimes found on the surface of desert rocks that have been exposed to the atmosphere for a long time.

DESERTIFICATION The conversion of a formerly fertile region into desert.

DESTRUCTIVE BOUNDARY See *convergent boundary*.

DEW POINT The temperature at which, under given conditions, liquid water in the atmosphere begins to condense. See *condensation*.

CREVASSE

A crevasse is a deep fissure in a glacier or other ice body. Crevasses usually form when some areas of ice move faster than others, consequently splitting the ice. This happens when glaciers change direction or pass over uneven topography, such as a rock ridge. Crevasses are usually wider at the top than at the bottom.

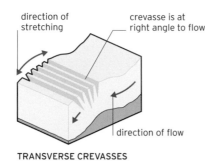

direction of stretching

crevasse is at right angle to flow

direction of flow

TRANSVERSE CREVASSES

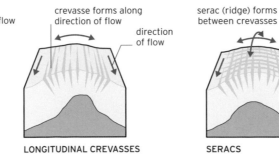

crevasse forms along direction of flow

direction of flow

LONGITUDINAL CREVASSES

serac (ridge) forms between crevasses

SERACS

DIKE A sheet of intrusive igneous rock inclined vertically or at a steep angle to the surface. A large number of dikes in an area are referred to as a dike swarm. Also known as a dyke. See also *basalt, igneous intrusion* (panel), *sill*.

DISTRIBUTARY A branch of a river that flows away from the main watercourse and does not reconnect with it later, for example in a delta.

DIVERGENT BOUNDARY The line along which two or more tectonic plates are moving away from each other, for example at a mid-ocean ridge. See also *boundary* (panel), *mid-ocean ridge* (panel).

DOLINE See *sinkhole*.

DOME VOLCANO A mound-shaped, steep-sided volcano. The shape is determined by the viscous nature of the lava, which cannot flow far from the vent. See also *volcano* (panel).

DRAINAGE BASIN The area within which all surface water from rain, melting snow, or ice is drained by a single river system. See also *catchment area, watershed*.

DROWNED COAST A stretch of coastline that has been inundated by a rise in sea level. Features of such coasts include drowned valleys (or rias). See also *isostasy, ria*.

DRUMLIN A streamlined mound of sediments left behind by a glacier after it has retreated. See also *glacier, moraine* (panel).

DRY FOREST Woodland growing in an area with a primarily dry climate.

DUNE See panel, below.

DYNAMIC METAMORPHISM The transformation of rocks brought about mainly by directed pressure or stress, rather than heat or chemical change.

E

ECOREGION A large area containing a distinctive group of species, communities of species, and environmental conditions.

ECOSYSTEM A community of interacting life forms and their environment. An ecosystem could be as small as a rotting log or as large as Earth.

EDDY A circular motion of any size and speed, in water or air. See also *vortex*.

EKMAN SPIRAL The tendency of deep ocean currents to change direction as they approach the surface, and of winds to change direction as they get closer to the ground. See also *Coriolis effect*.

EL NIÑO The phenomenon in which warm water from the western Pacific flows eastward, making the eastern Pacific warmer than usual and triggering worldwide changes to weather patterns. See also *La Niña*.

EMERGENT Describing forest trees that grow taller than the height of the canopy. See also *canopy*.

EMERGENT COAST A stretch of coastline along which formerly submerged features (for example, beaches and wave-cut platforms) have been exposed by a fall in sea level.

ENDEMIC A term for animals and plants native to one region and found nowhere else.

ENDORHEIC Describes a watershed or a lake that has no outflow to a larger body of water.

EPHEMERAL Temporary, as in a water body present only in the wet season.

EPIPHYTE A nonparasitic plant that grows on another plant.

ERG A large expanse of vegetation-free sand in a desert.

EROSION The processes by which water, moving ice, and wind, as well as the sand grains and other particles they carry, scrape and wear away rocks or soil from a land surface. See also *weathering*.

EROSIONAL COAST A coastline where there is net removal of sand, gravel, and other sediments.

ERRATIC A rock that has been transported, usually by a glacier, from its original location. See also *glacier*.

DUNE

A dune is a mound of sand in a desert, on a river bed, or close to the shore of a sea or lake, formed by the action of the wind or the flow of water currents. Dunes occur in a range of shapes and sizes, depending on how they have formed. With desert dunes, if the prevailing wind is usually from one direction, crescentic barchan or parabolic dunes will form. If the wind is variable, linear or star dunes will result. Dunes may migrate many kilometres or become stabilized by vegetation.

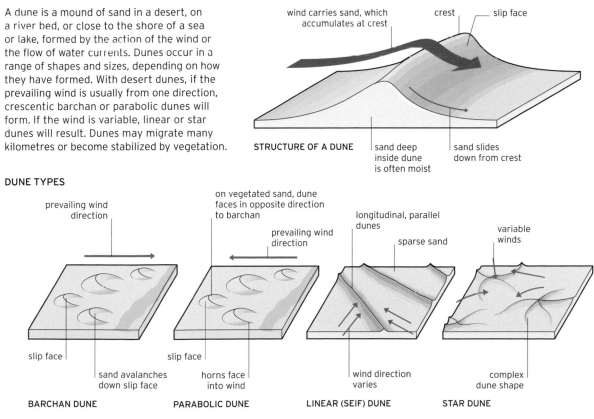

wind carries sand, which accumulates at crest — crest — slip face

STRUCTURE OF A DUNE — sand deep inside dune is often moist — sand slides down from crest

DUNE TYPES

prevailing wind direction — slip face — sand avalanches down slip face — **BARCHAN DUNE**

on vegetated sand, dune faces in opposite direction to barchan — prevailing wind direction — slip face — horns face into wind — **PARABOLIC DUNE**

longitudinal, parallel dunes — sparse sand — wind direction varies — **LINEAR (SEIF) DUNE**

variable winds — complex dune shape — **STAR DUNE**

ERUPTION

The discharge of gas, lava, ash, or volcanic bombs from a volcano is called an eruption. Different types of volcanic eruption include Hawaiian, Plinian, Strombolian, Surtseyan, and Vulcanian. Depending on the type of eruption, lava, dust, ash, or pyroclastic bombs are ejected from a main vent, while secondary vents, or fumaroles, may produce smoke and steam.

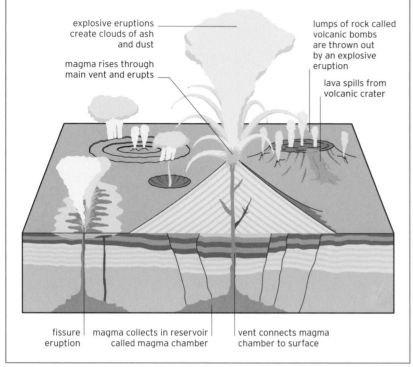

explosive eruptions create clouds of ash and dust

magma rises through main vent and erupts

lumps of rock called volcanic bombs are thrown out by an explosive eruption

lava spills from volcanic crater

fissure eruption

magma collects in reservoir called magma chamber

vent connects magma chamber to surface

ERUPTION See panel, above.

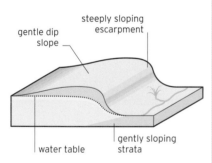

steeply sloping escarpment

gentle dip slope

water table

gently sloping strata

ESCARPMENT ▲ The steep slope at the edge of a plateau, or at the edge of an area of exposed, gently sloping strata, separating two relatively level areas of different elevation. Also called a scarp.

ESKER A long, typically winding, ridge of gravel or sand left behind by a glacier. Eskers mark the position of former meltwater channels beneath or within the ice. See also *glacier, meltwater channel*.

ESTUARY The wide lower course of a river from where it flows into the ocean. Estuaries experience tidal flows and are often zones of deposition.

EUSTASY A global change in sea level produced by a change in the amount of water in the oceans. This change usually results from a change in the volume of Earth's ice sheets.

EUTROPHICATION The process by which lakes and other water bodies are enriched with nutrients from fertilizers.

EVAPORITE A sediment formed when a salty lake or oceanic lagoon evaporates. Examples of evaporites include calcite, gypsum, and halite.

EVERGREEN Describing trees bearing leaves all year round. See also *deciduous*.

EXFOLIATION A weathering process that involves layers of rock peeling off in sheets instead of grain by grain.

EXFOLIATION DOME A large, dome-shaped mass or rock, such as granite, whose form has been produced by exfoliation. See also *exfoliation*.

EXTRUSIVE IGNEOUS ROCK Rock that forms when magma reaches Earth's surface and quickly cools and solidifies. For example, basalt and rhyolite. See also *intrusive igneous rock*.

FAST ICE Frozen water that is secured to the coast. If it rises more than 6 ft (2 m) above sea level, it is called an ice-shelf. See also *ice-shelf*.

FAULT See panel, below.

FEN A wetland that receives water mainly from groundwater seepage and where the soils are composed almost entirely of dead plant matter, such as peat.

FETCH The distance across open water that a wave or the wind has traveled.

FISSURE ERUPTION The emission of lava through a linear vent, which may be many miles long. See also *eruption* (panel).

FJORD A former glacial valley on a coast that has been submerged by a rise in sea level and has become an inlet of the sea.

FLASH POINT The temperature at which superheated water changes to steam.

FLASH-FLOOD A sudden, often destructive flood occurring after heavy rainfall.

FAULT

A fracture where the rocks on either side have moved relative to one another and often extend through the crust for many miles is called a fault. In normal and reverse faults, one section of the crust slips down, or rises up, in relation to the other. In strike-slip faults, one section slides past another, with little or no vertical movement. Great forces are involved in the formation of thrust faults, where older rocks are thrust over younger strata.

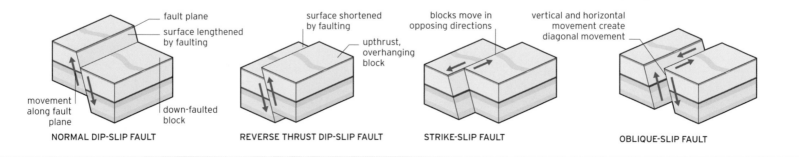

fault plane

surface lengthened by faulting

movement along fault plane

down-faulted block

NORMAL DIP-SLIP FAULT

surface shortened by faulting

upthrust, overhanging block

REVERSE THRUST DIP-SLIP FAULT

blocks move in opposing directions

STRIKE-SLIP FAULT

vertical and horizontal movement create diagonal movement

OBLIQUE-SLIP FAULT

FLOOD BASALT The result of a massive eruption that covers an extensive area with basalt. See also *basalt*.

FLOODPLAIN A flat area alongside a river that is liable to be covered with water when river volumes are particularly great. See also *alluvium*.

FLOWSTONE A coating of a carbonate mineral such as calcite on the walls or floor of a cave. The mineral precipitates out of flowing water. See also *calcite, precipitate*.

FOG A cloud of water droplets (or ice crystals) suspended in the air close to Earth's surface. See also *radiation fog*.

FOLD See panel, below.

FOLIATION Repetitive layering in metamorphic rocks where minerals become arranged in parallel bands. See also *lamination*.

FORESHORE The part of a shore lying between the low-water level and the average high-water level. See also *beach* (panel), *beach face*.

FORKED LIGHTNING A cloud-to-ground electrical discharge with a branched path in a storm. See also *sheet lightning, thunderstorm*.

FOSSIL The impression or remains of an ancient animal or plant buried in rock and preserved in petrified form.

FOSSIL FUEL An energy source such as coal, natural gas, or oil that is derived from the remains of long-dead plants and animals buried beneath the ground.

FOSSILIZATION The process of being transformed into a fossil.

FRINGING REEF A coral reef where either there is no lagoon between the reef and the shore or, if there is, it is entirely shallow. See also *atoll, coral reef*.

FRONT ▼ In meteorology, the forward-moving edge of an air mass. See also *air mass*.

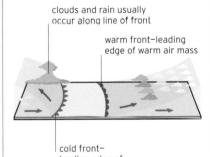

clouds and rain usually occur along line of front

warm front–leading edge of warm air mass

cold front– leading edge of cold air mass

FUMAROLE A small opening in the ground in volcanic regions through which steam and hot gases escape.

FUNNEL A tube of rapidly rotating air descending from a cloud. If its lower end touches the ground, it becomes a tornado. See also *tornado*.

G

GEODE A hollow cavity in a rock that is lined with crystals.

GEOTHERMAL Relating to the heat energy that is generated in Earth's interior. Geothermal energy is mainly produced by natural radioactive decay of uranium, thorium, and potassium.

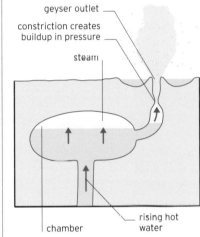

geyser outlet

constriction creates buildup in pressure

steam

rising hot water

chamber

GEYSER ▲ A jet of steam and boiling water that periodically erupts from the ground. Geysers are powered by hot rocks heating groundwater. See also *hot spring*.

GLACIER A slowly moving mass of ice formed by the accumulation and compaction of snow over a long period. There are many different types of glacier. See also *cirque glacier, piedmont glacier, surge-type glacier, valley glacier*.

GORGE A deep, narrow valley bounded by vertical or near-vertical cliffs. See also *canyon*.

GRABEN A block of Earth's crust lying between parallel faults and depressed relative to the blocks on either side. See also *rift valley*.

GRAIN The texture of a rock. For example, clay is made up of fine particles so is described as fine-grained, whereas conglomerate is coarse-grained.

GRANITE A coarse-grained igneous rock made up of the minerals quartz, feldspar, and mica.

GREASE ICE An early stage of sea-ice formation when tiny ice crystals called frazil join together to give the water surface the appearance of an oil slick. See also *sea-ice*.

GREENHOUSE EFFECT The tendency of the atmosphere to absorb a proportion of the Sun's radiation after it has been reradiated by Earth. As the amount of carbon dioxide in the atmosphere has increased, the greenhouse effect has become more marked. See also *greenhouse gas*.

GREENHOUSE GAS Any gas that promotes the greenhouse effect. The major greenhouse gases are carbon dioxide, methane, and water vapor. See also *greenhouse effect*.

FOLD

A geological structure in which once-horizontal layers of rock have been flexed and bent is called a fold. Folds develop when rocks are subjected to compression or tension. If the layers bend upward to form a ridge, the fold is an anticline; if they bend downward to form a trough, the trough is a syncline. Folds can be symmetrical or asymmetrical. Overfolds and recumbents are asymmetrical folds where the layers have folded beyond 90°. In an overfold, the layers have folded back on themselves. A fold's axis is its angle of inclination to Earth's surface.

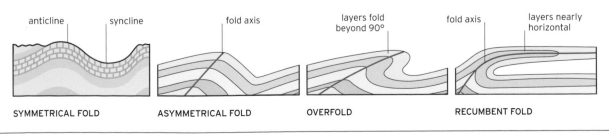

anticline   syncline   fold axis   layers fold beyond 90°   fold axis   layers nearly horizontal

SYMMETRICAL FOLD   ASYMMETRICAL FOLD   OVERFOLD   RECUMBENT FOLD

GROUNDMASS The fine-grained material in which larger crystals or fragments are embedded in igneous (and some sedimentary) rocks. See also *matrix*.

GROUNDWATER ▼ Water held in spaces (interstices) between the grains of rocks underground. The water table represents the top of the groundwater zone. See also *aquifer, water table*.

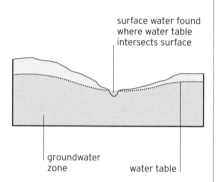

surface water found where water table intersects surface

groundwater zone   water table

GUYOT A flat-topped, undersea volcanic mountain, or seamount. The summit of a guyot is more than 660 ft (200 m) below sea level. See also *seamount*.

GYRE The large-scale rotation of an ocean current. See also *current*.

HABITAT Any area that can support a particular group of plants or animals.

HALO A hazy ring around the Moon or Sun, which is caused by the refraction of light as it passes through high clouds.

HAMADA A flat, rocky type of desert terrain, mainly devoid of sand. See also *desert pavement*.

HAMMOCK A stand of trees growing on raised ground above a wetland. Hammocks are common in the Florida Everglades.

HANGING VALLEY ▼ A tributary valley that joins a main valley high above the floor of the latter. Hanging valleys usually occur where a glacier has deepened the larger valley.

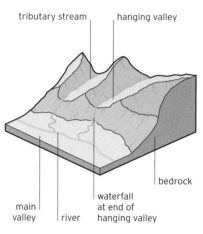

tributary stream   hanging valley

main valley   river   waterfall at end of hanging valley   bedrock

HAWAIIAN ERUPTION The least explosive type of volcanic eruption, with a relatively gentle flow of basaltic lava from a vent.

HEADLAND A narrow area of land projecting from a coastline. Headlands often exist because they are made of harder rocks than the coast on either side.

HEADWATER The upper portion of any stream or river, close to its source.

HOT SPRING The emergence of hot water and steam from the ground. Hot springs occur when hot rocks beneath the surface have heated groundwater. See also *geyser*.

HOTSPOT An area of volcanic activity thought to originate deep in Earth's mantle. Tectonic plates passing over hotspots are marked by lines of volcanoes that become older with increased distance from the hotspot. See also *mantle plume*.

HUMUS A dark-colored component of soil made up mainly of decomposed plant material.

HURRICANE A large weather system that develops and moves over tropical and subtropical oceans. Hurricanes feature destructive winds circulating rapidly around an area of low pressure, torrential rain, and storm surges. Also called a tropical cyclone or a typhoon. See also *cyclone, latent heat*.

HYDROTHERMAL VEIN Any concentration of minerals formed by the precipitation of solids from hot, mineral-laden water. See also *mineral, vein*.

HYDROTHERMAL VENT An outlet for water that has been heated by molten or hot rocks. Hydrothermal vents that emit hot, dark water from the ocean floor are called black smokers. See also *black smoker, geyser, hot spring, white smoker*.

I

ICE AGE A period of long-term reduction in global temperatures that results in the expansion of polar and continental ice-sheets and glaciers. See also *glacier, ice sheet*.

ICE CAP An ice mass covering a large area, but less than 19,000 square miles (50,000 square km). An ice cap's form overrides underlying landscape features. See also *glacier, icefield, ice sheet*.

ICE LEAD A large fracture in an expanse of sea-ice. See also *sea-ice*.

ICE SHEET A layer of ice covering an area greater than 19,000 square miles (50,000 square km) for

a prolonged period. See also *ice cap, icefield*.

ICE SHELF A large area of floating ice that is permanently attached to land. The Ross Ice Shelf, attached to Antarctica, is the world's largest.

ICE STREAM Part of an ice sheet that is moving faster than the surrounding ice. Ice streams are common in Antarctica. See also *ice sheet*.

ICE WEDGE A vertical, wedge-shaped mass of ice in the soil in a periglacial area. See also *ice-heave, periglaciation*.

ICE-HEAVE Disruption of the land surface due to the growth of ice within the soil. Ice-heave is one of the features of periglacial regions. See also *ice wedge, periglaciation, pingo*.

ICEBERG ▼ A floating mass of ice that has calved from a glacier, ice sheet, or ice shelf. See also *calving, glacier, ice shelf, ice sheet*.

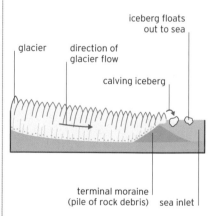

iceberg floats out to sea

glacier   direction of glacier flow

calving iceberg

terminal moraine (pile of rock debris)   sea inlet

ICEFIELD A large expanse of ice whose surface shape and extent is determined by underlying landscape features. Icefields are often interconnected valley glaciers from which higher mountain peaks rise. See also *glacier, ice cap, nunatak*.

IGNEOUS INTRUSION See panel, right.

IGNEOUS INTRUSION

An igneous intrusion is a mass of magma that forces its way into Earth's crust and cools and solidifies before reaching the surface. There are three main types: batholiths, dikes, and sills. Batholiths are large igneous bodies that may force the crust up into a dome. Dikes and sills are tabular sheets of igneous rock that have been intruded between layers of sedimentary, metamorphic, or older volcanic rocks. Dikes are vertical or have a high angle of elevation, whereas sills are horizontal or have a low angle of elevation. Granite, gabbro, diorite, and pegmatite are common intrusive igneous rocks. Over time, the erosion of softer surrounding rocks (referred to as country rock) may expose an intrusion at the surface.

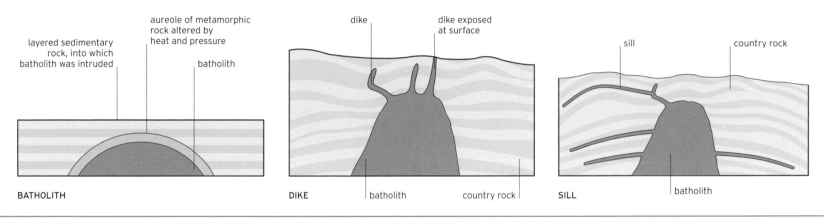

IGNEOUS ROCK Rock formed from the solidification of molten magma.

INFRARED Invisible radiation with a long wavelength, which we feel as heat. See also *ultraviolet*.

INSELBERG An isolated, steep-sided hill or small mountain rising up out of a flat plain. The Sugarloaf Mountain in Rio de Janeiro is an inselberg.

INTRUSIVE IGNEOUS ROCK A rock that formed due to the slow cooling of magma beneath Earth's surface. Examples include diorite and granite. See also *granite*, *igneous intrusion* (panel).

ISLAND ARC A chain of volcanic islands close to a subduction zone. Typically, there is a deep ocean trench on one side. See also *back arc*, *subduction*, *volcanic arc*.

ISOBAR An imaginary line connecting points with the same air pressure.

ISOSTASY The concept that continents rise high above the sea floor because continental crust is less dense than oceanic crust.

J

JET STREAM A ribbon of very strong, high-altitude winds, which move weather systems around Earth.

JOINT ▲ A fracture dividing rock into sections. The rocks on either side of a joint have not moved relative to each other, unlike the rocks either side of a fault. See also *fault* (panel).

K

KARST A landscape, usually found in limestone regions, characterized by caves, sinkholes, and underground watercourses. See also *limestone*, *sinkhole*.

KETTLE LAKE A water body occupying a kettle hole, a depression left after an ice block from a former glacier has melted.

L

LA NIÑA The phenomenon in which the waters of the eastern Pacific Ocean become unusually cold. This is the opposite of the El Niño effect. See also *El Niño*.

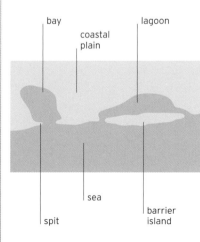

LAGOON ▲ An expanse of seawater that is almost cut off from the open ocean and consequently relatively sheltered. See also *atoll*.

LAHAR A flow of mud, water, ash, and other debris down the slopes of a volcano. See also *mass movement*.

LAMINATION Fine and generally parallel layers within rocks. It is usually less distinct than bedding. See also *bedding*.

LATENT HEAT The heat that is released, for example, when water vapor condenses into water droplets. This heat release is an important factor in the growth of storms. See also *condensation*.

LATERAL MORAINE The ridges of silt, sand, and gravel that build up along the sides of a glacier. They indicate the extent of ice before the glacier retreated. See also *moraine* (panel).

LAVA Molten rock that has reached Earth's surface as a result of an eruption. See also *basalt*, *pillow lava*.

LEACHING The process of rainwater washing minerals and nutrients through topsoil. Leached materials tend to be lost from the topsoil and deposited in the subsoil.

LEVEE ▼ A ridge of sediment deposited alongside a river when in flood, or an artificial embankment built to stop the river inundating its floodplain. See also *floodplain*.

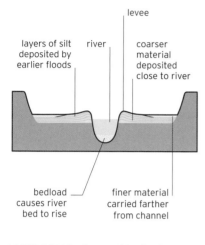

layers of silt deposited by earlier floods

river

levee

coarser material deposited close to river

bedload causes river bed to rise

finer material carried farther from channel

LIGHTNING The visible flash that accompanies electrical discharges from storm clouds. See also *forked lightning, sheet lightning*.

LIMESTONE A sedimentary rock composed mainly from the skeletons of marine life forms such as foraminifera and corals. Chemically, limestones are mostly calcium carbonate. See also *karst*.

LIMESTONE PAVEMENT An uneven, blocky surface layer of limestone rock with crevices or fissures (known as grikes) separated by blocks (or clints), created by rainwater dissolving the porous limestone.

LINEAR DUNE A long, narrow desert dune that is aligned parallel to the direction of the prevailing wind. These dunes usually form where wind from one direction dominates, and they often form parallel ridges many kilometres long. See also *dune* (panel).

LITHOSPHERE See panel, below.

LITTORAL Relating to the shoreline, especially between the low- and high-water marks. See also *beach* (panel), *tide*.

LONGSHORE BAR A ridge of sand, mud, or gravel parallel to the shore in the intertidal zone or just seaward of it. See also *beach* (panel).

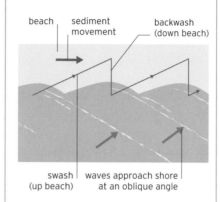

beach | sediment movement | backwash (down beach)

swash (up beach) | waves approach shore at an oblique angle

LONGSHORE DRIFT ▲ The transport of mud, sand, and gravel along a coastline by a current that flows roughly parallel to it. See also *current*.

MAAR A shallow but steep-sided volcanic crater formed when magma comes into contact with groundwater to produce a violent steam explosion.

MAGMA Molten, or semi-molten, rock that rises from parts of the mantle into the crust. It can cool and solidify beneath the surface or at the surface. See also *intrusive igneous rock, lava, mantle*.

MAGMA CHAMBER A large underground reservoir of molten rock. Sometimes, pressure forces the magma contained within to push its way to the surface in an eruption. See also *caldera, plug*.

MANGROVE SWAMP ▼ An intertidal area where salt-tolerant trees (halophytes) thrive. Mangrove swamps are most common on shallow coasts in the tropics and subtropics. See also *swamp*.

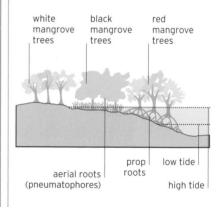

white mangrove trees | black mangrove trees | red mangrove trees

aerial roots (pneumatophores) | prop roots | low tide | high tide

MANTLE The region of Earth's interior lying between the core and the crust. It makes up 84 percent of Earth's volume. See also *core, crust*.

MANTLE PLUME A column of hot rock that rises through the mantle and into the crust, giving rise to a hotspot, which is likely to lead to volcanic activity at the surface. See also *hotspot*.

LITHOSPHERE

The solid, outer part of Earth, comprising the crust and the uppermost part of the mantle, is called the lithosphere. It lies above the asthenosphere, in which rocks are hot and viscous, and below the atmosphere and hydrosphere. The oceanic lithosphere is associated with oceanic crust and is slightly denser than the continental lithosphere, which is associated with continental crust and may extend to a depth of 120 miles (200 km).

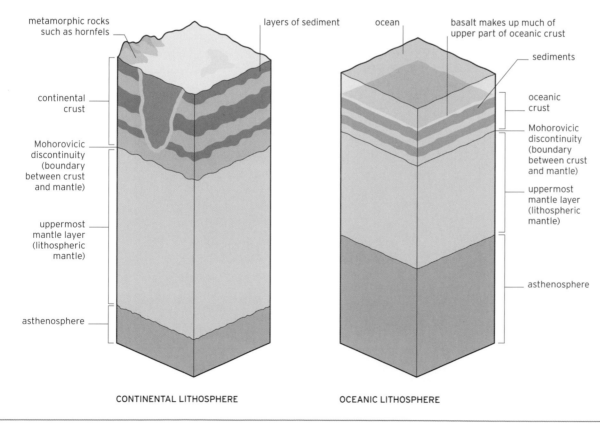

metamorphic rocks such as hornfels

continental crust

Mohorovicic discontinuity (boundary between crust and mantle)

uppermost mantle layer (lithospheric mantle)

asthenosphere

layers of sediment

ocean

basalt makes up much of upper part of oceanic crust

sediments

oceanic crust

Mohorovicic discontinuity (boundary between crust and mantle)

uppermost mantle layer (lithospheric mantle)

asthenosphere

CONTINENTAL LITHOSPHERE

OCEANIC LITHOSPHERE

MARBLE Metamorphosed limestone that is very hard and can be polished for use in construction and sculpture. See also *limestone, thermal metamorphism.*

MARSH A wetland dominated by plants such as grasses and reeds. Marshes are characterized by regular inundations, which can be of seawater or freshwater. See also *salt marsh* (panel), *swamp.*

MASS MOVEMENT The downslope movement of mud, soil, or rocks, as in a landslip. This process often takes place when soil and other surface debris are saturated with water. See also *transport.*

MASSIF A well-defined mountain mass or group of connected mountains whose rocks and landforms tend to be similar.

MASSIVE A term describing a rock without any obvious structures (such as bedding planes) or a mineral that is not visibly crystalline.

MATRIX The fine-grained material in which larger grains (in sedimentary rocks) or crystals (in igneous rocks) are embedded. See also *groundmass.*

MEANDER A loop in the course of a river. Meandering rivers gradually change their routes, as erosion takes place on the outside of the loop and deposition on the inside. See also *river valley.*

MEDIAL MORAINE A ridge of sediment running along the center of a glacial valley, and parallel with its sides. Medial moraines often form when two glaciers converge and two of their lateral moraines merge. See also *confluence, lateral moraine, moraine* (panel).

MELTWATER Water, from melted snow and ice, that flows within or from the snout of a glacier.

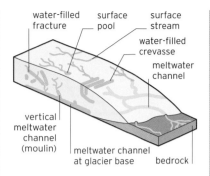

MELTWATER CHANNEL ▲ A channel cut by melted snow or ice flowing beneath, through, or near a glacier. The channel and its deposits usually remain after the glacier has retreated. See also *meltwater.*

MESA ▼ A "table-topped," steep-sided hill. Mesas are characterized by horizontal strata, capped by resistant top layers.

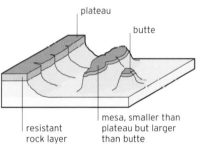

MESOSPHERE The layer of Earth's atmosphere between the stratosphere and thermosphere. It ranges from about 30 to 50 miles (50 to 80 km). See also *stratosphere, thermosphere.*

METAMORPHIC ROCK A rock that, under the influence of heat, pressure, or a combination of both, has had its texture or mineral makeup altered.

METAMORPHISM The processes— mainly heating and extreme pressure—by which rocks are transformed under ground. See also *contact metamorphism, dynamic metamorphism, regional metamorphism, thermal metamorphism.*

MICROCLIMATE The distinctive longterm weather conditions of a particular small place, such as a hilltop or a small valley.

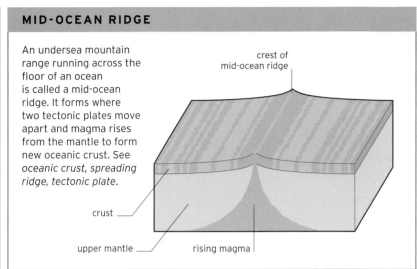

MID-OCEAN RIDGE

An undersea mountain range running across the floor of an ocean is called a mid-ocean ridge. It forms where two tectonic plates move apart and magma rises from the mantle to form new oceanic crust. See *oceanic crust, spreading ridge, tectonic plate.*

MID-OCEAN RIDGE See panel, above.

MINERAL A naturally occurring inorganic material that has a well-defined chemical composition. Most rocks are a mixture of several different minerals.

MIRE ▼ An area of swampy ground that retains wet, peaty soil. A mire often forms in a depression in impermeable rock.

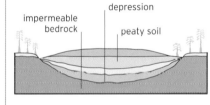

MONSOON Refers to a pattern of seasonal winds, especially in southern Asia. The term is also used to refer to the heavy rains these winds bring during the rainy, or monsoon, season. The southwest monsoon brings heavy rain to India, usually between June and September. These winds blow from one direction for about half the year, and from the opposite direction for the other half.

MONTANE Relating to mountains— for example, montane climate or montane fauna.

MORAINE See panel, below.

MORAINE

Ridges of silt, sand, gravel, and rock debris resulting from glacial action are called moraines. These features often remain after a glacier retreats. Medial moraines form where two glaciers converge, lateral moraines form along the sides of a glacier, and terminal moraines form at the furthest extent of the ice.

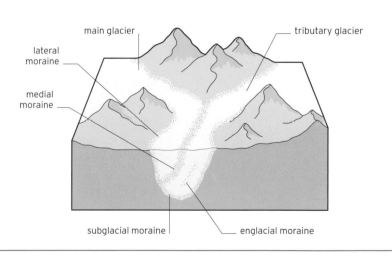

N

NATIVE Describing a plant or animal indigenous to a particular area. See also *endemic*.

NATIVE ELEMENT A chemical element that is found in its pure state in nature. Copper, gold, sulfur, and tin can all be found in their native form. See also *mineral*.

NEAP TIDE The tide with the least difference between low and high water. Neap tides occur twice a month when the gravitational pull of the Moon and Sun on Earth's oceans are working against each other. See also *spring tide*.

NEEDLE-LEAF A needle-shaped leaf typical of coniferous trees. See also *coniferous*.

NOCTILUCENT CLOUD The highest clouds in Earth's atmosphere, illuminated at night when the Sun is below the horizon and lower levels of the atmosphere are in shade. These ice clouds are usually at an altitude of 47 to 53 miles (76 to 85 km). See also *mesosphere*.

NORMAL FAULT An angled fault where one side (the hanging wall) has slipped down relative to the other side (the footwall). See also *fault* (panel).

NUNATAK A mountain peak rising out of an icefield or ice sheet. See *icefield, ice sheet*.

O

OASIS An isolated fertile area within a desert, usually fed by water from a spring. See also *aquifer*.

OCCLUSION In meteorology, the situation where a cold front overtakes a warm front, pushing a mass of warm air up and away from Earth's surface. There are two types: cold occlusions and warm occlusions. See also *front*.

OCEAN TRENCH A long, deep section of the ocean floor, usually lying above a subduction zone. See also *deep-sea trench, subduction*.

OCEANIC CRUST The outermost layer of Earth's lithosphere beneath the oceans. It is formed at spreading centers at oceanic ridges and is mainly made of basalt. See also *continental crust, crust, seafloor spreading, spreading ridge*.

OFFSHORE BAR A submerged or partly exposed ridge of sand or gravel away from and roughly parallel to the coast. Offshore bars are built up by waves and currents. See also *bar*.

OOZE The fine sediment covering a huge part of the deep ocean floor. Much of it is made up of the skeletal remains of microscopic marine life forms, such as foraminiferans and radiolarians.

ORE A rock from which a metal or valuable mineral can be profitably mined.

OROGENESIS The process of mountain-building that results from massive horizontal pressures as tectonic plates push together.

OROGENY A mountain-building event resulting from the collision of tectonic plates, when sections of Earth's crust are compressed, folded, and faulted. See also *tectonic plate*.

OUTLET GLACIER A mass of slow-moving ice that flows from an ice sheet, ice cap, or icefield before following a valley. See also *valley glacier*.

OXBOW LAKE A body of water that forms when a meander is completely cut off from the main channel of a river. See also *meander*.

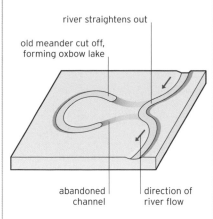

river straightens out

old meander cut off, forming oxbow lake

abandoned channel

direction of river flow

OVERTURNED FOLD A fold in which both limbs dip in the same direction. Overturned folds occur in areas of intense deformation of the rock layers. See also *fold* (panel).

P

PACK ICE A mass of ice floating on the sea, formed from many smaller pieces that have frozen together. See also *sea-ice*.

PAHOEHOE LAVA A type of basalt lava with a smooth or gently undulating surface. It often changes to a'a lava as it flows downslope. See also *a'a lava, block lava*.

PANCAKE ICE Small, flat, roughly circular areas of floating ice with curled margins where the "pancakes" have bumped into each other. Pancake ice is an early stage in the development of sea-ice. See also *sea-ice*.

PARABOLIC DUNE A U-shaped or V-shaped mound of fine to medium sand with long "arms" trailing upwind. Unlike barchans,

the steepest side (slip face) is on the convex side. See also *barchan, dune* (panel).

PARTIAL MELTING The process by which some minerals (in a rock being heated to great temperatures) melt, while others remain solid. This happens because some minerals have a lower melting point than others. See also *igneous rock*.

PEAT Partly decomposed plant material that accumulates in bogs, fens, and mires. Peat has a low mineral content.

PENINSULA An area of land protruding into a lake or sea.

PERIGLACIATION The processes that take place as a result of the regular freeze-thaw cycle in high-latitude or high-altitude regions where there is no permanent ice cover. Features include ice wedges and pingos. See also *ice wedge, permafrost, pingo, tundra*.

PERMAFROST Rock or soil that remains frozen for at least two years. Permafrost is characteristic of periglacial and glacial regions. See also *periglaciation, tundra*.

PHREATIC ERUPTION An explosive eruption that occurs when magma or hot rock come into contact with groundwater or surface water, turning it to steam.

PHYTOPLANKTON The tiny photosynthesizing life forms that float in the upper, sunlit layer of a lake or ocean and form the foundation of most aquatic food chains. See also *zooplankton*.

PIEDMONT GLACIER An area of slowly flowing ice that forms from the convergence of a number of valley glaciers at the foot of a mountain range. See also *valley glacier*.

PILLOW LAVA Molten rock that is extruded onto the ocean floor and cools quickly to form pillow-shaped lumps, usually of basalt. See also *lava*.

PINGO A periglacial landform comprising a dome-shaped hillock with a core of ice. Pingos form where groundwater—often from an old lake—rises into an area where it freezes and expands, pushing up the sediments above it. See also *periglaciation*.

PLATE BOUNDARY See *boundary* (panel).

PLATEAU ▼ A large area of flat or gently undulating land whose surface is significantly higher than the surrounding landscape.

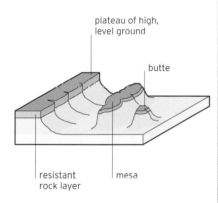

plateau of high, level ground

butte

resistant rock layer

mesa

PLAYA A flat-bottomed depression in a desert basin. Seasonally, a playa may be inundated to a shallow depth, but with no natural outlet, all the water evaporates.

PLINIAN ERUPTION The largest and most violent type of volcanic eruption, involving the release of enormous quantities of ash, lava, and gas. These eruptions are so violent that they can destroy large sections of the volcano itself. See also *eruption* (panel).

PLUG A mass of igneous rock filling the vent of an active volcano. If gas or magma is trapped under the plug, pressure will build beneath it, increasing the likelihood of the next eruption being particularly violent. See also *magma*.

PLUTON A large body of magma that solidified slowly underground to form a coarse-grained igneous rock, such as granite. See also *batholith*.

PLUTONIC Relating to igneous processes taking place, or igneous rocks formed, deep below Earth's surface. See also *batholith, igneous intrusion* (panel), *pluton*.

POINT BAR The area of sand, silt, or gravel on the inside of a river meander, where water flow is slowest. See also *bar, meander*.

POLDER An area of low-lying land reclaimed from the sea and enclosed within dikes.

POLYNA A stretch of open water in an expanse of sea-ice. See also *pack ice, sea-ice*.

POLYTHERMAL GLACIER A glacier whose base is partly "warm ice" (at 32°F, 0°C) and partly cold ice (below 32°F, 0°C). Typically, the central areas are warmer than the margins.

PRECIPITATE A substance that has come out of solution to form a solid deposit.

PRECIPITATION Water originating from the atmosphere that reaches Earth's surface, including rain, snow, hail, and drizzle.

PRESSURE GRADIENT The rate of change of atmospheric pressure with distance.

PRESSURE SYSTEM A weather pattern in which air circulates around an area of low pressure (a depression, or cyclone) or high pressure (an anticyclone). See also *anticyclone, cyclone, isobar*.

PREVAILING WIND The most common direction of airflow at a particular location.

PUMICE A light, porous volcanic rock, which solidified rapidly from gas-rich frothy magma. The numerous pores, or vesicles, were gas bubbles.

PYROCLASTIC DEPOSIT An accumulation of fragments of rock, including ash and volcanic bombs, erupted from a volcano.

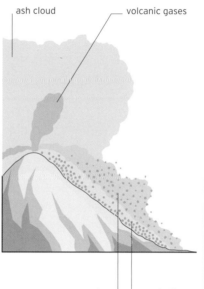

ash cloud

volcanic gases

lava fragments

pyroclastic flow

PYROCLASTIC FLOW ▲ A dense, fast-moving, and destructive cloud of hot ash, fragments of lava, and gas ejected violently from a volcano. See also *eruption* (panel).

R

RADIATION FOG The fog created by the night time cooling of air close to the ground until its content of water vapor condenses. See also *fog*.

RAIN FOREST A forest with high rainfall and humidity all year round. Most rain forests are in the tropics or subtropics, with some in temperate regions.

RAIN SHADOW The area of low rainfall downwind of a mountain range. This is produced as a result of the air shedding its moisture as it passes over the mountains.

RAINBOW A prismlike light effect caused by the Sun's rays being refracted into their constituent colors as they pass through raindrops.

RAPIDS Sections of a watercourse with a steep riverbed gradient, causing an increase in water velocity and turbulence.

RECUMBENT FOLD A fold of stratified rocks where the two limbs are almost parallel and close to the horizontal. See also *fold* (panel).

REGIONAL METAMORPHISM The processes of extreme heating and pressure by which the mineralogy and texture of rocks are transformed over a large area. See also *contact metamorphism, dynamic metamorphism, thermal metamorphism*.

REGOLITH The layer of soil and broken rock covering solid rock at Earth's surface.

RELATIVE HUMIDITY The amount of water vapor in a "parcel" of air, relative to the total amount that parcel can hold at that particular temperature. Warm air can hold more water vapor than cold air. See also *condensation, dew point*.

REMNANT ARC A formerly active volcanic island arc that has been moved away from where it was created by tectonic movement. See also *island arc, subduction*.

RESURGENCE The place where an underground watercourse emerges to become a surface stream or river. See also *karst, limestone*.

REVERSE FAULT An angled fault where one side has been pushed up relative to the other side. Reverse faults occur where two

blocks of rock have been forced together by compression. See also *fault* (panel).

RIA A former river valley that has been drowned by a rise in sea level to form an inlet of the sea.

RIDGE OF HIGH PRESSURE A wedge-shaped area of relatively high atmospheric pressure, an extension of an anticyclone. See also *anticyclone*.

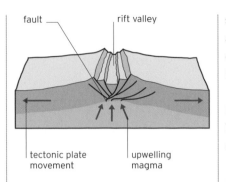

RIFT VALLEY ▲ A large block of land that has dropped vertically relative to the regions on either

side. A rift valley, or graben, forms as the result of horizontal extension of Earth's crust and normal faulting. See also *normal fault*.

RIVER BASIN A region drained by a river and its tributaries where sediment accumulates.

RIVER VALLEY See panel, left.

ROCK A natural material made up of one or more minerals. See also *mineral*.

RIVER VALLEY

A river valley is a depressed area, longer than it is wide, that has been carved by a river. A valley's shape depends on terrain, geology, and whether it has been modified by ice. Since a river has an upper, middle, and lower course, the shape of its valley reflects this. The upper section is V-shaped unless glacial ice has carved it into a U shape. Slopes are usually less steep in the middle course, and often very gentle in a valley's lower course, where the river may meander and deposit sediment on a floodplain, often dividing into channels called distributaries to form a delta. A river flows out to sea through its mouth and deposits more sediment on the seafloor.

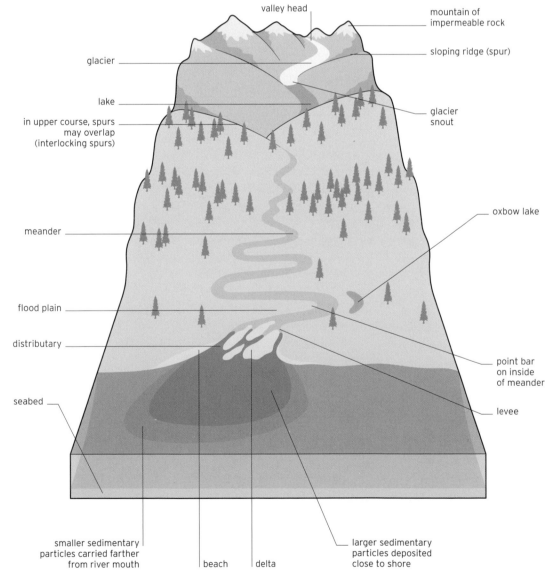

S

SALINITY The concentration of dissolved salts in water, usually expressed in parts per thousand.

SALT MARSH See panel, right.

SANDSTONE A sedimentary rock composed mostly of sand-sized grains of quartz or feldspar that have been cemented together.

SASTRUGI Sharp, irregular grooves or ridges on an ice or snow surface. Sastrugi, or zastrugi, patterns are formed by wind erosion, particularly the blasting of the surface by ice crystals.

SAVANNA A mixed grassland-woodland ecosystem of the tropics where the trees are not close enough to form a closed canopy. Often found between deserts and tropical forests, savannas are characterized by seasonal rainfall and regular wildfires. See also *cerrado*.

SEA ARCH A natural arch formed in a sea cliff by marine erosion.

SEA CAVE A cave eroded into a sea cliff by marine erosion.

SALT MARSH

Coastal wetlands that are flooded and drained by saltwater brought in by tides are called salt marshes. Salt-tolerant vegetation grows in mud and peat, which may be deep and characterized by very low oxygen levels. Many areas of salt marsh are bordered by tidal flats. See also *tidal flat*.

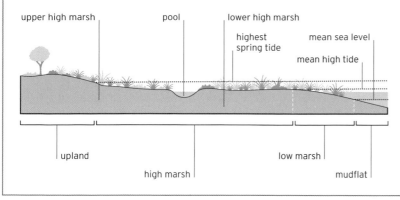

SEA STACK ▼ An offshore column of rock remaining after the erosion of adjacent sea cliffs, which are often of less-resistant rock.

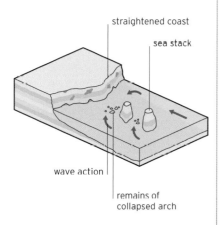

SEA-ICE Frozen ocean water. The sea freezes in several stages, first grease ice, then pancake ice, before forming continuous sheets of sea-ice. See also *grease ice, pancake ice*.

SEAFLOOR SPREADING A process that occurs at mid-ocean ridges where new oceanic crust is formed and gradually moves away from the ridge. See *mid-ocean ridge* (panel), *spreading ridge, tectonic plate*.

SEAMOUNT An undersea mountain, usually volcanic in origin, whose peak does not reach the surface. See also *guyot*.

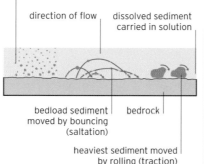

SEDIMENT ▲ Mud, sand, gravel, and other particles that have been transported by water, wind, gravity, or volcanic processes and later deposited. Moving water can transport its sediment load by several different processes.

SEDIMENTARY ROCK Material formed when mud, sand, gravel, or other particles are deposited in an ocean or lake, or on land, and are then hardened by the processes of compaction and cementation.

SEIF DUNE See *linear dune*.

SEISMIC WAVE A shock wave generated by an earthquake. There are three types of seismic wave: P-waves, S-waves, and surface waves.

SEMIDESERT An arid region that receives enough rainfall to support some plant life.

SERAC A pinnacle, or ridge, of glacial ice.

SHEET LIGHTNING An electrical discharge between two charged regions of one storm cloud. All or part of the actual discharge is obscured by cloud. It is also called intracloud lightning. See also *forked lightning, thunderstorm*.

SHIELD A large area of continental crust with ancient metamorphic rocks at the surface.

SHIELD VOLCANO The largest volcano type in terms of area, but with gently sloping sides. Shield volcanoes are the least violent, built mostly from liquid lava flows. See also *volcano* (panel).

SILL A thin, sheetlike igneous intrusion that forms when igneous rock forces its way between layers of sedimentary rocks. Most sills are roughly horizontal. See also *batholith, dike, igneous intrusion* (panel).

SINKHOLE A depression that forms where water has dissolved limestone bedrock. Sinkholes are characteristic of karst landscapes. See also *karst, limestone*.

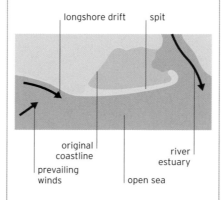

SPIT ▲ A peninsula of sand or shingle connected to the shore at one end. Spits are usually created by longshore drift, often where the coastline changes direction abruptly. See also *bar, longshore drift, tombolo*.

SPREADING RIDGE The raised area on the ocean floor where two tectonic plates are slowly moving apart. Molten rock from the mantle wells up along a spreading ridge to solidify into new oceanic crust. Situated between the Pacific and Nazca Plates, the East Pacific Rise contains one of Earth's fastest spreading ridge systems. See also *boundary* (panel), *mid-ocean ridge* (panel), *seafloor spreading, tectonic plate*.

SPRING ▼ A surface outlet for water that has flowed from porous rocks underground. See also *groundwater*.

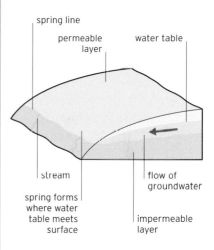

SPRING TIDE The tide with the greatest difference between low and high water. Spring tides occur twice a month when the gravitational pull on Earth's oceans of the Moon and Sun are working with each other. It falls much lower than a neap tide, exposing more of the shore, as well as rising much higher a few hours later. See also *neap tide*.

SPUR A sloping ridge projecting from the side of a mountain or valley. A truncated spur has a blunt end as a result of erosion by a glacier or faulting.

STALACTITE ▼ A deposit, usually of calcite, hanging from the roof of a cave. Stalactites are produced when minerals precipitate from dripping water. See also *calcite, karst, precipitate, stalagmite*.

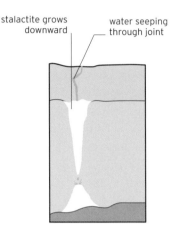

stalactite grows downward

water seeping through joint

STALAGMITE ▼ A deposit, usually of calcite, rising up from the floor of a cave. Like stalactites, stalagmites are produced when minerals precipitate from dripping water. See also *calcite, karst, precipitate, stalactite*.

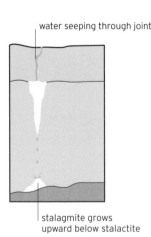

water seeping through joint

stalagmite grows upward below stalactite

STAR DUNE A pyramid-shaped mound of sand with three or more "arms" radiating from the center. Star dunes usually form where the wind blows from several directions, and they tend to grow up rather than spread out. See also *dune* (panel).

STEPPE A temperate grassland plain with very few trees, especially in regions with hot, dry summers and cold winters.

STORM SURGE ▼ An abnormal rise of sea level, higher than the predicted tide, generated by a storm. The surge is caused by the storm's winds pushing water onshore.

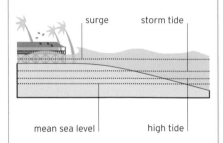

surge   storm tide

mean sea level   high tide

STRATOSPHERE The layer of the atmosphere between the top of the troposphere, at 5–10 miles (8–16km), and the bottom of the mesosphere, at about 30 miles (50km). See also *mesosphere, troposphere*.

STRATOVOLCANO A conical volcano composed of many layers of ash, pumice, and solidified lava. Unlike a shield volcano, its eruptions are usually violent. Stratovolcanoes are common near subduction zones. See also *pumice, subduction, volcano* (panel).

STRATUM (PL. STRATA) A layer of sedimentary rock. See also *bedding plane, sedimentary rock*.

STRIKE-SLIP FAULT A near-vertical fault with horizontal displacement of the rocks on either side. The San Andreas Fault in California is a strike-slip fault. See also *fault* (panel).

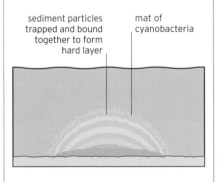

sediment particles trapped and bound together to form hard layer

mat of cyanobacteria

STROMATOLITE ▲ A layered mound, column, or sheet formed by layers of single-celled cyanobacteria in ancient seas. When cyanobacteria

became common about 2.5 billion years ago, the oxygen they produced through photosynthesis began to change the atmosphere in a way that enabled other life forms to develop.

STROMBOLIAN ERUPTION A short-lived explosive volcanic eruption characterized by the ejection of ash, lava, and volcanic bombs. See also *phreatic eruption*.

SUBDUCTION The convergence of two tectonic plates, as a result of which an oceanic plate descends beneath another plate. Subduction zones can be classified as either ocean–ocean or ocean–continent, depending upon the nature of the two converging plates. See also *boundary* (panel), *deep-sea trench, tectonic plate*.

SUBLIMATION The change in a substance, such as glacier ice, from a solid to a gas without passing through the liquid phase.

SUCCESSION The gradual change from one community of plants to another over time in a particular area. For example, grassland may change to forest if grazing animals are removed from it.

SUPERCELL CLOUD A large cloud characterized by the presence of a deep, persistently rotating updraft (a mesocyclone), lightning, torrential rain and hail, strong winds, and sometimes tornadoes. See also *thunderstorm, tornado*.

SURGE-TYPE GLACIER A glacier whose movement features periods of normal flow alternating with episodes of much more rapid flow, up to 100 times faster. The fast–slow cycle varies greatly between different surge-type glaciers. See *glacier*.

SURTSEYAN ERUPTION A volcanic eruption that takes place in relatively shallow water, with magma or lava interacting explosively with the

water. Surtseyan eruptions usually take place when an undersea volcano has finally grown large enough to break the water's surface. See also *phreatic eruption*.

SWAMP A freshwater or saltwater wetland that is dominated by trees. See also *mangrove swamp, marsh*.

SWASH The rush of seawater up a beach when a wave breaks.

SYMMETRICAL FOLD A geological structure in which once-horizontal layers of rock have been folded in such a way that the fold's axis is vertical. See also *asymmetrical fold, fold* (panel).

SYNCLINE A fold of stratified rocks in which the newest strata (layers) are at the core. Synclines result from massive horizontal compression. See also *anticline, fold* (panel).

T

TAIGA The moist coniferous forest (boreal forest) biome covering much of northern Europe, Asia, and North America, south of the tundra. See also *biome, coniferous, tundra*.

TECTONIC PLATE One of the large, rigid sections into which Earth's lithosphere is divided. Most earthquakes and volcanic activity are driven by the relative motion of plates. See also *lithosphere* (panel).

TEKTITE A glassy particle sometimes formed when a large meteorite strikes the Earth. Tektites are thought to have been created when a meteorite impact melted or vaporized surface rocks upon impact.

TEMPERATE Relating to those parts of Earth between the tropics and the polar regions. See also *tropical*.

TERMINAL MORAINE The ridge of silt, sand, or gravel that builds up at the snout (front end) of a glacier. It marks the maximum extent of the glacier's advance. See also *lateral moraine, medial moraine, moraine* (panel).

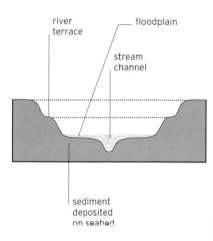

river terrace · floodplain
stream channel
sediment deposited on seabed

TERRACE ▲ A flat area in a river valley that is raised above the present floodplain. Terraces are remnants of former floodplains created when rivers flowed at a higher level. See also *floodplain*.

THERMAL METAMORPHISM The transformation of rocks brought about mainly by intense heat, rather than pressure. See also *contact metamorphism, dynamic metamorphism*.

THERMOCLINE A region in an ocean, a lake, or the atmosphere where temperature changes more rapidly with depth or height.

THERMOHALINE A term describing the temperature and salt content of water. Together, these properties determine water's density and influence ocean currents.

THERMOPAUSE The upper edge of the thermosphere, about 400 miles (640 km) above Earth's surface and below the exosphere. See also *thermosphere*.

THERMOSPHERE The layer of the atmosphere between the mesosphere and the exosphere (the layer separating the atmosphere from space). It extends between about 50–400 miles (80–640 km) above Earth's surface.

THRUST FAULT ▼ A type of reverse fault, which is an angled fault where one side (the hanging wall) has been pushed up relative to the other side (the footwall). In a thrust fault, the angle of the dip is less than 45°. See also *fault* (panel), *reverse fault*.

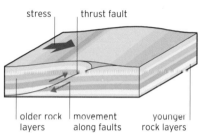

stress · thrust fault
older rock layers · movement along faults · younger rock layers

THUNDERSTORM A storm with discharges of lightning and its acoustic effect, thunder. Thunderstorms are often associated with massive cumulonimbus clouds and characterized by heavy rain, hail, strong but localized winds, and occasionally tornadoes. See also *lightning, supercell cloud*.

TIDAL BORE A single large wave sometimes created when a rising tide enters a narrowing channel such as a river estuary. See also *estuary*.

TIDAL FLAT An almost horizontal muddy or sandy area exposed at low tide and covered at high tide. Tidal flats are characteristic of sheltered areas such as estuaries. See also *estuary*.

TIDAL RACE A strong current created when a tide-generated water flow moves through a narrow channel.

TIDAL RANGE The vertical difference between a high tide and the succeeding low tide. At

any location, the tidal range varies according to the relative positions of the Sun and Moon.

TIDE ▼ The rise and fall (usually twice each day) of seawater on the shore caused by the gravitational pull of the Moon and the Sun and Earth's rotation. See also *neap tide, spring tide, tidal range*.

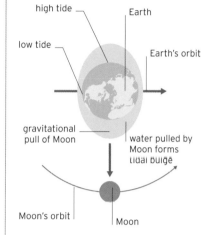

high tide · Earth
low tide
Earth's orbit
gravitational pull of Moon · water pulled by Moon forms tidal bulge
Moon's orbit · Moon

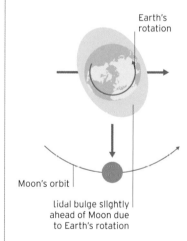

Earth's rotation
Moon's orbit
tidal bulge slightly ahead of Moon due to Earth's rotation

TILL A mixture of clay, sand, and rock fragments deposited by a glacier. Till that was left behind by glaciers in the last ice age covers many regions today. Also called boulder clay. See also *moraine* (panel).

TOMBOLO A spit that connects an island to the mainland. See also *spit*.

TORNADO A localized, violently destructive windstorm characterized by a tall funnel connecting a cloud with the ground. See *supercell cloud*.

TRANSFORM FAULT A type of fault in which two tectonic plates slide past each other. See also *boundary* (panel), *fault* (panel), *tectonic plate*.

TRANSPORT The movement of eroded and weathered material by water, ice, or wind. See also *weathering, erosion*.

TRAVERTINE A sedimentary rock, mostly calcium carbonate, that is precipitated around the edges of a hot spring. See also *precipitate, sedimentary rock*.

TREE LINE ▼ The altitude or latitude above which conditions are too harsh for trees to grow in a particular location.

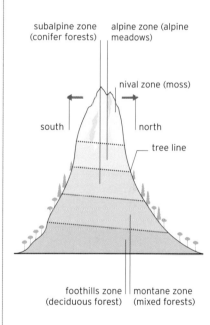

subalpine zone (conifer forests) · alpine zone (alpine meadows)
nival zone (moss)
south · north
tree line
foothills zone (deciduous forest) · montane zone (mixed forests)

TRENCH See *ocean trench*.

TRIBUTARY A river or stream that flows into a larger river.

TROPICAL Relating to those regions to the north and south of the equator, between the Tropic of Cancer (latitude 23.26°N) and the Tropic of Capricorn (latitude 23.26°S). See also *temperate*.

TROPICAL CYCLONE See *hurricane*.

TSUNAMI

Sometimes mistakenly called a tidal wave, a tsunami is not created by tidal action. It is a fast-moving ocean wave generated by an earthquake, volcanic eruption, or landslide above or below water. If a tsunami moves into shallow water it gains height rapidly, in extreme cases reaching a height of tens of feet and causing coastal destruction on a vast scale. The 2004 Indian Ocean tsunami killed at least 230,000 people in 14 countries.

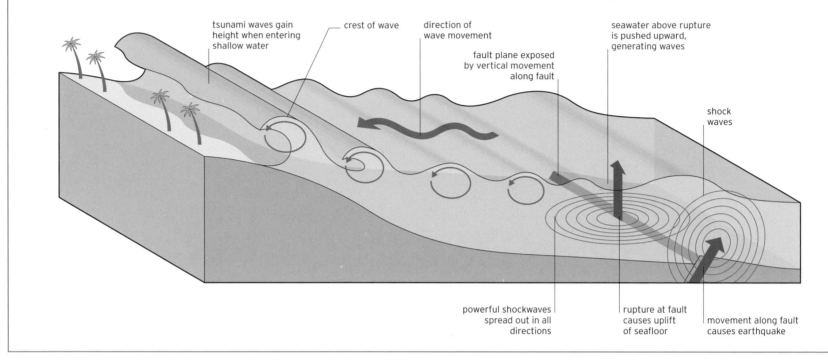

tsunami waves gain height when entering shallow water

crest of wave

direction of wave movement

fault plane exposed by vertical movement along fault

seawater above rupture is pushed upward, generating waves

shock waves

powerful shockwaves spread out in all directions

rupture at fault causes uplift of seafloor

movement along fault causes earthquake

TROPOPAUSE The boundary between the troposphere and the stratosphere, at altitudes of about 10 miles (16 km) at the equator and 5 miles (8 km) over the poles. Above the tropopause, the air gets warmer with height. See also *stratosphere, troposphere*.

TROPOSPHERE The lowest, most oxygen-rich, layer of the atmosphere, where most weather events take place. The upper limit of the troposphere is lower at the poles than at the equator. See also *tropopause*.

TROUGH In meteorology, a long, relatively narrow area of low atmospheric pressure. See also *front*.

TSUNAMI See panel, above.

TUFF An igneous rock made of fine-grained pyroclastic material, such as ash, produced by a violent volcanic eruption. See also *pyroclastic deposit*.

TUNDRA The treeless biome of low-growing, cold-tolerant

plants lying north of the taiga in northern Europe, Asia, and North America. See also *biome, taiga*.

TURBIDITE The sediment deposited by a turbidity current.

TURBIDITY CURRENT A rapidly moving flow of sediment-laden water, such as a flow down a continental slope onto the ocean floor.

TYPHOON See *hurricane*.

U

ULTRAVIOLET Invisible radiation with a short wavelength. Ultraviolet is a major component of sunlight. See also *infrared*.

UNDERSTORY The layer of smaller trees and shrubs growing mainly in the shade beneath a forest canopy. See also *canopy*.

UPWELLING The process whereby cold, and usually nutrient-rich, water from ocean depths rises toward the surface.

V

VALLEY FOG A type of fog that can form in low-lying areas, on cold, still nights. Heavy, chilled air settles in such places, and the water vapor it contains condenses when the temperature falls below the dew point. See also *radiation fog*.

VALLEY GLACIER A stream of flowing ice that is confined within the sides of a valley, often a V-shaped preglacial valley that is modified by the erosive action of the ice. See also *glacier*.

VEIN A fracture in rock containing mineral deposits. Veins are usually formed when minerals are

precipitated from a hot liquid in which they are dissolved. See also *hydrothermal vein*.

VENTIFACT A desert rock or pebble that has been polished by the action of windblown sand.

VERTICAL TRANSPORT The ascent (upwelling) or descent of nutrient-rich water in the oceans.

VOLATILE A term used to describe water, carbon dioxide, and some other gases dissolved in magma that form bubbles at low pressure near the surface. When the bubbles combine, they can make an eruption more explosive, showering sprays of lava. See also *eruption* (panel).

VOLCANIC ARC A chain of volcanoes formed above a subduction zone and arranged in an arc shape. The volcanoes may form a chain of islands in the ocean. See also *island arc, subduction, tectonic plate*.

VOLCANO

A volcano is a mountain with an opening, or vent, through which molten lava, volcanic ash, and gases are erupted. Eruptions range from violent andesitic eruptions to the gentle lava flows. Structurally, there are many types of volcanoes, their shapes being largely dependent on the type of eruptions and the quality of the magma. A dome volcano forms steep sides because the viscous lava it erupts cannot flow very far. Stratovolcanoes are formed of many layers of ash and solidified lava.

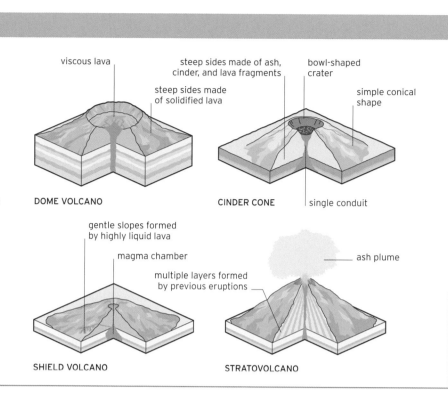

viscous lava

steep sides made of ash, cinder, and lava fragments

steep sides made of solidified lava

DOME VOLCANO

bowl-shaped crater

simple conical shape

CINDER CONE single conduit

gentle slopes formed by highly liquid lava

magma chamber

multiple layers formed by previous eruptions

SHIELD VOLCANO

ash plume

STRATOVOLCANO

VOLCANIC PLUG A mass of igneous rock filling the vent of an active volcano. If gas or magma is trapped beneath the plug, pressure will build up, increasing the likelihood of the next eruption being violent. See also *magma*.

VOLCANO See panel, above.

VORTEX A rapidly circulating mass of air or water. See *eddy, tornado, whirlpool*.

VULCANIAN ERUPTION An explosive type of volcanic eruption. Vulcanian eruptions occur when the pressure of trapped gases in magma becomes strong enough to blow off an overlying crust of solidified magma. See also *eruption* (panel), *magma*.

WADI A desert riverbed or valley that contains water only at times of heavy rainfall.

WATER TABLE ▼ The upper surface of the groundwater zone. Below the water table, the soil or rocks are permanently saturated, but its position fluctuates, often seasonally. See also *groundwater*.

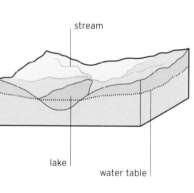

stream

lake

water table

WATERFALL ▼ A cascade of water formed when a stream or river flows over a cliff or steep incline.

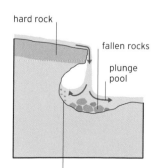

hard rock

fallen rocks

plunge pool

soft rock eroded by water and small swirling rocks

WATERSHED An imaginary line that separates neighboring drainage basins. See also *catchment area, drainage basin*.

WAVE A disturbance in the ocean, usually generated by wind. For a wave crossing the sea, the water itself does not move very much, except up and down as the wave passes. The high point of a wave is its crest, and the low point its trough.

WAVELENGTH The distance between successive crests (or troughs) of a wave.

WEATHERING The gradual breakdown of rocks on Earth's surface by chemicals in rainwater, temperature changes, and biological activity such as the growth of lichens and plants. Rocks are broken into small fragments, which are then transported away. See also *erosion*.

WHIRLPOOL A rapidly rotating body of water in a river or the ocean. See also *eddy, vortex*.

WHITE SMOKER A hydrothermal vent on the ocean floor emitting light-colored minerals such as those containing barium, calcium, and silicon. White smokers tend to be cooler than black smokers. See also *black smoker, hydrothermal vent*.

Y

YARDANG A wind-sculpted desert rock, eroded by windborne particles of sand and dust.

Z

ZOOPLANKTON Animal life, mostly microscopic, that drifts in the water column of oceans, lakes, and rivers. See also *phytoplankton*.

ZOOXANTHELLAE Single-celled algae that live in corals and supply them with vital nutrients to grow. See also *coral bleaching*.

W

INDEX

Page numbers in *italics* refer to photographs and illustrations.

ACKNOWLEDGMENTS

Dorling Kindersley would like to thank the following people for their work on this book: Professor Simon Lamb, Dr. Tony Waltham, and Tim Harris for fact-checking; Katie John for proofreading; Rob Houston for editorial help; Gregory McCarthy, Duncan Turner, and Adam Spratley for design assistance; Steve Crozier for Photoshop retouching; and Simon Mumford for cartographic work and advice.

DK India would like to thank Aisvarya Misra and Nisha Shaw for editorial assistance; Himshikha, Konica Juneja, Avinash Kumar, and Anjali Sachar for design assistance; and Deepak Negi for picture research assistance.

Smithsonian Enterprises:

President Christopher A. Liedel
Senior Vice President, Consumer and Education Products Carol LeBlanc
Vice President, Consumer and Education Products Brigid Ferraro
Licensing Manager Ellen Nanney
Product Development Manager Kealy Gordon

Data for the geological maps on the introductions to the continents is taken from the Generalized Geological Map of the World, published by the Geological Survey of Canada, and contains information licensed under the Open Government Licence—Canada.

Data and imagery for artworks has been used from the following sources: **pp.28-29, Crater Lake** USGS/USDA: NAIP Digital Ortho Photo Image and USGS NED; **pp.32-33, Yellowstone** Landsat 8 and USGS SRTM; **pp.44-45, the Kennicott Glacier** modified Copernicus Sentinel data (2016) and USGS NED; **pp.94-95, the Andes** Landsat 8 and USGS SRTM; **pp.108-109, the Amazon Basin** Blue Marble/NASA Earth Observatory and ETOPO1–NOAA; **pp.186-87, the Great Rift Valley** Landsat 8 and USGS SRTM; **pp.226-27, Mount Everest** NASA Earth Observatory image by Jesse Allen and Robert Simmon, using EO-1 ALI data from the NASA EO-1 team, archived on the USGS Earth Explorer and ASTER GDEM (a product of NASA and METI); **pp.238-39 Lake Baikal** Landsat 8 and USGS SRTM; **pp.286-87, the Great Barrier Reef** Landsat 8 and Deepreef Explorer, Beaman, R.J., 2010. Project 3DGBR: a high-resolution depth model for the Great Barrier Reef and Coral Sea. Marine and Tropical Sciences Research Facility (MTSRF) Project 2.5i.1a Final Report, MTSRF, Cairns, Australia, pp.13 plus Appendix 1; **pp.306-307 Antarctic Ice Sheet** British Antarctic Survey BedMap2 (www.bas.ac.uk/project/bedmap-2/); **pp.312-13 the Mid-Atlantic Ridge** Blue Marble/NASA's Earth Observatory and ETOPO1– NOAA; **pp.318-19, the Hawaiian Islands** Landsat 8 and Hawaii Mapping Research Group/School of Ocean and Earth Science and Technology—Main Hawaiian Islands Multibeam Bathymetry and Backscatter Synthesis: University of Hawai'i at Manoa; **pp.324-25 the Mariana Trench** Landsat 8 and Bathymetric Digital Elevation Model of the Mariana Trench—NOAA: National Geophysical Data Center (NGDC).

The publisher would like to thank the following for their kind permission to reproduce their photographs:

(Key: a-above; b-below/bottom; c-center; f-far; l-left; r-right; t-top)

Endpapers: AirPano Images **1 Getty Images:** DigitalGlobe. **2-3** Peter Franc. **4 Alamy Stock Photo:** National Geographic Creative (fcla). **Getty Images:** Mike Lanzetta (ca); G & M Therin-Weise (fcra). **Imagelibrary India Pvt Ltd:** Jinhu Wang (cla). **Philip Klinger (philip-klinger. photography):** (cra). **5 Getty Images:** Sue Flood / Oxford Scientific (ca); Buena Vista Images / DigitalVision (fcla); Wil Meinderts / Buiten-beeld / Minden Pictures (ca); James D. Morgan (fcra). **Matt Hutton:** (cla). **6-7 Alamy Stock Photo:** Steven Sandner. **8-9 Getty Images:** Yann Arthus-Bertrand. **11 Getty Images:** Danita Delimont (t); NOAA (b). **12 naturepl.com:** Alex Mustard (br). **13 Getty Images:** Education Images (bc); Mark Hannaford (bl); Douglas Peebles (br). **14 NASA:** JPL–Caltech (cl). **J. W. Valley, University of Wisconsin-Madison:** (tr). **14-15 Getty Images:** John Lund / Tom Penpark. **15 NASA:** (cr). **16 Alamy Stock Photo:** Science History Images (cra). **Allen Nutman, University of Wollongong:** (cl). **Science Photo Library:** (ca). **17 Alamy Stock Photo:** Life On White. **18-19 Getty Images:** Peter Adams. **22 Getty Images:** Ron Garnett (tc); Mike Grandmaison (tl). **22-23 Zack Frank:** (t). **23 Alamy Stock Photo:** Kip Evans (cra). **Science Photo Library:** W.K. Fletcher (ca). **24-25 Imagelibrary India Pvt Ltd:** Victor Aerden (t). **24 Carl Battreall / photographalaska.com:** (bl). **25 Copyright Tom Lussier Photography 2017:** (br). **26 Dorling Kindersley:** National Birds of Prey Centre, Gloucestershire (bc). **Imagelibrary India Pvt Ltd:** Mike Wilson (t). **27 Alamy Stock Photo:** Image Source (clb). **Science Photo Library:** USDA / Science Source (cr). **28 Dreamstime.com:** Maria Luisa Lopez Estivill (tl). **Getty Images:** Thomas Winz / Lonely Planet Images (bl). **29 123RF.com:** William Perry (tl). **Alamy Stock Photo:** Gerhard Zwerger-Schoner / Imagebroker (bl). **30 Getty Images:** Kevin Schafer (clb). **30-31 Alamy Stock Photo:** Christian Handl / Imagebroker (t). **National Geographic Creative:** Michael Nichols (b). **31 Brett Lange:** (bc). **32 Alamy Stock Photo:** robertharding (cra). **Getty Images:** Babak Tafreshi / National Geographic (cla, bc). **33 Alamy Stock Photo:** Gaertner (cb). **34 Alamy Stock Photo:** Brad Mitchell (tr). **Getty Images:** Wild Horizon / Contributor (b). **35 Getty Images:** DenisTangneyJr (br). **Imagelibrary India Pvt Ltd:** Josh Baker (tc). **36 Getty Images:** Carsten Peter / Speleoresearch & Films / National Geographic (bl). **37 Alamy Stock Photo:** Leonardo Díaz Romero / age fotostock (br). **Getty Images:** Manfred Gottschalk (cl). **38-39 Steve Morgan:** (t). **38 Getty Images:** Jason Edwards / National Geographic (br); Patrick Robert / Corbis Premium Historical (clb). **39 Dorling Kindersley:** Jerry Young (br). **40-41 Alamy Stock Photo:** NASA / Dembinsky Photo Associates (t). **40 Alamy Stock Photo:** Frans Lanting Studio (br). **Getty Images:** DEA / M. Santini / De Agostini (bl). **National Geographic Creative:** Design Pics Inc (clb). **41 Jason Hollinger:** (cb). **42 Alamy Stock Photo:** Marion Bull (crb). **David P. Reilander:** (cr). **43 Carl Battreall / photographalaska.com:** (c). **Imagelibrary India Pvt Ltd:** Lee Petersen (bl). **44 Getty Images:** Daniel A. Leifheit (tr). **45 123RF. com:** Galyna Andrushko (br). **Alamy Stock Photo:** John Schwieder (tr); Zoonar GmbH (crb). **46-47 Getty Images:** John Hyde. **46 Getty Images:** Sergey Gorshkov / Minden Pictures (clb). **47 Larry McCloskey:** (cb). **48-49 Getty Images:** Kevin Smith / Design Pics. **48 Alamy Stock Photo:** Fred Lord (clb). **49 Ardea:** Steffen & Alexandra Sailer (tc). **Getty Images:** Emmanuel Coupe / Photographer's Choice (br). **50 Imagelibrary India Pvt Ltd:** Jeff Moreau (br). **50-51 Dave Sandford:** (t). **51 Getty Images:** David Doubilet / National Geographic (crb); Rolf Hicker / All Canada Photos (bl). **52 Dorling Kindersley:** Neil Fletcher (br). **Imagelibrary India Pvt Ltd:** Jin Kim (t). **53 Alamy Stock Photo:** NASA / Landsat / Phil Degginger (tl). **Getty Images:** Cameron Davidson / Photographer's Choice (tr); Joel Sartore / National Geographic (crb). **54-55 Getty Images:** Paul Rojas / Moment Select. **54 123RF.com:** Tom Tietz (c). **56 Alamy Stock Photo:** Inge Johnsson (cb). **57 Alamy Stock Photo:** B.A.E. Inc. (bc); John Barger (tr). **Getty Images:** Pete Mcbride / National Geographic (ca). **58 Getty Images:** Joel Sartore / National Geographic (cla). **Imagelibrary India Pvt Ltd:** Cynthia Spence (tr). **SuperStock:** Keith Kapple (br). **59 Getty Images:** David Sieren / Visuals Unlimited (bc). **Diane Kirkland / dianekirklandphoto.com:** (t). **60 Getty Images:** Jupiterimages / Photolibrary (c). **iStockphoto.com:** Donyanedomam (clb). **Robert Harding Picture Library:** David Fleetham / Okapia (cra). **61 Courtesy of National Park Service, Lewis and Clark National Historic Trail:** G. Gardner. **62 Getty Images:** Nick Boren Photography / Moment (tr). **SeaPics.com:** Phillip Colla (bc). **63 Getty Images:** James P. Blair / Contributor / National Geographic (cb). **Imagelibrary India Pvt Ltd:** Clemens Ruehl (t). **64-65 Khanh Ngo, landscape photographer in Atlantic, Canada:** (b). **65 Getty Images:** Dale Wilson / Photographer's Choice (cla). **Imagelibrary India Pvt Ltd:** Arun Sundar (br). **66 Getty Images:** Thomas Kitchin & Victoria Hurst (cr). **Imagelibrary India Pvt Ltd:** 500px / Jakub Sisak (bl). **67 Alamy Stock Photo:** All Canada Photos (bc); Spring Images (bl). **Getty Images:** Konrad Wothe (t). **68 Alamy Stock Photo:** age fotostock (bc); Tierfotoagentur (c). **68-69 National Geographic Creative:** Micheal Nichols. **70 Dreamstime.com:** Betty4240 (bc). **Getty Images:** Danita Delimont (tl). **Stacey Putman Photography:** (br). **71 iStockphoto.com:** Gary R Benson (t). **William Neill Photography:** (cb). **72 Alamy Stock Photo:** FLPA (br). **Getty Images:** Steven Kazlowski / Science Faction (t). **73 Alamy Stock Photo:** Tom Bean (br). **Getty Images:** Antonyspencer / E+ (cr). **74 Getty Images:** Witold Skrypczak / Lonely Planet Images (cb). **iStockphoto.com:** Avatar Knowmad (bl). **74-75 iStockphoto.com:** Gary Kavanagh. **75 FLPA:** Mark Newman (bc). **Getty Images:** Andrew Kennelly / Moment Open (crb). **76 Getty Images:** Danita Delimont / Gallo Images (bl). **Imagelibrary India Pvt Ltd:** Jinhu Wang (tr). **76-77 Alamy Stock Photo:** Phil Degginger (b). **78 Getty Images:** Jeremy Duguid Photography / Moment (bc). **Katharina Winklbauer / www.winka-photography.de:** (t). **79 Getty Images:** Chris Saulit / Moment (cra). **Kris Walkowski / kriswalkowski.com:** (clb). **80 Imagelibrary India Pvt Ltd:** Michael T. Lim (r). **Alex E. Proimos:** (bc). **81 Dreamstime.com:** Eutoch (b). **FLPA:** Yva Momatiuk &, John Eastcott / Minden Pictures (cra). **82-83 Getty Images:** Mint Images—Frans Lanting. **86 Getty Images:** Apomares (tl). **86-87 Alamy Stock Photo:** Pulsar Images (t). **87 Getty Images:** EyeEm / André Reis (tc); Mint Images / Frans Lanting (cra). **iStockphoto.com:** Pxhidalgo (cla). **88 Getty Images:** traumlichtfabrik / Moment (c). **Rex Shutterstock:** WestEnd61 / REX (t). **89 Getty Images:** Andoni Canela / age fotostock (br). **Martin Rietze:** (tr). **90-91 AirPano images. 91 Ardea:** Adrian Warren (clb). **naturepl.com:** Luiz Claudio Marigo (c). **92 123RF.com:** Eric Isselee (br). **92-93 Ricardo La Piettra. 93 Alamy Stock Photo:** Francisco Negroni / Biosphoto (bl). **94 Alamy Stock Photo:** Kseniya Ragozina (tr). **95 Alamy Stock Photo:** Hemis (cra). **Getty Images:** Mint Images / Art Wolfe (br/flying flamingo). **iStockphoto.com:** DC_Colombia (br); Elisalocci (tl). **96-97 Getty Images:** Hans Neleman / Stone. **96 123RF.com:** Jarous (clb). **97 Getty Images:** Richard I'Anson / Lonely Planet Images (br). **Science Photo Library:** Bernhard Edmaier (tl). **98 Thanat Charoenpol:** (ca). **Getty Images:** Marisa L?pez Estivill / Moment Open (bl). **99 Getty Images:** Gina Pricope / Moment (c). **Imagelibrary India Pvt Ltd:** Craig Holden (b). **100 Alamy Stock Photo:** Minden Pictures (br). **Getty Images:** Bobby Haas / National Geographic (cra). **100-101 ESA:** KARI (t). **101 Alamy Stock Photo:** Tino Soriano / National Geographic Creative (bc). **Lucas Cometto / www.lucascometto.com:** (cra).